Biodiversity in Enclosed Seas and Artificial Marine Habitats

Developments in Hydrobiology 193

Series editor

K. Martens

Biodiversity in Enclosed Seas and Artificial Marine Habitats

Proceedings of the 39th European Marine Biology Symposium, held in Genoa, Italy, 21–24 July 2004

Edited by

G. Relini[1] & J. Ryland[2]

[1]*Laboratori di Biologia Marina ed Ecologia Animale, DIP.TE.RIS., Università di Genova, Corso Europa 26, 16132 Genova, Italy*

[2]*Emeritus Professor of Marine Biology, Biological Sciences, Wallace Building, University of Wales Swansea, Swansea SA2 8PP, Wales, UK*

Reprinted from Hydrobiologia, Volume 580 (2007)

Springer

Library of Congress Cataloging-in-Publication Data

A C.I.P. Catalogue record for this book is available from the Library of Congress.

ISBN-13: 978-1-4020-6155-4

Published by Springer,
P.O. Box 17, 3300 AA Dordrecht, The Netherlands

Cite this publication as Hydrobiologia vol. 580 (2007).

Cover illustration: Small red scorpionfish (*Scorpaena notata* Rafinesque 1810) on a module of the Monaco (Montecarlo) artificial reef. Photo credit: Prof. Roberto Pronzato, University of Genoa.

Printed on acid-free paper

Printed in the Netherlands

TABLE OF CONTENTS

OPEN SESSION

Hydrobiologia (2007) 580:1–5
DOI 10.1007/s10750-007-0574-0

Foreword

Giulio Relini · John Ryland

The 39th European Marine Biology Symposium, held from 21st to 24th July 2004 in Genoa (Italy), European Capital of Culture 2004, was hosted jointly by the University of Genoa and by the Italian Society of Marine Biology (S.I.B.M.), whose yearly congress was held on Monday and Thursday before the EMBS as a bridge between Italian and European marine biologists with the aim of fostering knowledge and cooperation. About 280 participants from 24 countries attended the Symposium; 6 keynote lecturers, 60 oral presentations and 130 posters were in the programme. All talks were presented in the Aula Magna while the poster sessions took place in the open gallery in front of the Aula Magna of the University, in a 16th-century monumental palace (via Balbi), within walking distance of the city centre. The traditional Yellow Submarine competitions were held in the "Porto Antico" area after the visit to the Aquarium of Genoa and before the Symposium Dinner, hosted by the Aquarium in the area in front of the Shark Tank.

There was a large participation of young and enthusiastic researchers, who had the opportunity to forge new friendships and research partnerships, to stimulate the exchange of scientific data and experiences and to have at the same time a link between generations of marine biologists. Five EMBS past-presidents (Bruno Battaglia, John Gray, Joerg Ott, John Ryland, Giulio Relini) were present and contributed strongly to the discussion of the papers presented.

The Symposium was convened under two main themes, partly reflecting the research interests at Genoa University:

1. Biodiversity in enclosed and semi-enclosed seas;
2. Artificial habitats and the restoration of degraded systems.

The organising Committee comprised members of the University of Genoa (DIP.TE.RIS., Dipartimento per lo Studio del Territorio e delle sue Risorse) and S.I.B.M.: G. Relini, R. Pronzato, C.N. Bianchi, C. Cima, S. Merello, E. Massaro, R. Simoni, S. Queirolo.

Financial and logistic support for the meeting was provided by University of Genoa, SIBM, Italian Ministry of Instruction, Univer-

Guest editors: G. Relini & J. Ryland
Biodiversity in Enclosed Seas and Artificial Marine Habitats

G. Relini (✉)
Laboratori di Biologia Marina ed Ecologia Animale, DIP.TE.RIS., Università di Genova, Corso Europa 26, 16132 Genova, Italy
e-mail: sibmzool@unige.it

J. Ryland
Emeritus Professor of Marine Biology, Biological Sciences, Wallace Building, University of Wales Swansea, Swansea SA2 8PP Wales, UK

sity and Research, ENEA, Fondazione Carige, Consorzio Nazionale Interuniversitario per le Scienze del Mare, Consiglio Nazionale Ricerche, Acquario di Genova—Costa Edutainment S.p.A., Erredi Grafiche Editoriali, Porto Antico S.p.A.

Many thanks to all sponsors, whose help contributed greatly to the success of the meeting.

After a long and difficult refereeing process, 43 of the 59 submitted papers were selected. Prof. John Ryland then carried out necessary revision of the texts as revised by authors following the referees' suggestions.

Each manuscript was reviewed by at least two internationally renowned scientists. It is not possible here to acknowledge individually the 135 referees involved, but we believe they contributed to enhancing the quality of papers. We would like to thank all colleagues who gave their time freely for referring manuscripts. We would like to thank also the authors, the editor-in-chief, and the publisher for the excellent cooperation throughout the numerous hours of work necessary to complete the publication of these proceedings.

A special word of recognition to Rossana Simoni and Maria Lombardo for their substantial help during the correspondence with referees and authors and in editing the manuscripts.

We have striven to maintain the high standard of previous EMBS Symposium volumes and hope that this volume will contribute to the further advancement of marine biology and bring back to the attendees of the 39th Symposium many happy memories of their stay in Genoa, in spite of a very hot July.

List of participants

Australia
Clynick, Brianna
Jelbart, Jane
Schlacher, Thomas
Schreider, Maria
Austria
Ott, Joerg
Chile
Duarte, Cristian
Jaramillo, Eduardo
Croatia
Travizi, Ana

continued

Cuba
Ramos Lachaise, Vladimir Alexander
Estonia
Lauringson, Velda
Martin, Georg
Möller, Tiia
Paalme, Tiina
Põllumäe, Arno
Finland
Bonsdorff, Erik
Nappu, Niko
Nordstrom, Marie
Ruuskanen, Ari
France
Braga De Mendonça Jr, Joel
Contino, Frédéric
Dupont, Lise
Maire, Olivier
Martin, Sophie
Richard, Joëlle
Santini, Francesco
Simon-Bouhet, Benoit
Susini, Marie-Lucie
Viard, Frédérique
Germany
Dannheim, Jennifer
Günther, Carmen-Pia
Kossak, Ute
Lenz, Mark
Molis, Markus
Reuter, Penpag
Rohde, Sven
Saborowski, Reinhard
Schatte, Jessica
Suck, Inken
Volkenborn, Nils
Wahl, Martin
Greece
Akoumianaki, Ioanna
Megalofonou, Persefoni
Nicolaidou, Artemis
Iran
Hajimoradloo, Abdolmajid
Israel
Benayahu, Yehuda
Perkol-Finkel, Shimrit
Italy
Abbiati, Marco
Airoldi, Laura
Aliani, Stefano
Amato, Alberto
Andaloro, Franco
Azzini, Francesca
Azzurro, Ernesto
Babbini, Lorenza
Badalamenti, Fabio
Balestri, Elena
Balzano, Raffaella

continued
Battaglia, Bruno
Beccornia, Eugenio
Belluscio, Andrea
Belmonte, Genuario
Beltrano, Anna Maria
Benedetti-Cecchi, Lisandro
Benfante, Mariagrazia
Bianchi, Carlo Nike
Blasutto, Oriana
Boero, Ferdinando
Boni, Laurita
Bressan, Guido
Brugnano, Cinzia
Bucci, Arianna
Cabiddu, Serenella
Canese, Simonepietro
Canestri Trotti, Giorgio
Capezzuto, Francesca
Caroppo, Carmela
Caruso, Tancredi
Castelli, Alberto
Cavallo, Rosa Anna
Cerrano, Carlo
Ceschia, Carlo
Chemello, Renato
Chessa, Lorenzo
Cibic, Tamara
Cilli, Elisabetta
Cima, Chantal
Cocito, Silvia
Corinaldesi, Cinzia
Coscia, Ilaria
Costantini, Federica
Curiel, Daniele
Cuttitta, Angela
Dalessandro, Santa
D'Anna, Giovanni
Danovaro, Roberto
De Luca, Massimo
Di Capua, Iole
Di Franco, Antonio
Di Nieri, Antonella
Di Stefano, Floriana
Faimali, Marco
Falace, Annalisa
Fanelli, Emanuela
Fauci, Anna
Ferretti, Cristina
Figus, Vincenza
Fiorentino, Fabio
Flagella, Maria Monia
Follesa, Maria Cristina
Fontani, Sonia
Frangipane, Gretel
Franzitta, Giulio
Galli, Paolo
Gallini, Alessandra
Gallizia, Ilaria
Gallo D'Addabbo, Maria

continued
Gambi, Maria Cristina
Garaventa, Francesca
Garibaldi, Fulvio
Garofalo, Germana
Giacalone, Vincenzo Maximiliano
Giacobbe, Mariagrazia
Giangrande, Adriana
Gianguzza, Paola
Giove, Agnese
Graziano, Mariagrazia
Greco, Silvestro
Irrera, Pia
Iuri, Vanessa
Lanteri, Luca
Lardicci, Claudio
Lattanzi, Loretta
Ledda, Fabio D.
Licandro, Priscilla
Ligas, Alessandro
Lipizer, Marina
Maggiore, Francesca
Manconi, Renata
Mangialajo, Luisa
Mannini, Alessandro
Mannino, Anna Maria
Marano, Giovanni
Marzialetti, Sara
Massaro, Elisabetta
Mastrototaro, Francesco
Masullo, Piero
Matarrese, Alfonso
Merello, Stefania
Miglietta, Annamaria
Milanese, Martina
Milazzo, Antonino
Mingazzini, Marina
Mo, Giulia
Montefalcone, Monica
Moreno, Mariapaola
Morizzo, Gaia
Morri, Carla
Mostarda, Edoardo
Mura, Marco
Nicoletti, Luisa
Occhipinti-Ambrogi, Anna
Orsi Relini, Lidia
Palandri, Giovanni
Pannacciulli, Federica G.
Pansini, Maurizio
Papetti, Chiara
Paravagna, Tatiana
Pasolini, Paola
Passaro, Romina
Patti, Carlo
Patti, Francesco Paolo
Peirano, Andrea
Penna, Antonella
Percopo, Isabella
Pesci, Paola

continued

Pessa, Giuseppe
Picollo, Andrea
Pinsino, Annalisa
Pipitone, Carlo
Prato, Ermelinda
Pronzato, Roberto
Puccio, Valentina
Pusceddu, Antonio
Queirolo, Sara
Ragonese, Sergio
Relini, Giulio
Relini, Marco
Riccardi, Manuela
Riggio, Silvano
Rismondo, Andrea
Rollandi, Lorenzo
Ruggiero, Emma
Russo, Giovanni
Sabatini, Andrea
Saggiomo, Maria
Sandulli, Roberto
Santangelo, Giovanni
Sarà, Antonio
Scardi, Michele
Serena, Fabrizio
Sidri, Marzia
Simoni, Rossana
Sinopoli, Mauro
Stabili, Loredana
Tagliapietra, Davide
Tanzarella, Simona
Tinti, Fausto
Tobbia, Valeria
Tunesi, Leonardo
Tursi, Angelo
Vallisneri, Maria
Vasapolli, Eleonora
Vaselli, Stefano
Vezzulli, Luigi
Vitale, Sergio
Volpi Ghirardini, Annamaria
Zanelli, Elisa
Zanon, Veronica
Lebanon
Bariche, Michel
Mexico
Ardisson, Pedro
Ferrara-Guerrero, Maria de Jesús
Signoret, Gisele
Signoret, Martha
Solis-Weiss, Vivianne
Norway
Ellingsen, Kari Elsa
Gray, John
Norderhaug, Kjell Magnus
Poland

continued

Bielecka, Luiza
Dzierzbicka-Glowacka, Lidia
Kosakowska, Alicja
Zmijewska, Maria Iwona
Portugal
Boaventura, Diana
Campos, Joana
Carvalho, Susana
Moura, Ana
Queiroga, Henrique
Ramos, Jorge H.
Santos, Miguel
Silva, Ana Catarina Ferriera
Spain
Abdulla, Ameer
Alonso Garcia, Carolina
Barcala-Bellod, Elena
Blanco Lizana, Gloria
Mancini, Agnese
Sabah Mazzetta, Sandra Carol
Sánchez Prado, José Antonio
Visauta, Eva
Sweden
Baden, Susanne P.
Eriksson, Susanne P.
Pihl, Leif
Wennhage, Hakan
The Netherlands
Hummel, Herman
Rossi, Francesca
United Kingdom
Beaumont, Jenny
Bell, James John
Dando, Margaret Ann
Dando, Paul
Davis, Martin
Gordon, John
Hughes, Adam
Ryland, Christine
Ryland, John S.
Sayer, Martin
Shelmerdine, Richard
Wilding, Thomas
Ukraine
Shulman, Georgiy
Tokarev, Yuriy
United States of America
Arena, Paul
Bailey, William
Bartholomew, Joy A.
Freeland, Rebecca
Quinn, T. Patrick
Schaffner, Linda
Seaman, William
Sherman, Robin
Spieler, Richard

Sponsors

Ministero dell'Istruzione, dell'Università e della Ricerca
ENEA
Fondazione Carige
CoNISMA, Consorzio Nazionale Interuniversitario per le Scienze del Mare
CNR, Consiglio Nazionale Ricerche
Acquario di Genova—Costa Edutainment S.p.A.
Erredi Grafiche Editoriali S.n.c.
Porto Antico di Genova S.p.A.

Hydrobiologia (2007) 580:7–21
DOI 10.1007/s10750-006-0469-5

BIODIVERSITY IN ENCLOSED SEAS

Biodiversity issues for the forthcoming tropical Mediterranean Sea

Carlo Nike Bianchi

Abstract Present-day Mediterranean marine biodiversity is undergoing rapid alteration. Because of the increased occurrence of warm-water biota, it has been said that the Mediterranean is under a process of 'tropicalization'. This paper analyses the main patterns of the Mediterranean Sea tropicalization and considers briefly its extent and consequences. As happened during previous interglacial phases of the Quaternary, Atlantic water, entering via the Straits of Gibraltar, carries into the Mediterranean species that are prevalently of (sub)tropical affinity. On the other side of the basin, Red Sea species penetrate through the Suez Canal, a phenomenon called lessepsian migration from the name of F. de Lesseps, the French engineer who promoted the cutting of the Canal. Also the many exotic species introduced by humans voluntarily or involuntarily are nearly always typical of warm waters. Climate change combines with Atlantic influx, lessepsian migration and the introduction of exotic species by humans to the establishment of tropical marine biota in the Mediterranean Sea. Present-day warming ultimately favours the spread of warm-water species through direct and indirect effects, and especially by changing water circulation. It is impossible at present to foresee to what extent the exuberance of warm-water species will affect the trophic web and the functioning of marine ecosystems in the Mediterranean Sea of tomorrow. While Mediterranean Sea communities are modifying their pattern of species composition, they do not seem to be acquiring a more marked tropical physiognomy: Mediterranean coastal marine ecosystems are still dominated by frondose algae (even if the species that are gaining ascendancy are of tropical origin) and not by corals as is normal in tropical seas.

Keywords Marine biodiversity · Marine biogeography · Climate change · Species distribution · Range extension · Mediterranean Sea

Guest editors: G. Relini & J. Ryland
Biodiversity in Enclosed Seas and Artificial Marine Habitats

C. N. Bianchi (✉)
DipTeRis, Dipartimento per lo studio del Territorio e delle sue Risorse, Università di Genova Corso Europa, 26 I-16132 Genova, Italy
e-mail: nbianchi@dipteris.unige.it

Introduction

The status of Mediterranean Sea biodiversity has been reviewed by Bianchi & Morri (2000) on the basis of information collected mostly in the mid 1990s (Bianchi, 1996). Their review was organised around six main points: (1) how many species are there in the Mediterranean Sea; (2) origins and causes of Mediterranean biodiversity; (3)

biodiversity and climate change; (4) the footprint of man; (5) protecting marine biodiversity; (6) the role of scientific research. Nearly ten years later all of these points need updating, as renewed research on Mediterranean Sea biodiversity has provided a wealth of new information on all of them. For instance, the number of macroscopic marine species inhabiting the Mediterranean, then reckoned as 8,565, is today estimated at about 12,000 (Boudouresque, 2004) and further increments may be expected by the study of 'inconspicuous' taxa or undersampled habitats such as submarine caves or the depths (Bianchi & Morri, 2002). The origins of the Mediterranean Sea biodiversity have been reanalysed by Taviani (2002), Boudouresque (2004), and Emig & Geistdoerfer (2004). A deficiency lamented by Bianchi & Morri (2000) was the derisory extent of marine protected areas (MPAs) in the Mediterranean Sea. Fortunately things have started to change, as numerous MPAs have been established in the last few years (Carrada et al., 2003); among them is the first international off-shore MPA in the world: the so-called 'whale sanctuary' of the Ligurian Sea (Diviacco, 2002).

In this paper, I will try to update knowledge about another point touched by Bianchi & Morri (2000): the rate of change that Mediterranean Sea biodiversity is presently facing under the action of climate and humans. While climate variation is apparently modifying the distribution patterns of Mediterranean Sea biodiversity (Bianchi & Morri, 1993, 1994; Francour et al., 1994; Bianchi, 1997; Morri & Bianchi, 2001; Bianchi & Morri, 2004a, 2004b), humans are altering the composition of Mediterranean marine biota by the introduction of exotic species (Zibrowius, 1991; Boudouresque & Ribeira, 1994; Occhipinti-Ambrogi, 2001; Occhipinti-Ambrogi & Savini, 2003; Streftaris et al., 2005). Since introduced species are nearly always typical of warm waters, anthropogenic and climatic actions combine to allow for an increased abundance and distribution of (sub)tropical species in the warm-temperate Mediterranean Sea. It is therefore often said that the Mediterranean Sea is heading towards a generalised phenomenon of 'tropicalization' (Bianchi & Morri, 2003). For example, out of 90 exotic fish species that entered the Mediterranean Sea in recent years, only three are boreal, all the others are tropical (Golani et al., 2002). I am inclined, however, to include in the phenomenon of Mediterranean tropicalization also the northward spread of Mediterranean indigenous species with (sub)tropical affinities that were confined in the southern parts of the basin until recently (Bianchi & Morri, 1993, 1994, 2004b). I believe, in fact, that the patterns of distribution change will be similar for all warm-water species, whether recently introduced in or native to the Mediterranean, and will be governed by the same climatic, hydrological and ecological factors.

This paper aims to analyse the main patterns of the Mediterranean Sea tropicalization and to consider briefly its extent and consequences. I will concentrate on the coastal benthos, for which I have more information, but the same phenomenon is observed in the pelagic realm (see Gómez & Claustre, 2003, for an example). Information on Mediterranean deep-water biota is still too scarce (Bellan-Santini et al., 1992; Emig & Geistdoerfer, 2004) for proper consideration.

The driving factors

The occurrence and spread of warm-water species in the Mediterranean Sea results from the action of four distinct causes, namely: Atlantic influx, lessepsian migration, introductions by humans, and present-day sea warming. The first is a natural cause, while the second and the third are clearly anthropogenic; the fourth may be considered natural only in part, as we are well aware that humans play a major role in planetary warming. These four causes act on very different time scales, but all have apparently accelerated in the last two decades or so.

The time scale of Atlantic influx is of the order of 10^4 years, i.e., since the beginning of the last interglacial. As happened during the Quaternary, Atlantic water, entering through the Straits of Gibraltar, carries into the Mediterranean species of prevalently (sub)tropical affinity (Bianchi et al., 2002). Many of these species originally established themselves exclusively in areas close to Straits of Gibraltar but, especially in recent years, some have penetrated farther east, reaching for instance the coasts of Sicily, i.e., the region that is traditionally taken as the boundary between western and eastern Mediterranean basins (but see below). A recent and well documented example is the crab *Percnon gibbesi* (H. Milne Edwards), of western tropical Atlantic origin (Relini et al., 2000; Pipitone et al., 2001; Mori & Vacchi, 2002). In addition to Sicily, this species is now found at the Pontine Islands, in the central Tyrrhenian Sea (Russo & Villani, 2004), in southern Sardinia (P. Panzalis, personal communication) and at Capo Rizzuto, on the Ionian coast of Calabria (I. Faccia, personal communication).

The phrase lessepsian migration was coined from the name of Ferdinand de Lesseps, the French engineer and diplomat who promoted the cutting of the Suez Canal, and was adopted to indicate the penetration of Red Sea species into the Mediterranean (Por, 1978). The time scale of lessepsian migration is of the order of 10^2 years, as it started soon after the opening of the Canal in 1869. However, it remained inconspicuous until the 1970s, when the penetration of lessepsian migrants increased because of the progressive reduction of salinity of the Bitter Lakes and the diminished outflow of the Nile at the northern end of the Canal, caused by the building of the Aswan dam. The spreading into the eastern Mediterranean of stenohaline Red Sea species was therefore facilitated (Galil, 1993). For a long period, the vast majority of these lessepsian migrants remained confined to the Levant Sea, where they now shape the coastal communities (Fishelson, 2000). However, many of them have now penetrated into the western Mediterranean (Galil et al., 2002; Golani et al., 2002; Ribera Siguan, 2002; Zenetos et al., 2003).

Lessepsian migration apart, the introductions of exotic species by humans have acted on a time scale of 10^3 years (Giaccone, 2002): well before the Christian era, Greek sailors travelled perhaps as far as Iceland, the Phoenicians circumnavigated Africa, and Punic merchants possibly reached Macaronesia, Brazil and the Maldives (Bianchi & Morri, 2000). We cannot say how many species the wooden ships of ancient times would carry with them. However, it seems indisputable that the amplitude of this phenomenon greatly increased since the World War II (Boudouresque & Ribeira, 1994). Species are being intentionally or accidentally introduced via ship fouling, ballast waters, aquaculture, trade in live bait, wrapping of fresh seafood with living algae, aquariology, and even scientific research (Bianchi & Morri, 2000).

Finally, sea warming has a time scale of 10^1 years: despite large cyclic fluctuations, a positive trend in Mediterranean temperatures is clearly seen after the mid 1980s: its effects include the northward extension of the range of warm-water species within the Mediterranean Sea (Bianchi & Morri, 1994; Astraldi et al., 1995; Bianchi, 1997, Vacchi et al., 2001).

A scope for biogeography

Biogeography is the study of the spatial and spatio-temporal patterns of biodiversity (Zunino & Zullini, 1995). Although long neglected in the recent past (Bianchi & Morri, 2000), Mediterranean marine biogeography is at present enjoying a certain revival (Koukouras et al., 2001; Arvanitidis et al., 2002; Bianchi & Morri, 2002; Baccetti, 2003; Harmelin, 2004; Logan et al., 2004). Tropicalization is said to be changing the pattern of Mediterranean Sea biodiversity, and changes in species distribution should be particularly obvious in those transitional areas that are close to biogeographic boundaries, i. e., at the limits of regions inhabited by a different biota (Bianchi & Morri, 2004b).

The Mediterranean Sea as a whole constitutes a distinctive province of the Atlantic-Mediterranean warm-temperate region. However, the Mediterranean is far from being homogenous biogeographically: its tormented geological history and the present-day variety of climatic and hydrologic situations that are found in the different areas of the basin have traditionally led to the recognition of ten distinct biogeographic sectors (Bianchi & Morri, 2000: 370, Fig. 2). Thanks to new knowledge (including the considerations below), I am now inclined to recognise at least two additional sectors (Bianchi, 2004), to distinguish the southern Tyrrhenian Sea from the Balearic-Sardinia area and the Ionian Sea from the Aegean Sea (Fig. 1). A thirteenth sector may well be recognised in the Straits of Messina: although small, this area harbours a wealth of biogeographic peculiarities, including Pliocene Atlantic remnants and local endemisms (Fredj & Giaccone, 1995).

While the core-zones of these biogeographic sectors are easily identifiable, tracing their boundaries on a map is difficult. A major boundary, often mentioned in Mediterranean literature, is that between the western and eastern basins of the Mediterranean Sea. In their highly influential '*Nouveau manuel*', Pérès & Picard (1964) placed this boundary somewhere in the mid Ionian Sea, thus including the whole Sicily, Calabria and the Gulf of Taranto in the western Mediterranean (Fig. 1, line i). Sarà (1968) expressed doubts about the placement of the Ionian coast of Calabria and the Gulf of Taranto in the western Mediterranean but nevertheless adopted the same scheme, which was later popularised in university text books (Cognetti & Sarà, 1974; Cognetti et al., 1999). The existence of a mid-Ionian boundary found confirmation in a recent study of the biogeography of Mediterranean Proseriata (a group of tiny interstitial flatworms supposedly provided with low dispersal capacity): species assemblages from the eastern Ionian Sea turned out more similar to those of other eastern Mediterranean localities, whereas assemblages from the western Ionian Sea grouped together with those from western Mediterranean localities (Curini-Galletti & Casu, 2003). A different picture was suggested by Giaccone & Sortino (1974) who, working on the algal flora, established the boundary between the western and the eastern Mediterranean in the middle of the Straits of Sicily, so that the island of Pantelleria should belong to the western Mediterranean, while the Pelagie islands and Malta should belong to the eastern Mediterranean (Fig. 1, line ii). Bianchi & Morri (2000) picked up the idea of excluding the Ionian coast of Calabria and the Gulf of Taranto from the western Mediterranean but still included in it the whole of Sicily, putting the Pelagie Islands in the eastern Mediterranean and leaving Malta on the border (Fig. 1, line iii). Despite minor differences, these last views agree in

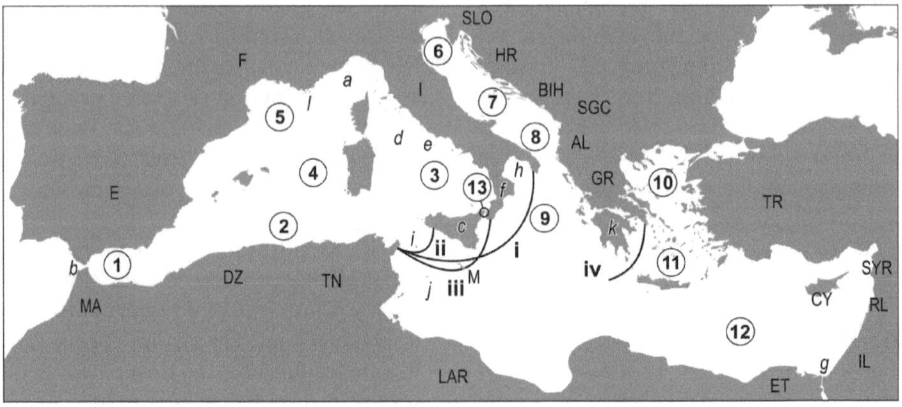

Fig. 1 Major biogeographic sectors within the Mediterranean Sea: (**1**) Alboran Sea; (**2**) Algeria and north Tunisia coasts; (**3**) southern Tyrrhenian Sea; (**4**) Balearic Sea to Sardinia Sea; (**5**) Gulf of Lions and Ligurian Sea; (**6**) northern Adriatic Sea; (**7**) central Adriatic Sea; (**8**) southern Adriatic Sea; (**9**) Ionian Sea; (**10**) northern Aegean Sea; (**11**) southern Aegean Sea; (**12**) Levant Sea; (**13**) Straits of Messina. Position of the boundary between the western and eastern Mediterranean according to different authors (see text): **i** Pérès & Picard (1964); **ii** Giaccone & Sortino (1974); **iii** Bianchi & Morri (2000); **iv** Costagliola et al. (2004). Mediterranean countries (clockwise): E Spain; F France; I Italy; SLO Slovenia; HR Croatia; BIH Bosnia-Herzegovina; SGC Serbia-Montenegro; AL Albania; GR Greece; TR Turkey; CY Cyprus; SYR Syria; RL Lebanon; IL Israel; ET Egypt; LAR Libya; M Malta; TN Tunisia; DZ Algeria; MA Morocco. Some additional localities mentioned in the text are also indicated: *a* Ligurian Sea; *b* Gibraltar Straits; *c* Sicily; *d* Tyrrhenian Sea; *e* Pontine Islands; *f* Calabria; *g* Suez Canal; *h* Gulf of Taranto; *i* Pantelleria Island; *j* Pelagie Islands; *k* Peloponnese; *l* Port-Cros Island

considering the area around the Straits of Sicily as the boundary zone between the western and eastern Mediterranean. This seems reasonable also in the light of geology and patterns of water circulation (Bianchi et al. 2002; Pinardi & Masetti, 2000). A radical departure from this established scheme was offered by population genetic studies on fish and invertebrates (Costagliola et al., 2004 and references therein), which revealed a strong genetic break between the Peloponnese and the Aegean and not at the Strait of Sicily (Fig. 1, line iv). However, in a recent population genetic study on the endemic seagrass *Posidonia oceanica* (L.) Delile, samples from the Gulf of Taranto showed greater genetic similarity to those from the Aegean Sea than to those from the Tyrrhenian Sea, thus supporting the traditional view that the Straits of Sicily represents the major biogeographic barrier separating western and eastern Mediterranean biota (Micheli et al., 2005).

Physical versus physiological barriers

The contrasting views about the position of the western/eastern Mediterranean boundary are of particular interest in considering the relatively recent colonisation of the Mediterranean from the Atlantic, after near extinction of the Mediterranean biota in the Messinian period, approximately 5.6 million years ago (Bianchi & Morri, 2000). It has been said that the Straits of Sicily acted as a filter to the recolonisation of the eastern Mediterranean, so that species richness should be lower there than in the western basin. This seems to be true for the Levant Sea: Taviani (2002) called it a 'Godot' basin, i. e., a basin waiting for Atlantic colonisers that were not arriving (colonisers are coming from the Red Sea, now!), but does not hold for the Aegean Sea, where recent research showed that its species richness is comparable to that of the western Mediterranean (Zenetos, 1997; Morri et al., 1999; Logan et al., 2002).

Mapping the surface isotherms of the Mediterranean Sea, averaged over a century of records and therefore representing the climatology of the basin (Brasseur et al., 1996), shows that the isotherm of 15°C for February (the coldest month in the year) crosses the Straits of Sicily, splits the Ionian Sea into a north-western and a south-eastern part, and finally separates the Peloponnese from the Aegean Sea (Fig. 2). In other words, the February 15°C surface isotherm follows quite closely all the biogeographic boundaries between the western and eastern Mediterranean proposed in turn! If temperature matters, that may be why the Aegean Sea biota is more similar to that of the western Mediterra-

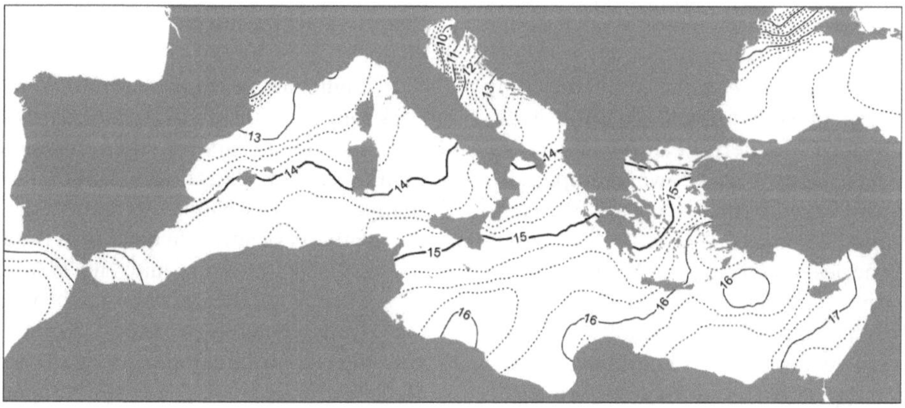

Fig. 2 Surface isotherms of February (traced every 0.25°C) of the Mediterranean Sea (climatological means from the historical data set 1906–1995). The 14°C and the 15°C 'divides' (see text) are highlighted by a thicker tract. Modified after MEDATLAS (Brasseur et al., 1996)

nean (both basins laying mostly to the north of the February 15°C surface isotherm) than to that of the Levant Sea (which remains to the south of that isotherm).

I am therefore tempted to suspect that the biotic differences between western and eastern Mediterranean are due more to differences in temperature regime, i.e., a physiological barrier, than to the physical barrier of the Straits of Sicily. This may have profound implications: sea warming will easily move isotherms but cannot change the morphology of straits. What evidence can species distribution provide with this respect?

If recent introductions (the already mentioned Atlantic and lessepsian migrants) are excluded, there are not many examples of well-known species occurring exclusively in only one of the two basins (Bianchi et al., 2002). One such example is found in the genus *Charonia*, one of the biggest Mediterranean gastropods, which is represented by *C. lampas lampas* (L.) in the western basin and by *C. tritonis variegata* (Lamarck) in the eastern basin; the two species meet in the Straits of Sicily (Russo et al., 1990). However, both species exist in the Atlantic Ocean, where the former exhibits a typical warm-temperate distribution (English Channel to West Africa) while the latter thrives in tropical and subtropical waters of both sides of the Atlantic. The February 15°C surface isotherm might well act as a divide between the ranges of two closely related species with distinct temperature requirements. Also the February 14°C surface isotherm may have a major biogeographic interest, as I will show later.

Charonia tritonis variegata is not the only species thriving in (sub)tropical Atlantic water and in the eastern Mediterranean but not in the area in between: other well-known examples are the ghost crab *Ocypode cursor* (L.) and the club-tipped anemone *Telmatactis cricoides* (Duchassaing). These species penetrated the Mediterranean during a warm interglacial period in the Quaternary and disappeared from the western basin when the climate got cooler. They have now a population 'trapped' in the warmer eastern basin, but sea-warming might join the Atlantic and Mediterranean populations again in the near future (Wirtz & Debelius, 2003).

Changing species ranges

A species once restricted to the eastern Mediterranean but now crossing the 15 °C divide is the parrotfish *Sparisoma cretense*: apart from two sightings in the Tyrrhenian Sea in summer 1991 (Bianchi & Morri, 1994), this species has become established since summer 2000 in the southern Adriatic (Guidetti & Boero, 2001, 2002).

The most striking example of a species that has recently expanded its range within the Mediterranean is the scleractinian coral *Astroides calycularis* (Pallas). This was the preferred example of a south-western Mediterranean species that 'does not go east' (Pérès & Picard, 1964; Zibrowius, 1980, 1983, 1995): its range before 1989 was confined to the south-western Mediterranean between the 15°C and 14°C divides (Fig. 3). It was explained that this species cannot go east because the life-span of its pelagic larva is too short to overtake the wide expanse of the Ionian Sea by means of the eastward flowing currents (Fig. 4). On the other hand, it could not cross this sea step by step along the coast: the northern coast would be too cool for this warm-water species, the southern coast is sandy and does not offer this rocky-bottom species place to settle (Pérès & Picard, 1964). Apparently, *A. calycularis* was not aware of these thoughtful explanations and went east anyway: it has recently been discovered along the coast of Croatia, in the Adriatic Sea (Kružić et al., 2002; Grubelić et al., 2004). It jumped at once over both the western/eastern Mediterranean boundary and the 14°C divide.

It might not be coincidental that the conspicuous range expansion of *A. calycularis* took place in the same years of the so-called Eastern Mediterranean Transient (EMT), a dramatic change in thermohaline circulation that involved the inversion of surface currents in the Ionian Sea (Briand, 2000). This may have provided *A. calycularis* with the carrier that brought it from the Straits of Sicily to the Adriatic (if so, the normal cyclonic circulation along the Ionian coast of Calabria and not the lower temperature was the obstacle to the eastward spreading of *A. calycularis*). It would be worth looking for this species along the Ionian coast of Calabria, but no

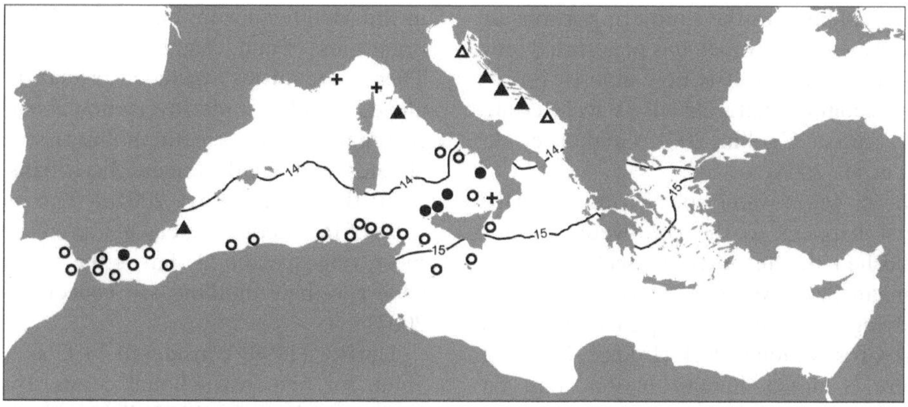

Fig. 3 Variations of the range of the scleractinian coral *Astroides calycularis* within the Mediterranean Sea. Open circles = confirmed distribution before 1980 (based on information in Zibrowius, 1980). Solid circles = later records within the known range (various sources). Open triangles = historical occurrences (1899 and 1904) not confirmed in recent decades (Grubelić et al., 2004). Solid triangles = recent records outside the known range: Cape Palos, Spain (Zibrowius, 1983); Giglio Island, Italy (Bianchi & Morri, 1994); islands and coast of Croatia (Kružić et al., 2002; Grubelić et al., 2004). + = Pleistocene fossil records (Zibrowius, 1995). The 14°C and 15°C 'divides' are also illustrated (see Fig. 2)

information is available at present. Changing surface circulation pattern in the Ionian Sea may also be invoked for the above-mentioned sudden appearance of the parrotfish *Sparisoma cretense* in the southern Adriatic. Galil & Kevrekidis (2002) attributed to the EMT the penetration of Indo-West Pacific crustacean species into the southeastern Aegean.

It is probable that the EMT has also influenced the western Mediterranean (Briand, 2000) but it cannot be said, at present, if the changes in species distribution observed in the latter basin hold any relationship with the transient. What can be said is that around 1990 many southern species moved northwards. The best studied case is that of the ornate wrasse *Thalassoma pavo* (L.), a

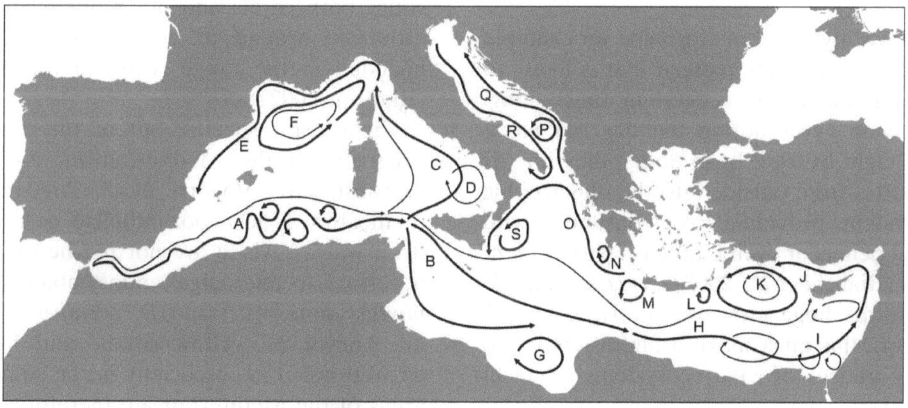

Fig. 4 A schematic summary of the major current and gyre systems of the Mediterranean Sea and their seasonal variability. Thick line = winter circulation; thin line = summer circulation. A: Algerian current and eddies; B: Branches of the Ionian stream; C: Tyrrhenian cyclonic current; D: summer antyciclone in the eastern Tyrrhenian Sea; E: Ligurian-Provençal current; F: Lions gyre; G: Syrte anticyclone; H: mid-Mediterranean jet; I: Shikmona and Mersa-Matruth gyres system; J: Cilician and Asia Minor current; K: Rhodes gyre; L: Iera-Petra gyre; M: western Cretan gyre; N: Pelops gyre; O: Ionian cyclonic current; P: southern Adriatic gyre; Q: eastern Adriatic coastal current; R: western Adriatic coastal current; S: western Ionian gyre. Modified after Pinardi & Masetti (2000)

species once confined to the southern portions of the Mediterranean Sea, which has penetrated into the Ligurian Sea, where it is now able to reproduce, thus becoming 'naturalised' (Vacchi et al., 1999, 2001; Sara & Ugolini, 2001; Guidetti et al., 2002; Sara et al., 2005).

If southern, warm-water species move northwards, what happens to the cold-water species long established in the northern sectors of the Mediterranean Sea? Are they at risk of extinction? Studying the marine decapod crustaceans of the Port-Cros National Park (France, Ligurian Sea), where a direct human action may be excluded, Noël (2003) related the increased rarity of the European lobster *Homarus gammarus* (L.), a northern species, to the increased abundance of the Mediterranean locust lobster *Scyllarides latus* (Latreille), a southern species. A well demonstrated case is that of two cave-dwelling mysids, the warm-water species *Hemimysis margalefi* Alcaraz, Riera and Gili and the cold-water species *H. speluncola* Ledoyer: in the submarine caves of the northern Mediterranean, the former is replacing the latter, which is therefore going extinct (Chevaldonné & Lejeusne, 2003).

Changing ecosystems

The two cave mysids above provide an example of change in biodiversity pattern that is likely to have great influence on ecosystem functioning. Mysids stay in caves during the day but move outside at night to feed. In so doing, they import organic matter from outside into the oligotrophic cave ecosystem, providing cave consumers with their faecal pellets or even falling prey to resident carnivores (Bianchi et al., 2003). As *H. speluncola* typically forms huge swarms and *H. margalefi* small groups, this must make a big difference to the energy budget of cave ecosystems (Bianchi et al., 1998).

Submarine caves are nothing but a very small portion of the Mediterranean Sea. Is there any indication that tropicalization is inducing major changes in marine ecosystems?

The most distinctive feature of tropical marine ecosystems are coral reefs. True reefs do not exist in the Mediterranean, but several Mediterranean organisms build significant bioconstructions. These organisms include primarily coralline algae, but also some invertebrates, such as scleractinian corals, vermetid molluscs, serpulid polychaetes and cheilostomate bryozoans (Bianchi, 1997; Bianchi & Morri, 2004b). The bioconstructions of the Mediterranean are monospecific or, at most, oligospecific, as far as the species responsible for their building are concerned (Bianchi, 2002).

Harriot (1999) considered 14°C as a threshold value for bioconstructional corals. In the Mediterranean Sea, the 14°C divide seems to represent the northern limit for the bioconstructional activity of the vermetid *Dendropoma petraeum* (Monterosato) (Antonioli et al., 1999) and of the scleractinian *Madracis pharensis* (Heller) (Morri et al., 2000a). During Quaternary phases warmer than at present, the bioconstructional activity of coralline algae in the Mediterranean Sea was more intense (Boudouresque et al., 1980; Sartoretto et al., 1996). Taken as a whole, these facts suggest that the present-day Mediterranean Sea represents a sort of hinge zone in space and time between a marine biota dominated by bioconstructors and one (nearly) deprived of them. The carbonate production by Mediterranean bioconstructors, taking into account both corals and other organisms, may be estimated around 10^3 $gCaCO_3$ $m^{-2}y^{-1}$, so being included in the range recorded for the tropics (Bianchi, 2002).

Eight coral species, out of the 37 presently occurring in the Mediterranean, are potential bioconstructors (Morri et al., 2000a). Five of them always lack zooxanthellae as they live in deep waters, which are not reached by the light necessary to microalgal endosymbionts. *Dendrophyllia ramea* (L.) and *D. cornigera* (Lamarck) are known as 'yellow corals' and live in the circalittoral zone, especially in the south-western areas of the Mediterranean, therefore showing a distribution typical of warm-water species. *Madrepora oculata* L., *Desmophyllum cristagalli* Milne Edwards and Haime and *Lophelia pertusa* (L.) are grouped under the name of 'white corals', live in the bathyal zone and have a strong affinity for cold waters. The three infralittoral species, obligatorily or facultatively zooxanthellate, are *Madracis pharensis*, *Oculina patagonica* De Angelis and *Cladocora caespitosa* (L.). *M. pharensis* occurs, without zooxanthellae, in submerged caves all over the Mediterranean, showing no significant bioconstruction capacity; however, in the south-eastern Mediterranean, beyond the 14°C divide, it may be found outside caves and with zooxanthellae, and in these situations it may play a significant bioconstructional role (Morri et al., 2000a). *O. patagonica*, probably originating from the south-western Atlantic, has been involuntarily introduced by humans to the Mediterranean: it is normally zooxanthellate and is able to build large colonies; although found also in the cold Ligurian Sea, it is especially abundant in western and eastern Mediterranean coasts to the south of the 14°C divide (Fine et al., 2001).

Cladocora caespitosa, the only species studied in some detail, is obligatorily zooxanthellate and may build banks more than one meter thick and several tens of meters wide (Morri et al., 1994, 2000b; Peirano et al., 1998, 2002; Kružić & Pozar-Domac, 2003). This species belongs to the family Faviidae, one of the most important in coral reef formation, and its calcification rates compare with those of tropical constructional corals (Rodolfo-Metalpa et al., 1999; Peirano et al., 2001). Growth of *Cladocora caespitosa* seems to be correlated with climate fluctuations (Morri et al., 2001; Rodolfo-Metalpa et al., 2002a, 2002b). Retrospective analysis, by X-radiography, on colonies older than 60 years, demonstrated that the highest growth rates coincided with the 'warm' period of the 1940s and the lowest with the 'cold' period of the 1970s (Peirano et al., 1999). This agrees with the palaeoecological information, indicating that *C. caespitosa* was more abundant—and its formations more conspicuous—during the warm periods of the Quaternary, and especially during the Tyrrhenian stage, when Mediterranean climate was subtropical (Peirano et al., 2004). It could therefore be supposed that, if the present sea-water warming continues, *Cladocora caespitosa* will play the role of constructional coral in a more and more 'tropical' Mediterranean Sea. In reality, in coincidence with positive anomalies of sea surface temperature recorded in these last few summers, this species underwent mass-mortality events, recalling those observed in the tropics (Rodolfo-Metalpa et al., 2000). Cases of mass-mortality correlated with high temperatures were recorded also in other organisms, such as sponges and gorgonians (Cerrano et al., 2000; Perez et al., 2000; Romano et al., 2000, Laubier, 2001; Garrabou et al., 2001, 2002; Sara et al., 2003).

Hypotheses, predictions and uncertainties

The tropicalization of the Mediterranean cannot be considered as a sort of improbable return of this sea to its ancient past of equatorial Mesozoic ocean, the Tethys. It is a completely new phenomenon that may rather be seen as the resultant of changes, not necessarily correlating among each other, induced by climate and human action.

Climate change combines with Atlantic influx, lessepsian migration and the introduction of exotic species by humans to favour the occurrence and establishment of warm-water species, whether exotic or native, in the Mediterranean Sea. While the latter three factors provide the 'raw material' (i. e., the warm-water species), the former is the 'mechanism' that ultimately favours the spread of these species through direct and indirect effects (Southward et al., 1995; Bianchi, 1997; Hiscock et al., 2004). Direct effects depends on temperature affecting survival rate, reproductive success and behaviour of organisms; indirect effects include those mediated by biotic interactions (e. g., conferral of competitive advantage to one of a pair of overlapping species, increased incidence of a parasite, or modified abundance of a predator) or by marine currents (climatic change may alter the emphasis of water flow and the pattern of water circulation, with great repercussions on the dispersal ability of marine organisms).

All these effects have been recognised in the Mediterranean biota (Morri & Bianchi, 2001; Chevaldonné & Lejeusne, 2003; Bianchi & Morri, 2004a). Perhaps, the change operated through marine currents has provided the most spectacular examples (Astraldi et al., 1995). In this paper, I am hypothesising that the surface current inversion in the Ionian Sea during the EMT is likely to have allowed species to cross the alleged boundary between western and eastern Mediterranean at the Straits of Sicily. If the hypothesis of a link between the EMT and the range extension of certain species would prove true, this might provide us with clues to previous occurrences of similar climatic events in the past. Records of *Astroides calycularis* in the Adriatic Sea at the turn of the XIX and XX centuries were later reputed erroneous by Zibrowius & Grieshaber (1977) but are now revalidated by Grubelić et al. (2004). Again at the turn of the XIX and XX centuries, *Thalassoma pavo* occurred in the Ligurian Sea (Vacchi et al., 1999). Was it a pure coincidence? Or should we read in that a proof that a climatic event similar to the recent EMT had happened a century earlier?

The two examples provided by the coral *Astroides calycularis* and the crab *Percnon gibbesi* show that warm-water species, whether they are native like the former, or exotic like the latter, have apparently followed the same route to cross the Straits of Sicily from west to east. On the other side, many lessepsian species have crossed the Straits of Sicily from east to west (see Pipitone et al., 2004, for a recent example). Thus, the traditional idea of a major biogeographic boundary between western and eastern Mediterranean should be abandoned in favour of a series of boundaries (or gradients, perhaps) in a south-north direction. Even the alleged difference in trophic status between eastern and western Mediterranean (the former being usually considered more oligotrophic than the latter) has to be questioned, as recent investigation in the south-western Mediterranean revealed a trophic regime similar to that of the eastern Mediterranean (D'Ortenzio, 2004). I am not therefore surprised that two recent thorough studies on within-Mediterranean distributions of sponges (Pansini & Longo, 2003) and pycnogonids (Chimenz-Gusso & Lattanzi, 2003) found higher faunal affinities between eastern and south-western Mediterranean localities than between north-western and south-western localities. I predict that while the southern portions of the Mediterranean will be more and more occupied by tropical exotic species, the northern portions will be invaded by warm-water native species that were once called 'southern'. Native cold-water species, typically confined to the northern portions of the basin, will probably rarefy and eventually be lost from the Mediterranean if sea-warming continues.

While there is no doubt that the biodiversity patterns are changing, it is impossible at present to foresee to what extent the exuberance of warm-water species in the Mediterranean Sea of tomorrow will affect the trophic web and, more in general, the functioning of marine ecosystems. The links between biodiversity and ecosystem functioning are elusive and perhaps insubstantial (Duarte, 2000; Price, 2001; Boero et al., 2004).

Tropical species are becoming more numerous in the Mediterranean Sea (Galil et al., 2002; Golani et al., 2002; Ribera Siguan, 2002; Zenetos et al., 2003), but the marine ecosystems do not seem yet to be acquiring a more marked tropical physiognomy. While the coastal seascape of tropical marine ecosystems is normally characterised by corals, the coastal seascape of the Mediterranean Sea is still dominated by frondose algae. Among these, however, the species that are gaining supremacy are introduced and exhibit a tropical affinity, such as *Stypopodium schimperi* (Buchinger ex Kützing) Verlaque and Boudouresque (Sartoni & De Biasi, 1999; Cocito et al., 2000) and the two species of *Caulerpa*, *C. taxifolia* (Vahl) C. Agardh (Meinesz et al., 2001) and *C. racemosa* (Forsskål) J. Agardh (Verlaque et al., 2000). Corals or other constructional organisms are not getting more abundant: on the contrary, the native constructional coral *Cladocora caespitosa* and other large invertebrates that 'shape' the submarine seascape of coastal Mediterranean ecosystems are perhaps going to face more frequent mass mortality events. The Mediterranean Sea biocoenoses might loose in the near future what have been called their 'peculiarities' (Bellan-Santini & Bellan, 2000) and acquire a different and unprecedented configuration and structure.

Acknowledgements This paper summarises the content of a plenary lecture at the 39th European Marine Biology Symposium. I wish to specially thank the President of the Organising Committee, Prof. Giulio Relini, for inviting me. C. Morri (Genoa) helped in many ways with the preparation of both the presentation and written contribution. K. Hiscock and A. Southward (Plymouth) kindly read an early version of the ms. N. Pinardi

(Bologna) and B. B. Manca (Trieste) drew my attention to the eastern Mediterranean climatic transient, and exchange of ideas with M. Astraldi and G. P. Gasparini (La Spezia) helped to let me thinking about its possible biological consequences. M. Sarà and the late E. Tortonese (Genoa) introduced me to marine biogeography, while continued correspondence with G. Giaccone (Catania), P. Wirtz (Madeira), P. Francour, A. Meinesz (Nice), C. F. Boudouresque, P. Chevaldonné, J. G. Harmelin, L. Laubier and H. Zibrowius (Marseille) was source of a wealth of information. I. Faccia (Isola Capo Rizzuto) and P. Panzalis (Carloforte) provided unpublished information. My research on the effects of climate change on Mediterranean marine ecosystems received financial support from the research projects SINAPSI (Seasonal, INterannual and decAdal variability of the atmosPhere, oceanS and related marIne ecosystems) and 'Ambiente Mediterraneo' (AdP MURST-CNR, L. 95/95).

References

Antonioli, F., R. Chemello, S. Improta & S. Riggio, 1999. *Dendropoma* lower intertidal reef formations and their palaeoclimatological significance, NW Sicily. Marine Geology 161: 155–170.

Arvanitidis, C., G. Bellan, P. Drakopoulos, V. Valavanis, C. Dounas, A. Koukouras & A. Eleftherious, 2002. Seascape biodiversity patterns along the Mediterranean and the Black Sea: lessons from the biogeography of the benthic polychaetes. Marine Ecology Progress Series 244: 139–152.

Astraldi, M., C. N. Bianchi, G. P. Gasparini & C. Morri, 1995. Climatic fluctuations, current variability and marine species distribution: a case study in the Ligurian Sea (north-west Mediterranean). Oceanologica Acta 18(2): 139–149.

Baccetti, B. (ed.), 2003. Marine biogeography of the Mediterranean Sea: patterns and dynamics of biodiversity. Part I. Biogeographia 24: 1–328.

Bellan-Santini, D. & G. Bellan, 2000. Distribution and peculiarities of Mediterranean marine biocoenoses. Biologia marina mediterranea 7(3): 67–80.

Bellan-Santini, D., G. Fredj & G. Bellan, 1992. Mise au point sur les connaissance concernant le benthos profond méditerranéen. Oebalia suppl 17: 21–36.

Bianchi, C. N., 1996. The state of marine biodiversity. In Carrabba, P., L. M. Padovani & F. Mauro (eds), International symposium on Mediterranean biodiversity. Ministero dell'Ambiente and Enea, Rome: 51–61.

Bianchi, C. N., 1997. Climate change and biological response in the marine benthos. Proceedings of the Italian Association for Oceanology and Limnology 12(1): 3–20.

Bianchi, C. N., 2002. Bioconstruction in marine ecosystems and Italian marine biology. Biologia marina mediterranea (2001) 8(1): 112–130.

Bianchi, C. N., 2004. Proposta di suddivisione dei mari italiani in settori biogeografici. Notiziario della Società Italiana di Biologia Marina 46: 57–59.

Bianchi, C. N., M. Abbiati, L. Airoldi, M. Alvisi, L. Benedetti-Cecchi, A. Cappelletti, F. Cinelli, P. Colantoni, P. R. Dando, C. Morri, I. Niccolai, P. Picco, A. Southward & E. Southward, 1998. Hydrology and water budget of a submarine cave with sulphur water spring: the Grotta Azzurra of Capo Palinuro (Southern Italy). Proceedings of the Italian Association for Oceanology and Limnology 12(2): 285–301.

Bianchi, C. N., F. Boero, S. Fraschetti & C. Morri, 2002. La fauna del Mediterraneo. In Argano, R., G. Chemini, S. La Posta, A. Minelli & S. Ruffo (eds), La fauna in Italia. Touring Club Italiano, Milan and Ministero dell'Ambiente e della Tutela del Territorio, Rome: 247–335.

Bianchi, C. N. & C. Morri, 1993. Range extension of warm-water species in the northern Mediterranean: evidence for climatic fluctuations? Porcupine Newsletter 5(7): 156–159.

Bianchi, C. N. & C. Morri, 1994. Southern species in the Ligurian Sea (northern Mediterranean): new records and a review. Bollettino dei Musei e degli Istituti biologici dell'Università di Genova (1992–1993) 58–59: 181–197.

Bianchi, C. N. & C. Morri, 2000. Marine biodiversity of the Mediterranean Sea: situation, problems and prospects for future research. Marine Pollution Bulletin 40(5): 367–376.

Bianchi, C. N. & C. Morri C., 2002. La biogeografia marina del Mediterraneo (XXXIV Congresso della Società Italiana di Biogeografia). Notiziario della Società Italiana di Biologia Marina 42: 78–79.

Bianchi, C. N. & C. Morri, 2003. Global sea warming and "tropicalization" of the Mediterranean Sea: biogeographic and ecological aspects. Biogeographia 24: 319–327.

Bianchi, C. N. & C. Morri, 2004a. Uomo, clima e biodiversità marina: esempi dal Mar Ligure. Uomo e Natura 9(1): 15–23.

Bianchi, C. N. & C. Morri, 2004b. Climate change and biological response in Mediterranean Sea ecosystems – a need for broad-scale and long-term research. Ocean Challenge (2003) 13(2): 32–36.

Bianchi, C. N., C. Morri & G. F. Russo, 2003. Deplezione trofica. In Cicogna, F., C. N. Bianchi, G. Ferrari & P. Forti (eds), Grotte marine: cinquant'anni di ricerca in Italia. Ministero dell'Ambiente e della Tutela del Territorio, Rome, 297–305.

Boero, F., G. Belmonte, S. Bussotti, G. Fanelli, S. Fraschetti, A. Giangrande, C. Gravili, P. Guidetti, A. Pati, S. Piraino, F. Rubino, O. D. Saracino, J. Schmich, A. Terlizzi & S. Geraci, 2004. From biodiversity and ecosystem functioning to the roots of ecological complexity. Ecological Complexity 1(2): 101–109.

Boudouresque, C. F., 2004. Marine biodiversity in the Mediterranean: status of species, populations and communities. Scientific Reports of Port-Cros National Park 20: 97–146.

Boudouresque, C. F. & M. A. Ribeira, 1994. Les introductions d'espèces végétales et animales en milieu marin: conséquences écologiques et économiques et problèmes législatifs. In Boudouresque, C. F., A. Meinesz & V. Gravez (eds), First international workshop on *Caulerpa taxifolia*. GIS Posidonie, Marseille, 29–103.

Boudouresque, C. F., J. Thommeret & Y. Thommeret, 1980. Sur la découverte d'un bioconcrétionnement fossile intercalé dans l'herbier à *Posidonia oceanica* de la Baie de Calvi (Corse). In Journées d'étude sur la systématique et la biogéographie de la mer Méditerranée, Cagliari. CIESM, Monaco, 139–142.

Brasseur, P., J. M. Beckers, J. M. Brankart & R. Schoenauen, 1996. Seasonal temperature and salinity fields in the Mediterranean Sea: climatological analyses of an historical data set. Deep Sea Research 42(2): 159–192.

Briand, F. (ed.), 2000. The Eastern Mediterranean climatic transient, its origin, evolution and impact on the ecosystem. CIESM, Monaco, Workshop Series n° 10, 86 pp.

Carrada, G. C., P. Coiro & G. F. Russo (eds), 2003. Le Aree Marine Protette del Mediterraneo. Centro Studi Tegea, Electa, Napoli, 174 pp.

Cerrano, C., G. Bavestrello, C. N. Bianchi, R. Cattaneo-Vietti, S. Bava, C. Morganti, C. Morri, P. Picco, G. Sara, S. Schiaparelli, A. Siccardi & F. Sponga, 2000. A catastrophic mass-mortality episode of gorgonians and other organisms in the Ligurian Sea (North-western Mediterranean), summer 1999. Ecology Letters 3: 284–293.

Chevaldonné, P. & C. Lejeusne, 2003. Regional warming-induced species shift in NW Mediterranean marine caves. Ecology Letters 6: 371–379.

Chimenz-Gusso, C. & L. Lattanzi, 2003. Mediterranean Pycnogonida: faunistic, taxonomical and zoogeographical considerations. Biogeographia 24: 251–262.

Cocito, S., C. N. Bianchi, C. Morri & A. Peirano, 2000. First survey of sessile communities on subtidal rocks in an area with hydrothermal vents: Milos Island, Aegean Sea. Hydrobiologia 426: 113–121.

Cognetti, G. & M. Sarà, 1974. Biologia marina. Calderini, Bologna, 439 pp.

Cognetti, G., M. Sarà & G. Magazzù, 1999. Biologia marina. Calderini, Bologna, 596 pp.

Costagliola, D., D. R. Robertson, P. Guidetti, S. Stefanni, P. Wirtz, J. B. Heiser & G. Bernardi, 2004. Evolution of coral reef fish *Thalassoma* spp. (Labridae). 2. Evolution of the eastern Atlantic species. Marine Biology 144: 377–383.

Curini-Galletti, M. & M. Casu, 2003. Species composition and biogeographical relationships of the Proseriata (Platyhelminthes) of the Mediterranean Sea. Biogeographia 24: 115–130.

Diviacco, G. (ed.), 2002. Il santuario dei cetacei: stato attuale e prospettive. Quaderni del Centro Studi sulle Aree protette marine e costiere 1: 1–134.

D'Ortenzio, F., 2004. Space and time occurrence of algal blooms in the Mediterranean: their significance for the trophic regime of the basin. PhD Thesis, Joint program Stazione Zoologica 'A. Dohrn' – Open University of London, 298 pp.

Duarte, C. M., 2000. Marine biodiversity and ecosystem services: an elusive link. Journal of experimental marine Biology and Ecology 250: 117–131.

Emig, C. C. & Geistdoerfer P., 2004. The Mediterranean deep-sea fauna: historical evolution, bathymetric variations and geographical changes. Carnets de Géologie 2004(1): 1–10.

Fine, M., H. Zibrowius & Y. Loya, 2001. *Oculina patagonica*: a non-lessepsian scleractinian coral invading the Mediterranean Sea. Marine Biology 138: 1195–1203.

Fishelson, L., 2000. Marine animal assemblages along the littoral of the Israeli Mediterranean seashore: the Red-Mediterranean Seas communities of species. Italian Journal of Zoology 67: 393–415.

Francour, P., C. F. Boudouresque, J. G. Harmelin, M. L. Harmelin-Vivien & J. P. Quignard, 1994. Are the Mediterranean waters becoming warmer? Information from biological indicators. Marine Pollution Bulletin 28(9): 523–526.

Fredj, G. & G. Giaccone, 1995. Particularités des peuplements benthiques du détroit de Messine. In Guglielmo, L., A. Manganaro & E. De Domenico (eds), The Strait of Messina ecosystem. Dipartimento di Biologia animale ed Ecologia marina, University of Messina, 119–128.

Galil, B. S., 1993. Lessepsian migration: new findings on the foremost anthropogenic change in the Levant basin fauna. In Della Croce, N. F. R. (ed.), Symposium Mediterranean Seas 2000. Istituto di Scienze Ambientali Marine, University of Genova, 307–323.

Galil, B. S. & K. Kevrekidis, 2002. Exotic decapods and stomatopods off Rhodes Island (Greece) and the Eastern Mediterranean Transient. Crustaceana 75(7): 925–930.

Galil, B., C. Froglia & P. Noel, 2002. CIESM Atlas of Exotic Species, 2 - Crustaceans. CIESM, Monaco, 192 pp.

Garrabou, J., L. Laubier & T. Perez, 2002. Sessile benthic invertebrates as indicators of climate change – mass mortalities in the north-west Mediterranean in summer 1999. Ocean Challenge 12(1): 17–23.

Garrabou, J., T. Perez, S. Sartoretto & J. G. Harmelin, 2001. Mass mortality event in red coral (*Corallium rubrum*, Cnidaria, Anthozoa, Octocorallia) populations in the Provence region (France, north-western Mediterranean). Marine Ecology Progress Series 217: 263–272.

Giaccone, G., 2002. Disseminazione antropica di organismi marini nel Mediterraneo durante gli ultimi 3000 anni. Bollettino dell'Accademia Gioenia di Scienze naturali di Catania 35(361): 95–108.

Giaccone, G. & M. Sortino, 1974. Significato biogeografico della vegetazione marina della Sicilia e delle isole minori nell'area del Mare Mediterraneo. Bollettino di Studi ed Informazioni del Giardino coloniale di Palermo 26: 130–146.

Golani, D., L. Orsi-Relini, E. Massutí & J. P. Quignard, 2002. CIESM Atlas of Exotic Species, 1 - Fishes. CIESM, Monaco, 256 pp.

Gómez, F. & H. Claustre, 2003. The genus *Asterodinium* (Dinophyceae) as a possible biological indicator of warming in the western Mediterranean Sea. Journal of the marine biological Association of the UK 83: 173–174.

Grubelić, I., B. Antolić, M. Despalatović, B. Grbec & G. B. Paklar, 2004. Effect of climatic fluctuations on the distribution of warm-water coral *Astroides calycularis* in the Adriatic Sea: new records and review. Journal of the marine biological Association of the UK 84: 599–602.

Guidetti, P., C. N. Bianchi, G. La Mesa, M. Modena, C. Morri, G. Sara & M. Vacchi, 2002. Abundance and size structure of *Thalassoma pavo* (Pisces: Labridae) in the western Mediterranean Sea: variability at different spatial scales. Journal of the marine biological Association of the UK 82: 495–500.

Guidetti, P. & F. Boero, 2001. The occurrence of the Mediterranean parrotfish *Sparisoma cretense* (L.) (Perciformes, Scaridae) in south-eastern Apulia (SE Italy). Journal of the marine biological Association of the UK 81: 717–719.

Guidetti, P. & F. Boero, 2002. Spatio-temporal variability in abundance of the parrotfish, *Sparisoma cretense*, in SE Apulia (SE Italy, Mediterranean Sea). Italian Journal of Zoology 69: 229–232.

Harmelin, J. G., 2004. Environnement thermique du benthos côtier de l'île de Port-Cros (parc national, France, Méditerranée nord-occidentale) et implications biogéographiques. Scientific Reports of Port-Cros National Park 20: 97–146.

Harriot, V. J., 1999. Coral growth in subtropical eastern Australia. Coral Reefs 18: 281–291.

Hiscock, K., A. Southward, I. Tittley & S. Hawkins, 2004. Effects of changing temperature on benthic marine life in Britain and Ireland. Aquatic Conservation: marine and freshwater Ecosystems 14: 333–362. .

Koukouras, A., E. Voultsiadou, M. S. Kitsos & S. Doulgeraki, 2001. Macrobenthic fauna diversity in the Aegean Sea, affinities with other Mediterranean regions and the Black Sea. Bios 6: 61–76.

Kružić, P. & A. Pozar-Domac, 2003. Banks of the coral *Cladocora caespitosa* (Anthozoa, Scleractinia) in the Adriatic Sea. Coral Reefs 22(4): 536.

Kružić, P., H. Zibrowius & A. Pozar-Domac, 2002. Actiniaria and Scleractinia (Cnidaria, Anthozoa) from the Adriatic Sea: first records, confirmed occurrences and significant range extensions of certain species. Italian Journal of Zoology 69: 345–353.

Laubier, L., 2001. Climatic changes and Mediterranean marine invertebrates: observation network and ecophysiology. Proceedings of the Italian Association for Oceanology and Limnology 14: 15–23.

Logan, A., C. N. Bianchi, C. Morri & H. Zibrowius, 2004. The present-day Mediterranean brachiopod fauna: diversity, life habits, biogeography and paleobiogeography. Scientia marina 68(Suppl. 1): 163–170.

Logan, A., C. N. Bianchi, C. Morri, H. Zibrowius & G. Bitar, 2002. New records of Recent brachiopods from the eastern Mediterranean Sea. Annali del Museo civico di Storia naturale G. Doria, Genova 94: 407–418.

Meinesz, A., T. Belsher, T. Thibaut, B. Antolic., K. Karim Ben Mustapha, C. F. Boudouresque, D. Chiaverini, F. Cinelli, J. M. Cottalorda, A. Djellouli, A. El Abed, C. Orestano, A. M. Grau, L. Ivesa, A. Jaklin, H. Langar, E. Massuti-Pascual, A. Peirano, L. Tunesi, J. de Vaugelas, N. Zavodnik & A. Zuljevic, 2001. The introduced green alga *Caulerpa taxifolia* continues to spread in the Mediterranean. Biological Invasions 5: 201–210.

Micheli, C., P. Paganin, A. Peirano, G. Caye, A. Meinesz & C. N. Bianchi, 2005. Genetic variability of *Posidonia oceanica* (L.) Delile in relation to local factors and biogeographic patterns. Aquatic Botany 82(3): 210–221.

Mori, M. & M. Vacchi, 2002. On a new occurrence of the alien flat crab, *Percnon gibbesi* (H. Milne Edwards), in the southern Sicily (central Mediterranean Sea) (Crustacea, Brachyura, Grapsidae). Annali del Museo civico di Storia naturale G. Doria, Genova 94: 295–301.

Morri, C. & C. N. Bianchi, 2001. Recent changes in biodiversity in the Ligurian Sea (NW Mediterranean): is there a climatic forcing? In Faranda, F. M., L. Guglielmo & G. Spezie (eds), Structure and processes in the Mediterranean ecosystems. Springer Verlag, Milan, 375–384.

Morri, C., C. N. Bianchi, S. Cocito, A. Peirano, A. M. De Biasi, S. Aliani, M. Pansini, M. Boyer, F. Ferdeghini, M. Pestarino & P. Dando, 1999. Biodiversity of marine sessile epifauna at an Aegean island subject to hydrothermal activity: Milos, Eastern Mediterranean Sea. Marine Biology 135(4): 729–739.

Morri, C., A. Peirano & C. N. Bianchi, 2001. Is the Mediterranean coral *Cladocora caespitosa* an indicator of climatic change? Archivio di Oceanografia e Limnologia 22: 139–144.

Morri, C., A. Peirano, C. N. Bianchi & R. Rodolfo-Metalpa, 2000b. *Cladocora caespitosa*: a colonial zooxanthellate Mediterranean coral showing constructional ability. Reef Encounter 27: 22–25.

Morri, C., A. Peirano, C. N. Bianchi & M. Sassarini, 1994. Present-day bioconstructions of the hard coral, *Cladocora caespitosa* (L.) (Anthozoa, Scleractinia), in the eastern Ligurian Sea (NW Mediterranean). Biologia marina mediterranea 1(1): 371–372.

Morri, C., D. Vafidis, A. Peirano, C. C. Chintiroglou & C. N. Bianchi, 2000a. Anthozoa from a subtidal hydrothermal area of Milos Island (Aegean Sea), with notes on the construction potential of the scleractinian coral *Madracis pharensis*. Italian Journal of Zoology 67: 319–325.

Noël, P., 2003. Les crustacés du Parc National de Port-Cros et de la région des îles d'Hyères (Méditerranée), France. État actuel des connaissances. Scientific Reports of Port-Cros National Park 19: 135–306.

Occhipinti-Ambrogi, A., 2001. Transfer of marine organisms: a challenge to the conservation of coastal biocoenoses. Aquatic Conservation: marine and freshwater Ecosystems 11(4): 243–251.

Occhipinti Ambrogi, A. & D. Savini, 2003. Bioinvasions as a component of global change in stressed marine ecosystem. Marine Pollution Bulletin 46(5): 542–551.

Pansini, M. & C. Longo, 2003. A review of the Mediterranean Sea sponge biogeography with, in appendix, a list of the demosponges hitherto recorded from this sea. Biogeographia 24: 251–262.

Peirano, A., C. Morri & C. N. Bianchi, 1999. Skeleton growth and density pattern of the temperate, zooxanthellate scleractinian *Cladocora caespitosa* from the Ligurian Sea (NW Mediterranean). Marine Ecology Progress Series 185: 195–201.

Peirano, A., C. Morri, C. N. Bianchi, J. Aguirre, F. Antonioli, G. Calzetta, L. Carobene, G. Mastronuzzi & P. Orrù, 2004. The Mediterranean coral *Cladocora caespitosa*: a proxy for past climate fluctuations? Global and planetary Change 40: 195–200.

Peirano, A., C. Morri, C. N. Bianchi & R. Rodolfo Metalpa, 2001. Biomass, carbonate standing stock and production of the Mediterranean coral *Cladocora caespitosa* (L.). Facies 44: 75–80.

Peirano, A., C. Morri, G. Mastronuzzi, C. N. Bianchi, 1998. The coral *Cladocora caespitosa* (Anthozoa, Scleractinia) as a bioherm builder in the Mediterranean Sea. Memorie descrittive della Carta geologica d'Italia (1994) 52: 59–74.

Peirano, A., R. Rodolfo Metalpa, C. Morri & C. N. Bianchi, 2002. The Mediterranean coral *Cladocora caespitosa* (Anthozoa, Scleractinia). Biologia marina mediterranea (2001) 8(1): 259–263.

Pérès, J. M. & J. Picard, 1964. Nouveau manuel de bionomie benthique de la mer Méditerranée. Recueil des Travaux de la Station marine d'Endoume 31(47): 1–137.

Perez, T., J. Garrabou, S. Sartoretto, J. G. Harmelin, P. Francour & J. Vacelet, 2000. Mortalités massives d'invertébrés marins: un événement sans précédent en Méditerranée nord-occidentale. Comptes-rendus de l'Académie de Sciences de Paris, Sciences de la Vie 323: 853–865.

Pinardi, N. & E. Masetti, 2000. Variability of the large scale general circulation of the Mediterranean Sea from observations and modelling: a review. Palaeogeography, Palaeoclimatology, Palaeoecology 158: 153–174.

Pipitone, C., F. Badalamenti F. & A. Sparrow, 2001. Contribution to the knowledge of *Percnon gibbesi* (Decapoda, Grapsidae), an exotic species spreading rapidly in Sicilian waters. Crustaceana 74: 1009–1017.

Pipitone, C., G. D'anna, M. Coppola, G. Di Stefano & F. Badalamenti, 2004. First record of the lessepsian fish *Fistularia commersonii* in the Western Mediterranean. Biologia marina mediterranea 11(3): 327.

Por, F. D., 1978. Lessepsian migrations: the influx of Red Sea biota into the Mediterranean by way of the Suez Canal. Ecological Studies, 23. Springer, Heidelberg, 228 pp.

Price, A. R. G., 2001. The marine food chain in relation to biodiversity. The Scientific World 1: 579–587.

Relini, M., L. Orsi, V. Puccio & E. Azzurro, 2000. The exotic crab *Percnon gibbesi* (H. Milne Edwards, 1853) (Decapoda, Grapsidae) in the Central Mediterranean. Scientia marina 64: 337–340.

Ribera Siguan, M. A., 2002. Review of non-native marine plants in the Mediterranean Sea. In Leppäkoski, E., S. Gollasch & S. Olenin (eds), Invasive aquatic species of Europe: distribution, impacts and management. Kluwer Academic Publishers, Dordrecht: 291–310.

Rodolfo-Metalpa, R., M. Abbate, C. N. Bianchi, A. Peirano, G. Cerrati & V. Di Fesca, 2002a. Cambiamenti climatici e ambiente marino: effetti della temperatura e della salinità sulla crescita in acquario del corallo *Cladocora caespitosa*. Enea, Roma, RT/2002/22/Clim, 27 pp.

Rodolfo Metalpa, R., M. Abbate, A. Peirano & C. N. Bianchi, 2002b. Crescita in acquario di coralli mediterranei. Energia, Ambiente e Innovazione 48(2): 86–87.

Rodolfo-Metalpa, R., C. N. Bianchi, A. Peirano & C. Morri, 2000. Coral mortality in NW Mediterranean. Coral Reef 19(1): 24.

Rodolfo-Metalpa, R., A. Peirano, C. Morri & C. N. Bianchi, 1999. Coral calcification rates in the Mediterranean scleractinian coral *Cladocora caespitosa* (L. 1767). Proceedings of the Italian Association for Oceanology and Limnology 13(1): 291–299.

Romano, J. C., N. Bensoussan, W. A. N. Younes & D. Arlhac, 2000. Anomalie thermique dans les eaux du golfe de Marseille durant l'été 1999. Une explication partielle de la mortalité d'invertébrés fixes? Comptes-rendus de l'Académie de Sciences de Paris, Sciences de la Vie 323: 415–427.

Russo, G. F., G. Fasulo, A. Toscano & F. Toscano, 1990. On the presence of triton species (*Charonia* spp.) (Mollusca Gastropoda) in the Mediterranean Sea: ecological considerations. Bollettino malacologico 26(5–9): 91–104.

Russo, G. F. & G. Villani, 2004. Spreading of the alloctonous species *Percnon gibbesi* in the central Tyrrhenian Sea (Decapoda, Grapsidae). Biologia marina mediterranea 11(3): 341.

Sara, G., C. N. Bianchi & C. Morri, 2005. Mating behaviour of the newly-established ornate wrasse *Thalassoma pavo* (Osteichthyes, Labridae) in the Ligurian Sea (north western Mediterranean). Journal of the marine biological Association of the UK 85: 191–196.

Sara, G., G. P. Gasparini, C. Morri, C. N. Bianchi & F. Cinelli, 2003. Reclutamento di gorgonie dopo un episodio di moria di massa in Mar Ligure orientale. Biologia marina mediterranea 10(2): 176–182.

Sara, G. & U. Ugolini, 2001. Recent naturalization of the warm-water wrasse *Thalassoma pavo* (Pisces, Osteichthyes) in the Ligurian Sea. Proceedings of the Italian Association for Oceanology and Limnology 14: 249–258.

Sarà, M., 1968. La biogeografia marina. In Enciclopedia della natura. G. Casini, Bologna, 709–742.

Sartoni, G. & A. M. De Biasi, 1999. A survey of the marine algae of Milos Island, Greece. Cryptogamie et Algologie 20: 271–283.

Sartoretto, S., M. Verlaque & J. Laborel, 1996. Age of settlement and accumulation rate of submarine "coralligène" (–10 to –60 m) of the northwestern Medi-

terranean Sea; relation to Holocene rise in sea level. Marine Geology 130: 317–331.
Southward, A. J., S. J. Hawkins & M. T. Burrows, 1995. Seventy years' observations of changes in distribution and abundance of zooplankton and intertidal organisms in the western English Channel in relation to rising sea temperature. Journal of thermal Biology 20(1/2): 127–155.
Streftaris, N., A. Zenetos & E. Papathanassiou, 2005. Globalisation in marine ecosystems: the story of non-indigenous marine species across European seas. Oceanography and Marine Biology: An Annual Review 43: 419–453.
Taviani, M., 2002. The Mediterranean benthos from late Miocene up to present: ten million years of dramatic climatic and geologic vicissitudes. Biologia marina mediterranea 9: 445–463.
Vacchi, M., C. Morri, M. Modena, G. La Mesa & C. N. Bianchi, 2001. Temperature changes and warm-water species in the Ligurian Sea: the case of the ornate wrasse *Thalassoma pavo* (Linnaeus, 1758). Archivio di Oceanografia e Limnologia 22: 149–154.
Vacchi, M., G. Sara, C. Morri, M. Modena, G. La Mesa, P. Guidetti & C. N. Bianchi, 1999. Dynamics of marine populations and climate change: lessons from a Mediterranean fish. Porcupine marine natural History Society Newsletter 3: 13–17.
Verlaque, M., C. F. Boudouresque, A. Meinesz & V. Gravez, 2000. The *Caulerpa racemosa* complex (Caulerpales, Ulvophyceae) in the Mediterranean Sea. Botanica marina 43: 49–68.
Wirtz, P. & H. Debelius, 2003. Mediterranean and Atlantic invertebrate guide. ConchBooks, Hackenheim, Germany, 305 pp.
Zenetos, A., 1997. Diversity of marine bivalvia in Greek waters: effects of geography and environments. Journal of the marine biological Association of the UK 77: 463–472.
Zenetos, A., S. Gofas, G. Russo & J. Templado, 2003. CIESM Atlas of Exotic Species, 3 - Molluscs. CIESM, Monaco, 376 pp.
Zibrowius, H., 1980. Les scléractiniaires de la Méditerranée et de l'Atlantique nordoccidental. Mémoires de l'Institut océanographique, Monaco 11: 1–284.
Zibrowius, H., 1983. Nouvelles données sur la distribution de quelques scléractiniaires méditerranéens à l'est et à l'ouest du détroit de Gibraltar. Rapports et procès-verbaux des réunions de la Commission international pour l'Exploration scientifique de la Mer Méditerranée 28(3): 307–309.
Zibrowius, H., 1991. Ongoing modification of the Mediterranean marine fauna and flora by the establishment of exotic species. Mésogée 51: 83–107.
Zibrowius, H., 1995. The "southern" *Astroides calycularis* in the Pleistocene of the northern Mediterranean – an indicator of climatic changes (Cnidaria, Scleractinia). Geobios 28(1): 9–16.
Zibrowius, H. & A. Grieshaber, 1977. Scléractiniaires de l'Adriatique. Tethys 4: 375–384.
Zunino, M. & A. Zullini, 1995. Biogeografia: la dimensione spaziale dell'evoluzione. Casa Editrice Ambrosiana, Milano, 310 pp.

Hydrobiologia (2007) 580:23–33
DOI 10.1007/s10750-006-0468-6

BIODIVERSITY IN ENCLOSED SEAS

Biodiversity in the Black Sea: effects of climate and anthropogenic factors

Yuriy Tokarev · Georgiy Shulman

Abstract The Black Sea ecosystem and diversity underwent dramatic adverse changes during the 1960s and, especially, 1970s and 1980s of the last century. Anthropogenically-induced eutrophication increased through greater biogenic flow, dumping and pollutant discharge, in turn causing red tides, fish kills and oxygen depletion over the northwestern shelf.

Anthropogenic pressures, associated with the economic situation of the Black Sea countries, has decreased during the last decade, allowing some improvement in the state and biodiversity of the ecosystem. The abundances of several native species have increased. However, mediterranization—the invasion by species from the adjacent basin and beyond—has continued.

The conclusion is grounded, that biodiversity is not only inter- and intra-species diversity but also spatial–temporal variability, abundance and productivity dynamics, differences of the metabolic strategies providing sustainable existence in the changing environment. Biodiversity at the intra-species level expresses itself in spatial and temporal variations of the Black Sea biota.

It has been shown, that preservation of the Black Sea ecosystem's biodiversity must be based on the measures which should be undertaken in national and social spheres, and be directed to the recreation, stabilization and conservation of this unique sea basin.

Measures must be implemented nationally to conserve, stabilize or recreate the ecosystem biodiversity of this unique sea basin.

Keywords Plankton · Benthos · Ichthyofauna · Ecological situation · Adaptation strategies

Guest editors: G. Relini & J. Ryland
Biodiversity in Enclosed Seas and Artificial Marine Habitats

Y. Tokarev (✉) · G. Shulman
Institute of Biology of Southern Seas, Nakhimov Av. 2, Sevastopol 99011 Crimea, Ukraine
e-mail: tokarev@ibss.iuf.net

Introduction

The Black Sea, which was formed 5–7 thousand years ago, is the youngest among the semi-enclosed seas in the Atlantic basin. The comprehensive Black Sea ecosystem formation began 1–1.5 thousand years ago, in other words in the new era (Zaitsev & Mamaev, 1997). This ecosystem is both the youngest and the most dynamic among all semi-enclosed seas ecosystems.

The Black Sea can be characterized as a water body with very low ecological environmental capacity, because it has a very thin aerobic biotic layer, stretching to the depth of 150–200 m and comprising less than 13% of the water basin volume. Furthermore it has poor water exchange with the adjacent seas through the narrow straits and poor vertical exchange with the deep hydrosulphide layer.

Change of the global climatic factors and increase of anthropogenic pressure on marine ecosystems (increase of the biogenic flow, dumping and the discharge of pollutants of different kinds), especially in near-coastal zones is possibly leading to changes in the functional and structural characteristics of these ecosystems (Keodjyan et al., 1990). Eutrophication increased many times and caused a number of adverse processes (red tides, fish kill) and, as a consequence, oxygen depletion in the wide shelf zone in the northwestern part.

The purpose of this work is to show, the first, that the biodiversity of the Black Sea ecosystem is determined by both anthropogenic and biological factors, which act in different ways, and, the second, that preservation of the Black Sea ecosystem's biodiversity is based on the measures which should be undertaken in national and social spheres, and be directed to the recreation, stabilization and conservation of this unique sea basin.

Materials and methods

The present evaluation of the Black sea marine ecosystem condition is based, first of all, on the data of many Ukrainian cruises carried out between 1960 and 2003. For example, Fig. 1 shows the ichthyoplankton sampling stations in the Black Sea during 1988–1996 with the R/V "Professor Vodyanitsky" (IBSS, Ukraine) and R/V "Bilim" (Institute of Marine Sciences, Middle East Technical University (IMS-METU), Turkey) (Gordina et al., 1998). The phyto-, zoo- and ichthyoplankton were collected also by the R/V "Academik A. Kovalevsky", "Prof. Vodyanitsky" (IBSS), "Mikhail Lomonosov", "Academic Vernadsky" (Marine Hydrophysical Institute, Ukrainian Academy of Science) and R/V "Bilim" (IMS-METU, Turkey).

The spatial structure of the bioluminescence field, temperature and electrical conductivity of the water were studied by multiple bathyphotometric soundings (Gitelson et al., 1992; Tokarev et al., 1999) using the bathyphotometer "SALPA" (Vasilenko et al., 1997). A two-measure matrix of the small-scale distribution of the bioluminescent and hydrological fields, together with their statistical characteristics, was determined.

Plankton was sampled by vertical hauls using Hensen, Bogorov-Rass and Juday nets, and with 5 l water bottles for the qualitative and quantitative determination of plankton organisms. Additionally, horizontal tows were also carried out with the Melnikov's trawl (Melnikov, 1993). During the collection period more than 40,000 phytoplankton, 60,000 zooplankton and 8,000 ichthyoplankton samples were analyzed. The total plankton biomass (wet weight) was determined in the laboratory and certain samples were

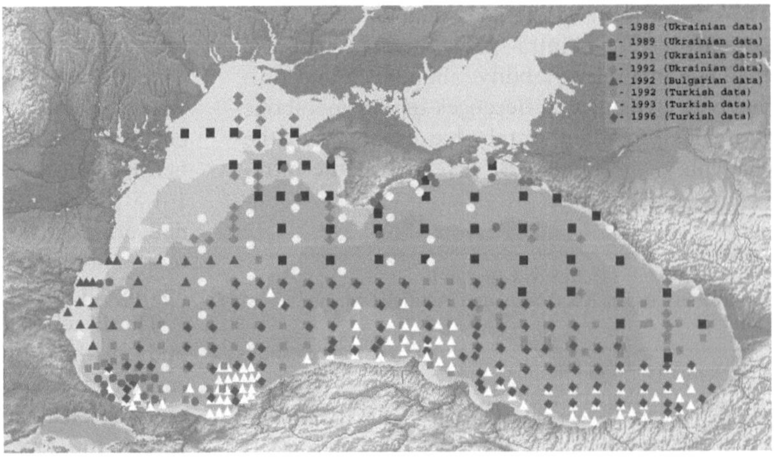

Fig. 1 The ichthyoplankton sampling stations in the Black Sea during 1988–1996 (Gordina et al., 1998)

processed to species level with corresponding recalculation for 1 m^3. Details of nets, vessels, and other information concerning the sampling procedure are summarized by Niermann et al. (1994) and Tokarev et al. (2003).

Biologic and abiotic environment parameters underwent different forms of mathematical processing with the help of univariate and multivariate statistics methods (Jenkins & Vatts, 1972; Marple, 1990). Average number (*N*, individuals m^{-2}), average biomass (B, g m^{-2}), occurrence (P, %) were determined for each species.

Results

The Black Sea ecosystem and its biodiversity underwent several different stages in its development during the last century, there are two most important of them:

(1) Dramatical alterations in the 60s and, especially, in the 70s and 80s of the XX century;
(2) Its stabilization with the elements of recreation from 1994 up to the present time.

It was reflected, first of all, at the most vulnerable elements of ecosystem—benthic and plankton communities. For example, the sudden disappearance of the phyllophore "Zernov field"—a unique accumulation of unattached Phyllophora—was the greatest ecological catastrophy in the Black Sea basin according to Milchakova (2003) (Fig. 2). Most researchers think that destructive influence of the long-term eutrophication of the northwestern Black Sea ecosystem, and uncontrolled removal of the Phyllophora for agaroid production are the reasons of this.

Less but still significant changes occurred in the macrozoobenthic communities in the coastal zone of Crimea (Mazlumyan et al., 2004). This demonstrated itself at an example of the *Chamelea galena* community in Lisya Bay: due to the increase in eutrophication and considerable decrease of the biodiversity in this period, the average abundance and biomass of this species increased about 20 times in the period from 1973 to 1998 (Mazlumyan et al., 2004). Many planktonic species disappeared, kills of macrobenthic Crustacea became more frequent (*Crangon crangon*, *Palaemon adspersus*) and meiobenthos degradation began.

Ctenophora invasions (*Mnemiopsis leidyi* first and then *Beroe ovata*) occurred because the stability of the Black Sea ecosystem decreased. The first invasion deformed trophic links and sharply decreased mesozooplankton abundance, including the copepod *Calanus euxinus*—an important component of the pelagic fish diet, especially for the anchovy (*Engraulis encrasicolis ponticus*), the scad (*Trachurus mediterraneus ponticus*) and the sprat (*Sprattus sprattus phalericus*). The number of copepod species in the Sevastopol region decreased from 1976 to 1990 from 13 to 7 and their total abundance fell 12 times (Gubanova et al., 2001) (Fig. 3).

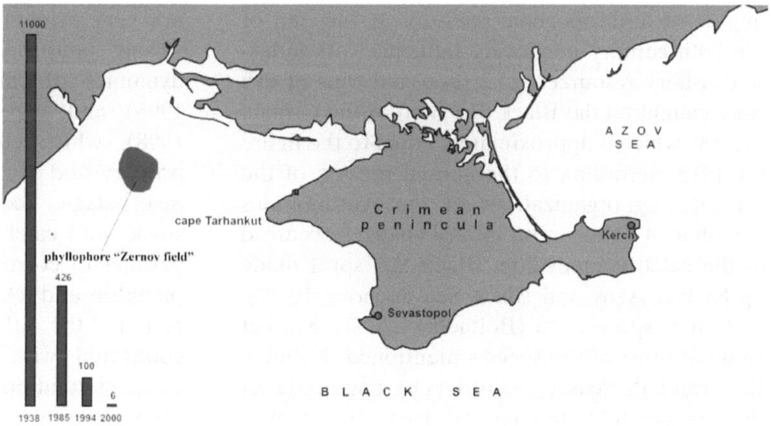

Fig. 2 Long-term changes in Phyllophora field biomass at the Black Sea (from Milchakova, 2003)

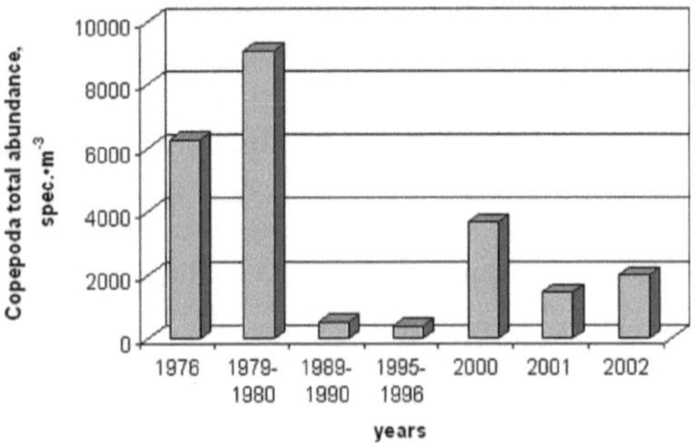

Fig. 3 Long-term changes in Copepoda total abundance (from Gubanova et al., 2001)

That led to the undermining of the ichthyofaunal food source, growth of constraints in the food relationships, decreases in species diversity, abundance and biomass of the ichthyoplankton, and in spawning and adult populations of the pelagic fishes (including size–weight composition and level of the accumulated energy resources). As a result, in the Sevastopol region, of 40 species listed in the literature only 27 remained by 1989–1990.

The commercial fish fauna of the Crimean neritic zone was diverse and included 50 species of industrial value at the beginning of the 20th century (Boltachev, 2003). Grey mullet (4 spp.), mackerel, herring (3 spp.), anchovy, beluga and sturgeon, which made 60% of the total catch, were the most important. Flounder, mullet, scad, stellate sturgeon, goby and bullhead together contributed more than 20% of the catch. Other species were of less importance. Radical changes in the Azov-Black Sea basin ecosystems under the impact of anthropogenic pressure at the end of the 20th century negatively influenced its industrial fishery resources. 23.2 thousand tons of fish were caught on the Black Sea shelf of the Crimea in 2000, which is approximately equal to the figure for 1913, according to the annual reports of the fish-catching organizations of the Autonomous Republic of Crimea. But radical changes occurred in the catch composition. Black Sea sprat made up 88.3%, Azov and Black Sea anchovy 10.7%, and other species 1% (Boltachev, 2003). Market value of most of the species mentioned declined, their catch decreased 1–3 orders of magnitude, or they completely disappeared from the coastal Crimean zone in comparison with the beginning of the last century (Fig. 4).

By the middle of 1990s a number of researchers registered stabilization and obvious signs of improvement in the Black Sea ecosystem's condition. Frequency of occurrence of the native species, abundance and biomass of phyto-, zoo- and ichthyoplankton, fishes and benthic animals increased. For example, average annual abundance of planktonic crustacea in 2000 increased 6.4 times in comparison with 1998 and 4.3 times in comparison with 1999 (Gubareva et al., 2004) (Fig. 5). This is connected with a decrease of the ctenophore *M. leidyi* population, and its influence on the ecosystem, related to the occurrence of another ctenophore species—*Beroe ovata*, and the decrease of environmental pollution with biogenic elements, heavy metals, pesticides in 1998–1999.

These conclusions were confirmed by studies on the stomach fullness of fish larvae, conducted in the last quarter of the 20th century in the IBSS (Gordina et al., 2004). This permitted, in particular, an evaluation of changes in the Black Sea fish food base (Table 1), and also revealed the main trends of the ichthyofauna state in the Black Sea: worsening larval survival conditions from the end of 80s to the middle of 90s followed by step-by-step improvement of these conditions at the beginning of the 21st century (Fig. 6).

Discussion

The role of climatic regional and global factors is not very evident against the background of the strong anthropogenic pressure. Abundance dynamics of dinoflagellate (Bryantseva et al., 1996) and copepod biomass (Kovalev et al., 1998), which contribute considerably both to primary and secondary production of the Black Sea pelagic ecosystem, and to oscillations of stocks and catches of pelagic fish, are observed parallel to corresponding changes in the phytoplankton and ichthyofauna in the Mediterranean Sea, and the Atlantic and Pacific Oceans. This is connected with global atmospheric processes, water circulation and temperature fluctuations at the world scale (Efimov, 2000).

Fig. 4 Comparison of the food fish species proportion in the catches of the Crimea coastal area at different time (from Boltachev, 2003)

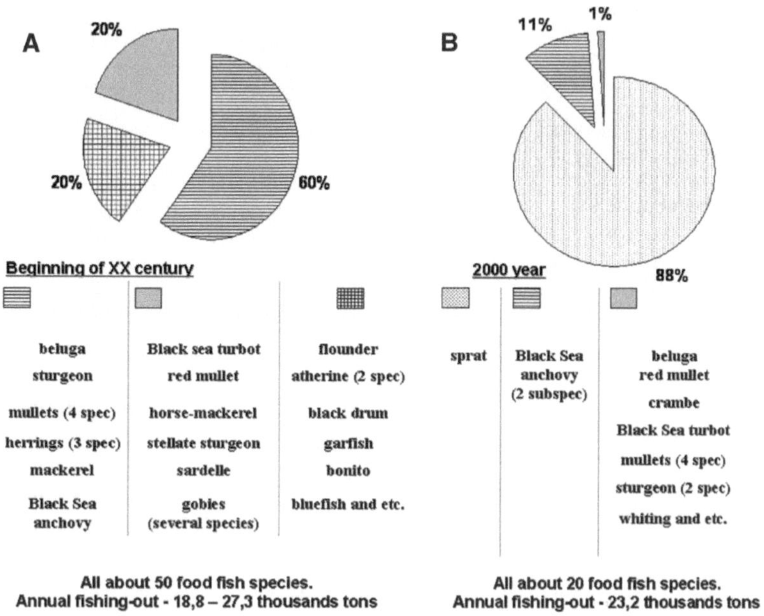

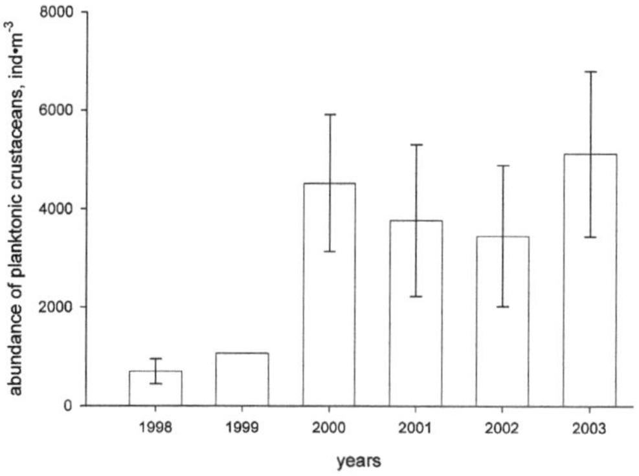

Fig. 5 Average abundance of planktonic crustaceans during spring–autumn period (May–September) from 1998 to 2003 (from Gubareva et al., 2004)

Average global temperature of the Earth has increased by ~0.6°C over the past 100 years. Numerous assessments, based on the data of averaged indirect and direct instrumental measurements, show that this rise in temperature is exceptional, at least in the period of the Holocene, which began 10–12 thousand years ago (Efimov, 2000). This general tendency is broken at regional levels, where local exclusions from this rule are observed. The Black Sea is such a region (Efimov, 2000). Change of the average winter temperature in the period of 1959–1998 in the Black Sea is negligible but for the summer months temperature increased by 0.8°C according to the re-analysis of data of NCEP/NCAR (National Center for Environmental Prediction/ National Center of Atmospheric Research) (Kalnay et al., 1996).

The difference between surface water temperature in the Black Sea and temperature in the adjacent regions of dry land seems to have been important during the past 30–40 years, in other words for the period of the most considerable global warming. The lack of change in the water winter temperature of the sea despite considerable rises in both the summer water temperature and the winter air temperature over the seaside dry land region is clearly apparent, in spite of significant inter-annual variations. This tendency is abnormal, because the temperatures of the air and surface layer water through the seasons are close and their difference does not exceed 2–3°C, with a different sign for winter and summer, in the temperate latitudes of the earth (Efimov, 2000).

Changes in the hydrochemical characteristics of the Black Sea waters over 10-year periods are

Table 1 Abundance change of zooplankton organisms of <0.5 mm size, which are the food for fish larvae in the northern part of the Black Sea (average for July in the layer of 0–10 m) (Gordina et al., 2004)

Years	Nauplii (ind m^{-3})	Other organisms (ind m^{-3})	Data sources
1960–1969	10,74,000	13,210	Greze et al. (1971)
1988	47,500	No data	Ostrovskaya et al. (1993)[a]
1989	21,500	2092	Ostrovskaya et al. (1993)[a] and own data
1990	29,670	708	Own data
1992–1993	11,670	No data	Gruzov et al. (1994)
1998	10,700	200	Own data
1999	86,000	2,250	Own data
2000	59,670	2,353	Own data
2001	382,330	4,870	Own data

[a] Data of the direct calculation (bathometer)

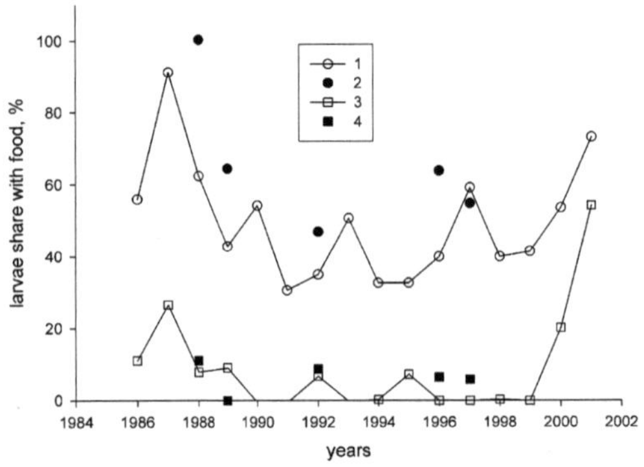

Fig. 6 Percentages of fish larvae with food in the intestines in the Black Sea for the period of 1986–2001. Northern sea half: 1—goby and striped blenny, 3—anchovy; southern sea half: 2—goby and striped blenny, 4—anchovy (from Gordina et al., 2004)

less studied. Konovalov & Murray (2001) pointed out, in particular, the decrease of the average oxygen concentrations from the middle of seventies to the beginning of the 90s. At the same time no marked decrease in biogenic element discharge has been observed in the shore run-off, while the published data reveal some stabilization (Konovalov & Murray, 2001). First of all, decrease of anthropogenic pressure on the Black Sea ecosystem for the past 10 years (Konovalov, 1995), connected with the economical situation in the Black Sea countries, led to some improvement in the state of the ecosystem and its biodiversity.

Gubareva et al. (2004) analyzed seasonal dynamics of the zooplankton community structure in 1999–2003 as an example of the need to consider climatic factors. Authors had shown the tendency of the Black Sea ecosystem to self-restore, which demonstrated itself in the increase in food zooplankton biomass during the second half of the 90s from the eastern to western shores of the Black Sea. This is bound up with superposition of the following factors: decrease in the anthropogenic pressure on the ecosystem, change of water mass temperature and balanced alternation of the Ctenophora (*M. leidyi*, *B. ovata*) (Fig. 7).

Certainly, any final conclusion about climatic effects can be drawn in future only after further detailed studies, but we believe that anthropogenic pressure is the main factor influencing the Black Sea ecosystem's condition. Having increased many times during the 70s and 80s, under the influence of anthropogenic pressure, eutrophication strongly affects the ecosystem—its structure and functioning, productivity and biodiversity (Keodjyan et al., 1990).

A process of mediterranization, or the invasion of new species from the adjacent basin (Zaitsev, 2000; Boltachev & Yurakhno, 2002) and other regions of the World Ocean, is also taking place. Oysters *Ostrea gigas* from the Pacific Ocean, the bivalve mollusc *Scapharca enaequivalvis* (*Cunearca cornea*), the gastropod *Rapana thomasiana*, the Far-Eastern "mullet haarder" (*Mugil soiuy*) and others have already acclimatized in the Black Sea. Acclimatizants increased sea food productivity on one hand but also became rivals of the native biota, disturbed nature balance and introduced new parasites, as it happened with the Pacific oyster. Building water channels, mariculture and aquarium facilities, together with aspects

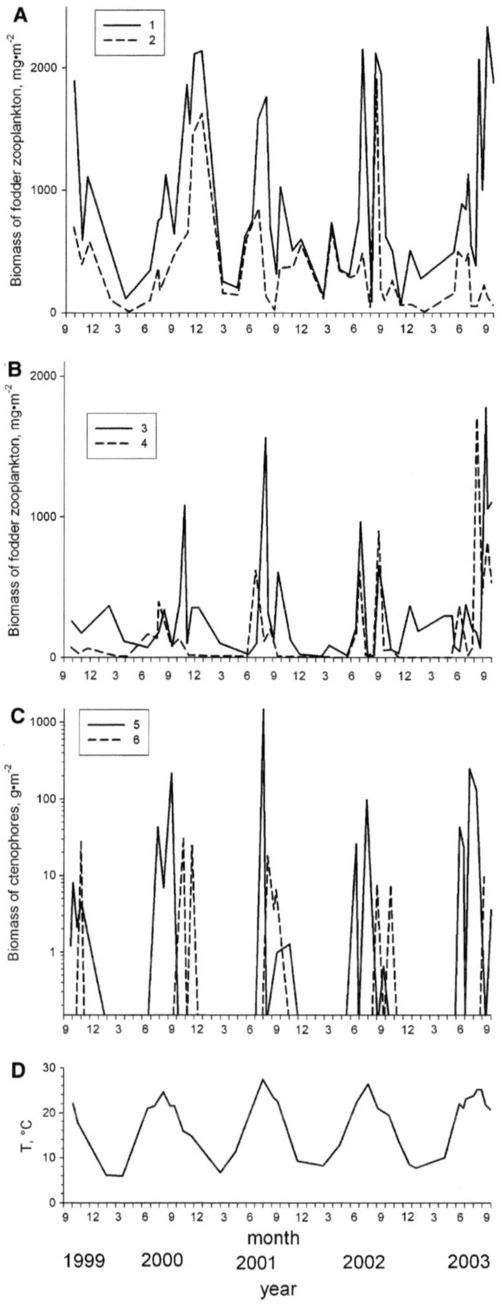

Fig. 7 The biomass of fodder zooplankton (A, B), ctenophores (C) and surface water temperature (D) 1—total fodder zooplankton; 2—meroplankton; 3—Copepoda; 4—Cladocera; 5—*Mnemiopsis leidyi;* 6—*Beroe ovata* (from Gubareva et al., 2004)

of shipping, including transport of the organisms on the vessels' hulls and in water ballast were the main sources of the invasions (Alexandrov, 2003).

Contemporary data of the pilot service in the Bosporus record the volume of shipping in the Black Sea. 47–51 thousand vessels pass through the Bosporus every year; of them 2–7 thousand had a length of >200 m during 1995–2000 (Alexandrov, 2003). 57% of the total number of cargo transportations are by vessels of countries bordering the Black Sea, 11% of which belong to Ukraine (Alexandrov, 2003). More than 11 million tons of ballast water were discharged in Ukrainian ports alone in 2001. The list of species invading the Black Sea increased more than five times between 1995 and 2001 (Table 2).

Not every invasion of exotic organisms results in significant ecological consequences and economical problems. However, such events became more frequent and the scale of their consequences became more serious as the invasions intensified as a result of the expansion of ballast water transport. Invasion of the Black Sea by the north American ctenophora *Mnemiopsis leidyi* at the beginning of eighties reduced anchovy stocks, with total economical losses of 240 million US dollars per year.

Consequences of introductions and invasions by alien species pointed to the insufficiency of our knowledge about the functioning of the Black Sea ecosystem. A discovery, made in the last 4–5 years, points to this circumstance: many aerobic forms of multicellular animals, most of which are unknown to science, were repeatedly observed in

Table 2 Number of species-invaders registered in the Black Sea in different years (Alexandrov, 2003)

Year of the analysis	Organisms groups					
	PA	MP	IV	F	M	Total
1995		1	15	10		26
1999	3	2	29	5		39
2000	7	3	30	13	5	58
2001	7	4	40	9	5	65
2001	29	38	53	15	5	140

PA, plankton algae; MP, macrophytes; IV, invertebrates; F, fishes; M mammals

the hydrosulphide zone in bottom samples from depths of 700 to 2,200 m (Sergeeva, 2003). Their presence can be connected with the occurrence of oxygen-saturated water masses, the origin of which is still uncertain, in the deep-water zone. Sergeeva proposed a new and considerably changed conception of zonation of the biota distribution in the Black Sea.

These new results were, at the same time, a confirmation of the postulate, that biodiversity should be understood as applying not only to species but also to the temporal-spatial variability of biota, its abundance and productivity dynamics, and differences in metabolic strategies, providing stable development in the changing environment.

So, sharp stratification, reflected in the formation of a layer or layers of higher concentrations of organisms in the photic layer, is a characteristic peculiarity of the Black Sea plankton vertical distribution (Tokarev et al., 2003). General conjunction of plankton biomass vertical distribution and the planktonic bioluminescence field is observed parallel to this (Fig. 8). The superposition of physical and biological parameters (stratification and movement of waters, light level, food availability, elements of animals social behavior, anthropogenic pressure, etc.) causes spatial heterogeneity (patchiness) in pelagic populations, and the role of biological (reproductive, migrational, ethological) mechanisms increases considerably with decrease of the studied scales of space and time (Tokarev & Sokolov, 2001) (Table 3).

All these factors have an influence upon the Black Sea ecosystem: its structure and functioning, productivity and biodiversity. So, to our opinion, biodiversity is not only of structural but of functional character too. We think, that differences in metabolic strategies provide sustainable existence in the changing environment (Table 4). Alternative metabolic features lie at the base of these strategies. First of all, these are different levels of energy metabolism, illustrated in Table 5, where rate of oxygen consumption, Q, in the Black Sea fishes (ml g^{-1} h^{-1}) is shown. Coefficient "a" is calculated from equation:

$$Q = aW^k$$

where W is the weight of fish.

Table 3 The wave length (m), corresponding to maxima of the spatial spectra of plankton (P), bioluminescence (B) and temperature (T) in the Black Sea surface layer (Tokarev & Sokolov, 2001)

Number of max.	P		B		T	
	Day	Night	Day	Night	Day	Night
1	450	818	345	1,500	450	1,800
2	300	600	214	643	250	562
3	188	428	132	450	173	392
4	155	321	100	346	141	251
5	137	250	76	290	122	200
6	86	200	–	225	99	164
7	–	184	–	183	80	–
8	–	164	–	171	–	–

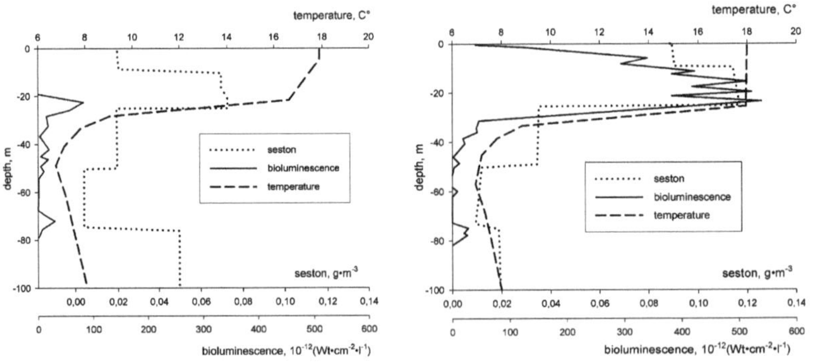

Fig. 8 The typical vertical structure of plankton biomass (1), bioluminescence (2) and temperature (3) at the day (a) and night (b) time in the center of the Black Sea western halistase (from Tokarev & Sokolov, 2001; Tokarev et al., 2003)

Table 4 Alternative life strategies providing biodiversity and biological progress in the Black Sea (Shulman & Urdenko, 1989)

Strategies	
Expansion	Specialization
Development of enormous areas	Occupation of narrow ecological niches
High biomass and productivity	
High inter- and intra-species differentiation	High inter- and intra-species differentiation

Table 5 Rate of oxygen consumption in Black Sea fishes (ml g^{-1} h^{-1}) (Belokopythin, 1993)

Species	Standard metabolism (coefficient *a*)
Anchovy	0.97
Horse-mackerel	0.700
Mullet	0.572
Pickerel	0.572
Red mullet	0.247
Whiting	0.276
Scorpion fish	0.084

The difference of metabolic strategies is well shown in the intensity of food consumption and the efficiency of its utilization (or assimilation). However, metabolic variability is shown not only at inter-species but also at intra-species levels. For instance, even in the same school fishes differ in their swimming "capacity" and concentration of neutral lipids (triacylglycerols), the main energy substrate for swimming (Shulman et al., 1976) (Table 6).

It is important to notice that biodiversity is not only inter- and intra-species diversity but also spatial–temporal variability, abundance and productivity dynamics, differences of the metabolic

Table 6 Lipid characteristics (triacylglycerols, mg g^{-1}) in good and bad swimmers of horse-mackerel (Shulman et al., 1976)

	Good swimmers	Bad swimmers
Red muscles	102	56
While muscles	96	32

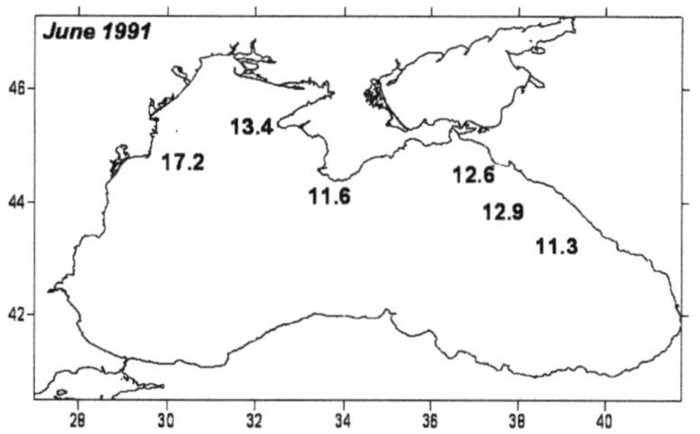

Fig. 9 Fat content (% WW) in sprat populations differences are due to differences in food supply (from Minyuk et al., 1997)

strategies providing sustainable existence in the changing environment. Biodiversity at the intra-species level expresses itself in spatial and temporal variations of the Black Sea biota. Examples of spatial variability are fat content (% WW) in sprat populations due to differences in food supply (Fig. 9) (Minyuk et al., 1997). This variability is related to the degree of preparedness for migration, caused by different food supply. The same influence of food supply is seen by comparing the main energy substrates: total lipid content of copepod *Calanus euxinus* and glycogen content of ctenophora *Pleurobrachia rhodopis* were different in cyclonic and anticyclonic zones (Yuneva et al., 1999).

As for temporal diversity, we pay attention first of all to ontogenesis (life history). Large variability is expressed during annual cycles (we mean first of all seasonal changes of metabolism). We note diurnal variability too when finishing the observation of terminal diversity (Shulman et al., 1976; Shulman & Urdenko, 1989). Most of our data correspond to phenotypic diversity in the Black Sea biota. We should also pay attention to genotypic diversity (Dobrovolov, 1976, 2000), but it is unfortunately less studied (Table 7). These data show the place of the Black and Azov Sea anchovies within the species *Engraulis encrasicolus*. We see that the Black Sea anchovy is very close to the Marmara Sea one (closer than to the Azov Sea one). This is not occasional because there is close gene transfer between the Black Sea and the Marmara Sea subspecies.

Table 7 Indices of genetic identity (under diagonal) and genetic distance (above diagonal) in *Engraulis encrasicolus* L. (Dobrovolov, 1976)

Populations	1	2	3	4	5	6	7
Black sea	*	0.0014	0.0004	0.0083	0.0032	0.0023	0.0706
Azov sea	0.9986	*	0.0020	0.0083	0.0066	0.0039	0.0602
Marmara sea	0.9996	0.9980	*	0.0048	0.0030	0.0021	0.0623
Aegean sea	0.9917	0.9923	0.9952	*	0.0066	0.0061	0.0440
Adriatic	0.9968	0.9932	0.9970	0.9934	*	0.0008	0.0612
Portuguese	0.9977	0.9961	0.9972	0.9939	0.9992	*	0.0620
Cape Blanc	0.9318	0.9416	0.9396	0.9570	0.9406	0.9399	*

The research cited testifies to the fact that the biodiversity of the Black Sea ecosystem is determined by both anthropogenic and biological factors, which act in different ways. Conservation of Black Sea biodiversity is connected with the operation of complex state and social actions, directed on reconstruction, stabilization and conservation of this unique marine basin. Creation of the reserve and national parks can be one of the ways to save the genofund of this unique sea. Creating the conserved and reserved aquatic environment is the solution for very important ecological problems, fulfillment of which will promote preservation of the existing communities and the genetic diversity of rare and vanishing species.

Improvement of international scientific cooperation within the framework of international and national projects, and also the coherence of national legislative norms in the sphere of environmental protection, rational use of its resources and biodiversity conservation are all very important.

Acknowledgements We present our sincere gratitude to all scientific IBSS staff for active discussion of the materials obtained, and for valuable remarks.

References

Alexandrov, B. G., 2003. Problem of aquatic organisms by ships and some approaches for risk assessment of the new invasions. Marine Ecological Journal. 3: 5–17 (in Russian).

Belokopythin, Yu. S., 1993. Energy Metabolism in Marine Fishes. Naukova Dumka, Kiev, 128 pp (in Russian).

Boltachev, A. R., 2003. Diversity of the food fishes. In Eremeev, V. N. & A. V. Gaevskaya (eds), Modern Condition of Biological Diversity in Near-Shore Zone of Crimea (the Black sea sector). Ecosi-Gidrofyzika, Sevastopol: 409–424 (in Russian).

Boltachev, A. R. & V. M. Yurakhno, 2002. New evidence of ongoing mediterranization of the Black Sea ichthyofauna. Journal of Ichthiology. 42: 713–719 (in Russian).

Bryantseva, Y. V., V. A. Bryantsev, L. A. Kovalchuk & E. Z. Samyshev, 1996. For the question of long-term changes of biomass in diatom and peridinien algae of the Black sea in relation with atmospheric activity. Ekologiya morya 45: 13–18 (in Russian).

Dobrovolov, I. S., 1976. Multiple forms of lactate dehydrogenase in anchovy (*Engraulis encrasicolus*) from the Black Sea, the Sea of Azov and the Atlantic Ocean. Comptes rendu de L'Academie Bulgare des Sciences 29: 877–880.

Dobrovolov, I. S., 2000. Genetic divergence between the scad subspecies *Trachurus mediterraneus* from the Black Sea and Mediterranean. Mediterranean Marine Sciences 1: 133–139.

Efimov, V. V., 2002. The development of the Black Sea area. In Rolston, S. (ed), Proceedings of the PACEM in MARIBUS International Conference, Hamburg, 3–6 December, 2000. IOI, Halifax, 74 pp.

Gitelson, I. I., L. A. Levin, R. N. Utyushev, O. A. Cherepanov & Yu. N. Chugunov, 1992. Ocean Bioluminescence. S.-Petersburg, Nauka, 284 pp (in Russian).

Gordina, A. D., U. Niermann, A. E. Kideys, A. A. Subbotin, Yu. G. Artyomov & F. Bingel, 1998. State of summer ichthyoplankton in the Black Sea. In Ivanov, L. I. & T. Oguz (eds), Ecosystem Modeling as a Management Tool for the Black Sea, Vol. 1. Kluwer Academic Publishers, the Netherlands, 367–378.

Gordina, A. D., L. P. Salekhova & T. N. Klimova, 2004. Fish specific composition of the Black Sea near Sevastopol as indicators of current condition of coastal ecosystem of south-western part of Crimean shelf. Marine Ecological Journal 3: 15–24 (in Russian).

Gubanova, A. D., I. Yu. Prusova & N. V. Shadrin et al. 2001. Dramatic change in the Copepod community in Sevastopol Bay (Black Sea) during two Decades (1976–1996). Senckenbergiana maritime 31: 17–27.

Gubareva, E. S., L. S. Svetlichny & Z. A. Romanova et al. 2004. Zooplankton community state in Sevastopol Bay after the invasion of ctenophore Beroe ovata into the Black Sea (1999–2003). Marine Ecological Journal 3: 39–46 (in Russian).

Jenkins, G. & D. Vatts, 1972. Spectral Analysis and its Appendixes. Mir, Moscow, 317 pp (in Russian).

Kalnay, E., M. Kanamitsu, R. Kustler, W. Collins, D. Deaven, L. Gandin, M. Iredell, S. Saha, G. White, J. Woollen, Y. Zhu, M. Chelliah, W. Ebisuzaki, W. Higgins, J. Janowiak, K. C. Mo, C. Kopelewski,

J. Wang, A. Leetmaa, R. Reynolds, R. Jenne & D. Joseph, 1996. The NCEP/NCAR 40-year reanalysis project. Bulletin of the American Meteorological Society 77: 437–471.

Keodjyan, V. P., A. M. Kudin & Yu. V. Terekhin (eds), 1990. Applied Ecology of Sea Region. The Black Sea. Naukova Dumka, Kiev, 252 pp (in Russian).

Konovalov, S. M. 1995. Anthropogenic impact and ecosystems of the Black Sea. In Briand, I. (ed), Les mers tributaires de Mediterranee. CIESM Science series, Vol. 1: 53–83.

Konovalov, S. K. & J. W. Murray, 2001. Variations in the chemistry of the Black Sea on a time scale of decades (1960–1995). Journal of Marine Systems 31: 217–243.

Kovalev, A., U. Niermann, V. V. Melnikov, Z. Uysal, A. E. Kideys, M. Unsal & D. Altukhov, 1998. Long-term changes in the Black Sea zooplankton: the role of natural and anthropogenic factors. In Ivanov, L. I. & T. Oguz (eds), Ecosystem Modeling as a Management Tool for the Black Sea, Vol. 1. Kluwer Academic Publishers, the Netherlands, 221–234.

Marple, S. L. Jr., 1990. Numerical Spectral Analysis and its Appendixes. Mir, Moscow, 582 pp.

Mazlumyan, S. A., N. A. Boltacheva & E. A. Kolesnikova, 2004. Analysis of long-term changes of benthos diversity in the Lisya bay (southern-eastern Crimea). Marine Ecological Journal 3: 59–72 (in Russian).

Melnikov, V. V., 1993. New Models of Selfclosing Macroplankton Trawls. Second Polish-Soviet Antarctic Symposium "Arctowski 89". Institute of Ecology Polish Academy of Science, Warszavva, 141–143.

Milchakova, N. A., 2003. Mycrophytobentoc. In Eremeev, V. N. & A. V. Gaevskaya (eds), Modern Condition of Biological Diversity in Near-shore Zone of Crimea (the Black Sea sector). EKOSI-Gidrophizika, Sevastopol, 152–208 (in Russian).

Minyuk, G. S., G. E. Shulman, V. Ya. Shchepkin & T. V. Yuneva, 1997. Black Sea Sprat: The Relationship Between Lipid Dynamics, Biology and Fishery. Ekosi-Hydrophysica, Sevastopol, 139 pp (in Russian).

Niermann, U., F. Bingel, A. Gorbun, A. D. Gordina, A. C. Giicu, A. E. Kideys, A. Konsulov, G. Radu, A. A. Subbolin & V. E. Zaika, 1994. Distribution of anchovy eggs and larvae (*Engraulis encrasicolus Cuv.*) in the Black Sea in 1991–1992. ICES Journal of Marine Sciences 51: 395–406.

Sergeeva, N. G. 2003. Meiobentos of deep-water anoxic hydrogen sulphide zone of the Black Sea. In Proceeding of the "Second International Conference on Oceanography of the Mediterranean and Black Sea. Similarities and differences of two interconnected basins", 14–18 October, 2002. Tubitak Publishers, Ankara, 880–887.

Shulman, G. E. & S. Yu. Urdenko, 1989. Productivity of Fishes of the Black Sea. Naukova Dumka, Kiev, 188 pp (in Russian).

Shulman, G. E., V. Ya. Shchepkin, K. K. Yakovleva & T. V. Khotkevich, 1976. Lipids and their utilization during fish swimming. In Shulman, G. E. (ed.), Elements of Physiology and Biochemistry of Total and Active Metabolism in Fish. Naukova dumka, Kiev, 100–121 (in Russian).

Tokarev, Yu. N. & B. G. Sokolov, 2001. Effect of physical and biological factors on forming of small-scale of bioluminescent and acoustic fields in the Black and Mediterranean seas. Gidrobiologicheskii Jurnal 31: 3–13 (in Russian).

Tokarev, Yu. N., E. P. Bityukov, R. Williams, V. I. Vasilenko, S. A. Piontkovski & B. G. Sokolov, 1999. The bioluminescence field as an indicator of the spatial structure and physiological state of the planktonic community at the Mediterranean sea basin. In Malanotte-Rizzoli, P. & V. N. Eremeev (eds), The Eastern Mediterranean as a Laboratory Basin for the Assessment of Contrasting Ecosystems. Kluwer Academic Publishers, The Netherlands, 407–416.

Tokarev, Yu. N., E. P. Bityukov, V. I. Vasilenko, B. G. Sokolov & I. M. Serikova, 2003. Bioluminescence from the Black Sea to the eastern Mediterranean: The spatial structure and functional connection with the characteristics of plankton in the two interconnected basins. In Yilmaz, A. (ed.), Oceanography of Eastern Mediterranean and Black Sea: Similarities and Differences of Two Interconnected Basins. Tubitak, Turkey, 785–793.

Vasilenko, V. I., E. P. Bityukov, B. G. Sokolov & Yu. N. Tokarev, 1997. Hydrobiophysical Device "SALPA" of Institute of Biology of the Southern Seas Used for Bioluminescent Investigation of the Upper Layers of the Ocean // Bioluminescence and Chemiluminescence. Molecular Reporting with Photons. J. Wiley & Sons, N.Y., 549–552.

Yuneva, T. V., L. S. Svetlichny, O. A. Yunev, Z. A. Romanova, A. E. Kideys, F. Bingel, A. Yilmaz, Z. Uysal & G. E. Shulman, 1999. Nutritional condition of female Calanus euxinus from cyclonic and anticyclonic regions of the Black Sea. Marine Ecology Progressive Series 189: 195–204.

Zaitsev, Yu. P., 2000. The Black Sea: Ecosystem State and Ways of its Improvement. Molodezhnyi ecologicheskii tsentr im. V.I. Vernadskogo, Odessa, 48 pp (in Russian).

Zaitsev, Yu. & V. Mamaev, 1997. Biological Diversity in the Black Sea. A Study of Change and Decline. United Nations Publications, New York, 208 pp.

Hydrobiologia (2007) 580:35–41
DOI 10.1007/s10750-006-0467-7

Measuring change of Mediterranean coastal biodiversity: diachronic mapping of the meadow of the seagrass *Cymodocea nodosa* (Ucria) Ascherson in the Gulf of Tigullio (Ligurian Sea, NW Mediterranean)

Mattia Barsanti · Ivana Delbono · Ornella Ferretti · Andrea Peirano · Carlo Nike Bianchi · Carla Morri

Abstract Monitoring the extension of seagrass meadows over time is of primary importance for the surveillance of marine coastal biodiversity. Here, we analyse the evolution of the meadow of *Cymodocea nodosa* in the Gulf of Tigullio, a coastal tract of naturalistic interest but subjected to high anthropogenic pressure. Historical maps at a scale 1:25,000 of *C. nodosa* meadow drawn in 1986, 1991 and 2001 were processed with GIS (Geographical Information System), using overlay vector methods. Diachronic analyses allowed the measurements of temporal changes, in term of percentage gain or loss of meadow extension, through concordance and discordance maps. A general increase in the extension of the meadow from 1986 to 1991 was evidenced, but the disparity of mapping methods (SCUBA diving in 1986, Side Scan Sonar in 1991) in the two surveys imposes caution when interpreting this result. On the other hand, the comparison of 1991 and 2001 maps, both derived from Side Scan Sonar surveys, showed a regression of the meadow of about 60% in the northern area, and modifications in the upper and lower limits of the meadow due to the impact of coastal works. *C. nodosa* meadow showed the only enlargement in front of the mouth of the Entella River, due to the increase in nutrient contents for rainfall in the period 1988–1994. The overall analysis evidenced a net decrease in seagrass meadow extension, an early warning of risk for marine coastal biodiversity in the Gulf of Tigullio.

Keywords Biodiversity · *Cymodocea nodosa* · Diachronic analysis · Geographic Information System

Guest editors: G. Relini & J. Ryland
Biodiversity in Enclosed Seas and Artificial Marine Habitats

M. Barsanti (✉) · I. Delbono · O. Ferretti · A. Peirano
ENEA S. Teresa, Marine Environment Research Centre, P.O. BOX 224, La Spezia I-19100, Italy
e-mail: mattia.barsanti@santateresa.enea.it

C. N. Bianchi · C. Morri
Dip.Te.Ris, University of Genoa, Corso Europa 26, Genova I-16132, Italy

Introduction

Biodiversity is a cluster of concepts that encompasses many interrelated levels, from genes to whole biological communities and habitats (Bianchi & Morri, 2000). Traditionally, the attention to biodiversity problems is mostly directed toward species richness but experience has shown that species are effectively preserved if attention is paid to habitats (Bianchi, 2002).

Seagrasses form the most productive autotrophic communities on the planet (Duarte & Chiscano, 1999) and are major foundation species (sensu Bruno et al., 2003), shaping submarine seascape and generating habitats for the associated mobile invertebrates and fishes. Seagrass meadows are therefore ecosystems of primary importance for coastal biodiversity (Duarte, 2000, 2002).

Among the 50 species of seagrass described world-wide (Short et al., 2001), five species inhabit the marine coasts of the Mediterranean Sea, *Posidonia oceanica* and *Cymodocea nodosa* being the most widely distributed (Den Hartog, 1970; Furnari et al., 2003; Buia et al., 2004). A majority of studies has focused on the former (Pergent et al., 1995; Buia et al., 2000), while the latter has received comparatively less attention (Cebrián et al., 2000; Cancemi et al., 2002). Recent investigations have shown important structural and functional differences between the meadows of *P. oceanica* and those of *C. nodosa*, especially in terms of their response to coastal hydrodynamics, sediment transport, nutrient content and anthropogenic disturbance (Duarte & Sand-Jensen, 1990a, b, 1996; Marbà et al., 1994; Peirano & Bianchi, 1997).

It is therefore important to monitor *C. nodosa* meadows as well as those of *P. oceanica*. This is particularly true for the coasts of Liguria, an Italian region along the North Western Mediterranean Sea where the extension of *C. nodosa* is comparable to that of *P. oceanica* (Bianchi & Peirano, 1995). Methods should be chosen in order to get an estimation of change in status and extension of seagrass meadows with time. The approach presented in this paper uses the diachronic analysis of cartographic data of *C. nodosa* meadows taken in 1986, 1991 and 2001: meadow extensions are herein compared and differences with time are critically discussed in relation both to different mapping systems and coastal evolution.

Study area

The *C. nodosa* meadow considered in this study is located in the Gulf of Tigullio (Eastern Ligurian Sea, NW Italy). This area shows high naturalistic interest (a Marine Protected Area has been established at its western end) but is subject to high anthropogenic pressure due especially to tourism and coastal development (Morri et al., 1988). The sandy coast in the eastern part of the Gulf has been experiencing erosion phenomena since the second half of the 20th century (Cortemiglia, 1987). The *C. nodosa* meadow covers large areas in front of this sandy coast (nearly 4 km), from 5 to 20 m depth approximately (Barsanti et al., 2003). Observations by SCUBA in 2001 showed that this meadow harbours a rich and diverse associated fauna including the epiphytic hydroid *Laomedea angulata*, the snail *Smaragdia viridis* and numerous species of crabs, gastropods and fish. Among the latter, must be underlined the abundance of the seahorse *Hippocampus hippocampus*, which has been included in a list of protected species by UNEP (Relini, 2000) and the Convention on International Trade in Endangered Species of wild flora and fauna (Wabnitz et al., 2003).

Materials and methods

Surveys on the *C. nodosa* meadow were carried out in 1986, 1991 and 2001 by using different methods. The first data on the extension of the meadow derived from the biocenotic map produced at a scale 1:30,000 by Morri et al. (1988), who used information collected by SCUBA in 1986; the map was subsequently processed in GIS format by Tunesi et al. (2002). A further mapping of the meadow was performed through Side Scan Sonar (SSS) in May 1991 at a scale 1:25,000 (Bianchi & Peirano, 1995). In June 2001, a new survey with SSS was undertaken, integrated with underwater observations and samplings by Remotely Operated Vehicle (ROV) and by SCUBA (Barsanti et al., 2003; Delbono et al., 2003). The 2001 SSS survey resulted in three mosaic raster-format maps in scale 1:5,000; the conversion from raster to vector format was performed in the nominal scales of both 1:5,000 and 1:25,000.

Since the three surveys considered do not cover exactly the same area, a common area was located by marking the limits of the 2001 SSS sonograms. To make the diachronic analyses of the *C. nodosa* meadow easier, this common area was divided into three sectors (Fig. 1): (I) northern sector, from Punta Chiappe to the

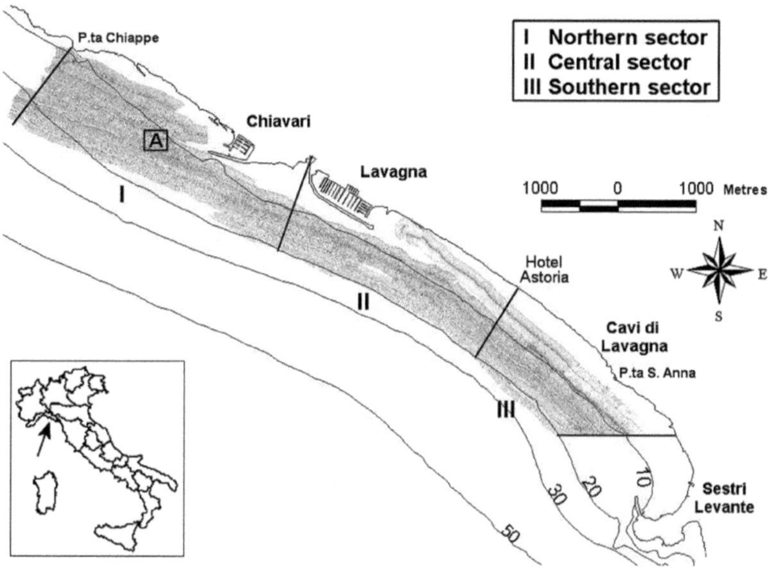

Fig. 1 Study area and the three sectors described in text. The mosaic map of the Side Scan Sonar sonograms taken in 2001 are superimposed

mouth of the Entella river (located between the two harbours of Chiavari and Lavagna); (II) central sector, from the mouth of the Entella River to Hotel Astoria of Cavi di Lavagna; (III) southern sector, from Hotel Astoria to the beginning of the seafront area of Sestri Levante. For each sector, the extension in hectares (Table 1) of *C. nodosa* meadow was calculated through a Geographical Information System bringing the three maps to a common nominal scale of 1:25,000. All vector models of the *Cymodocea nodosa* meadow are of polygonal type and are georeferenced at WGS84 datum with UTM metric co-ordinates.

Concordance and discordance maps for the periods 1986–1991 and 1991–2001 were produced using vector overlay methods. For each sector,

Table 1 Areas in hectares of the *Cymodocea nodosa* meadow in the three sectors

	Northern Sector	Central Sector	Southern Sector	Total
Hectares in 1986	86.4	50.1	93.5	230.0
Hectares in 1991	201.1	85.8	120.5	407.4
Hectares in 2001	77.7	86.3	115.7	279.7

the percentage differences in extension of the *C. nodosa* meadow for the period under consideration were calculated through the following formulae:

$[A/(A + B + C)]$ %: percentage value of the portion of the meadow present in the year X only;

$[B/(A + B + C)]$ %: percentage value of the portion of the meadow present in the year Y only;

$[C/(A + B + C)]$ %: percentage value of the portion of the meadow common to the two years.

where X and Y are two generic years; A is the value in hectares of meadow present in the X year and not in the Y year; B is the value in hectares of meadow present in the Y year and not in the X year; C is the value in hectares of the common portion in the years X and Y.

Results

Concordance and discordance maps for the period 1986–1991 (Fig. 2) showed that *Cymodocea nodosa* spread out in sector I expanding both westward and eastward and reaching a slightly deeper lower limit and a shallower upper limit;

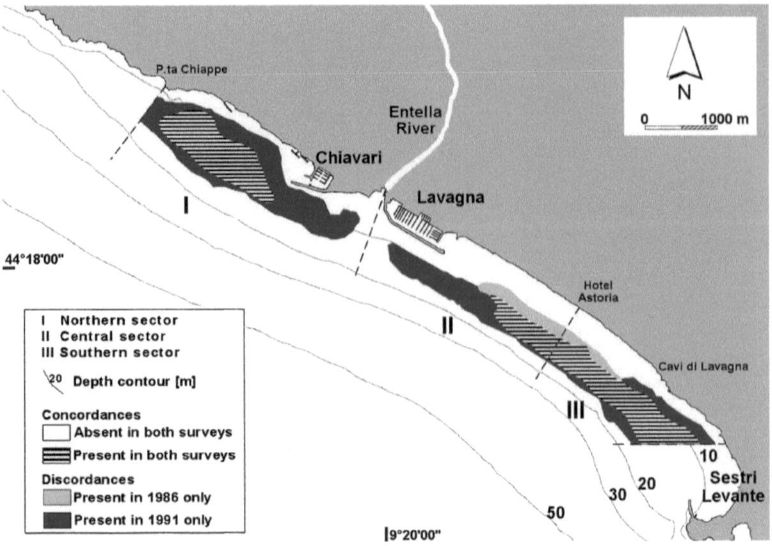

Fig. 2 Concordance and discordance map of the meadow of *Cymodocea nodosa* for the period 1986–1991

the common part is about 43% while the increased part is about 57%. In sector II, the meadow extended westward by about 53%; the lower limit of the meadow remained stable, while the upper limit deepened with a decrease of shallow beds of about 19%. In sector III, about 65% of the meadow kept steady, while an increase of about 28% was evident towards the coast and offshore.

Concordance and discordance maps for the period 1991–2001 (Fig. 3) showed that in sector I the meadow split in two parts: one in front of Chiavari, the other at the mouth of the Entella River; as a consequence, the meadow extension decreased of about 63%. In sector II, the meadow enlarged towards the river mouth and onshore by about 26% but the lower limit regressed in a similar proportion; 48% remained stable. In

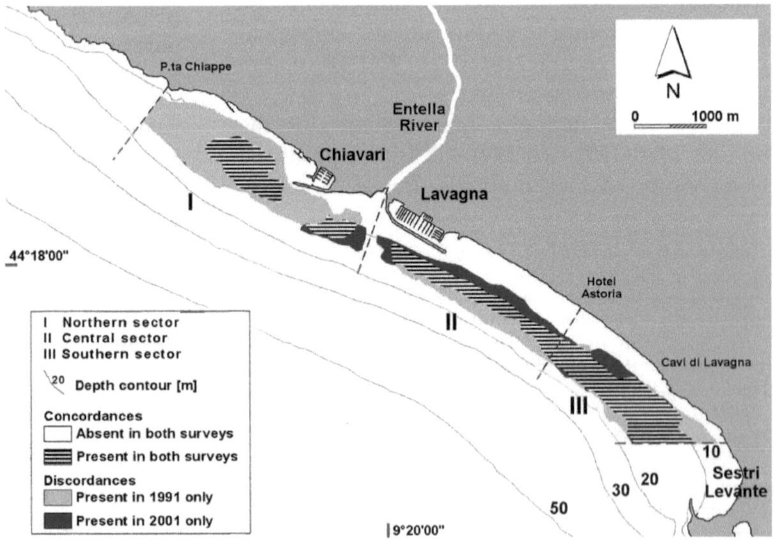

Fig. 3 Concordance and discordance map of the meadow of *Cymodocea nodosa* for the period 1991–2001

sector III there was no relevant change: around 77% of the meadow kept steady and the upper and lower limits remained basically the same.

Discussion

Comparing data from different sources and methods makes it possible that some differences do not reflect change over time, as pointed out by Leriche et al. (2004). This may be the case for the temporal comparison made in this work, which had to take into account different methods used to map the *Cymodocea nodosa* meadow: SCUBA diving in 1986 and Side Scan Sonar in 1991 and 2001.

The apparent enlargement of the meadow between 1986 and 1991, particularly in sectors I and II, may depend to a great deal on the disparity of methods. The interpolation point data obtained by SCUBA in 1986 might have led to great inaccuracy, compared to the complete coverage assured by Side Scan Sonar in 1991. However, location of upper and lower limits of the meadow by SCUBA diving should be the most reliable.

In both 1991 and 2001, the same mapping technique (SSS) was used. Positioning at sea was different in the two surveys: Loran C in 1991, differential GPS in 2001. Once again, however, errors in georeferencing should be negligible at the nominal scale of 1:25,000. On the whole, the maps produced in these two years can be considered fully comparable.

Between 1991 and 2001, the extension of the meadow regressed significantly in sector I, with complete disappearance of *C. nodosa* in front of the pier of Chiavari harbour. SCUBA diving and ROV observations suggested that the defensive works of Chiavari harbour made in previous years exerted a negative impact on the meadow: increased water turbidity prevented sufficient light penetration and the deposition of silt and clay suffocated the plant.

SCUBA diving and ROV observations in 2001 also showed *C. nodosa* plants partially or completely embedded in sand dunes in sector II: the high wave energy in this tract of coast (Cortemiglia, 1978) is likely to cause the migration of sand dunes and the presence of scattered *C. nodosa* patches (Marbà et al., 1994).

The extension of the meadow in both sector I and II towards the mouth of Entella River has to be related to a net acquisition of nutrients (Duarte & Sand-Jensen, 1996; Ceccherelli & Cinelli, 1997). Hydrological monitoring conducted in the area between 1996 and 1997 (ENEA, 1997) showed greater concentration of nutrients (NO_3 and silicates) during rainy winter months (January to April) at the Entella mouth. The continued enlargement of the meadow in the proximity of the river mouth from 1986 to 2001 is in accordance with the positive trend in annual rainfall recorded in Liguria from 1988 to 1994 (ISTAT, 2000).

Despite the unhomogeneity in the methods used to produce the original maps in different years, diachronic map analysis proved effective in showing that the *C. nodosa* meadow along this tract of coast is subject to variations in its extension. Even though the meadows of *C. nodosa* are known to be intrinsically more variable than those of *P. oceanica* (Buia et al., 2000; Guidetti et al., 2001), the influence of both climate (increase in rainfall) and human activity (harbour defences) is indisputable, the latter being cause of local meadow disappearance. *C. nodosa* is considered a tolerant species with high capacity of colonisation (Aliani et al., 1998) but long-established meadows have greater biomass and play a significant role in entrapping detritus in the sediment (Cebrián et al., 2000). The loss of 'old' *Cymodocea nodosa* meadows individuated throughout the years in the Tigullio Gulf may therefore be detrimental to ecosystem functioning and has to be interpreted as an early warning of risk for marine coastal biodiversity.

Acknowledgement The authors wish to thank Olivia Vannello (Genoa) for her contribution in GIS maps production and analysis.

References

Aliani, S., C. N. Bianchi, S. Cocito, P. R. Dando, R. Meloni, C. Morri, A. Niemeyer, A. Peirano & W. Ziebis, 1998. A map of seagrass meadows in Palaechori Bay (Milos Island, Greece), a marine area with hydrothermal activity. Rapports et Procès-Verbaux des Réunions de la Commission Internationale pour l'Exploration Scientifique de la Mer Méditerranée 35(2): 512–513.

Barsanti, M., A. Peirano, S. Sgorbini, S. Cocito, C.N. Bianchi & C. Morri, 2003. Rilevamento dei prati di *Cymodocea nodosa* mediante Side Scan Sonar, ROV ed immersioni subacquee: area costiera-marina tra Chiavari e Sestri Levante. In Studi per la creazione di strumenti di gestione costiera: Golfo del Tigullio. ENEA, Centro Ricerche Ambiente Marino, La Spezia—Ministero Ambiente e Tutela del Territorio, 141–155.

Bianchi, C. N., 2002. Il monitoraggio della biodiversità nelle Aree Marine Protette: considerazioni scientifiche e metodologiche. Notiziario della Società Italiana di Biologia Marina 41: 89–95.

Bianchi, C. N. & C. Morri, 2000. Marine biodiversity of the Mediterranean Sea: situation, problems and prospects for future research. Marine Pollution Bulletin 40(5): 367–376.

Bianchi, C. N. & A. Peirano, 1995. Atlante delle fanerogame marine della Liguria: *Posidonia oceanica* e *Cymodocea nodosa*. ENEA. Centro Ricerche Ambiente Marino, La Spezia, 1–146.

Bruno, J. F., J. J. Stachowicz & M. D. Bertness, 2003. Inclusion of facilitation into ecological theory. Trends in Ecology and Evolution 18: 119–125.

Buia, M. C., M. C. Gambi & M. Dappiano, 2004. Seagrass system. Biologia Marina Mediterranea 10(suppl.): 133–183.

Buia, M. C., M. C. Gambi & V. Zupo, 2000. Structure and functioning of Mediterranean seagrass ecosystems: an overview. Biologia Marina Mediterranea 7(2): 167–190.

Cancemi, G., M. C. Buia & L. Mazzella, 2002. Structure and growth dynamics of *Cymodocea nodosa* meadows. Scientia Marina 66(4): 365–373.

Cebrián, J., M. F. Pedersen, K. D. Kroeger & I. Valiela, 2000. Fate of production of the seagrass Cymodocea nodosa in different stages of meadow formation. Marine Ecology Progress Series 204: 119–130.

Ceccherelli, G. & F. Cinelli, 1997. Short-term effects of nutrient enrichment of the sediment and interactions between the seagrass *Cymodocea nodosa* and the introduced green alga *Caulerpa taxifolia* in a Mediterranean bay. Journal of Experimental Marine Biology and Ecology 217: 165–177.

Cortemiglia, G. C., 1978. Le modificazioni dell'assetto strutturale del litorale di Lavagna quale fattore erosivo della spiaggia. Memorie della Società Geologica Italiana 19: 369–380.

Cortemiglia, G. C., 1987. Evoluzione della piana di Sestri Levante (Liguria Orientale) dal secolo XVIII d.c. alla situazione attuale. Geografia Fisica e Dinamica del Quaternario 10: 122–131.

Delbono I., C. N. Bianchi, C. Morri & A. Peirano, 2003. Influenza della tessitura dei sedimenti e dei fenomeni erosivi nella distribuzione del prato di Cymodocea nodosa nel tratto di litorale compreso fra Chiavari e Sestri Levante. In Studi per la creazione di strumenti di gestione costiera: Golfo del Tigullio. ENEA, Centro Ricerche Ambiente Marino, La Spezia—Ministero Ambiente e Tutela del Territorio, 121–129.

Den Hartog C., 1970 The Seagrasses of the World. Amsterdam, Verhandelingen der Koninklijke Nederlandse Akademie van Wetenschappen Afdeling Natuurkunde, 1–275.

Duarte, C. M., 2000. Marine biodiversity and ecosystem services: an elusive link. Journal of Experimental Marine Biology and Ecology 250: 117–131.

Duarte, C. M., 2002. The future of seagrass meadows. Environmental Conservation 29(2): 192–206.

Duarte, C. M. & C. L. Chiscano, 1999. Seagrass biomass and production: a reassessment. Aquatic Botany 65: 159–174.

Duarte, C. M. & K. Sand-Jensen, 1990a. Seagrass colonisation: patch formation and patch growth in Cymodocea nodosa. Marine Ecology Progress Series 65: 193–200.

Duarte, C. M. & K. Sand-Jensen, 1990b. Seagrass colonisation: biomass development and shoot demography in *Cymodocea nodosa*. Marine Ecology Progress Series 67: 93–103.

Duarte, C. M. & K. Sand-Jensen, 1996. Nutrient constraints on the establishment from seed and on vegetative expansion of the Mediterranean seagrass *Cymodocea nodosa*. Aquatic Botany 54: 279–286.

ENEA, 1997. Realizzazione di un sistema integrato di difesa dall'inquinamento delle acque marine costiere del Golfo Tigullio e del Golfo Paradiso. ENEA, La Spezia Rapporto Tecnico Ambiente.

Furnari, G., G. Giaccone, M. Cormaci, G. Alongi & D. Serio, 2003. Biodiversità marina delle coste italiane: catalogo del macrofitobenthos. Biologia Marina Mediterranea 10(1): 1–482.

Guidetti, P., M. C. Buia, M. Lorenti, M. B. Scipione, V. Zupo & L. Mazzella, 2001. Seasonal trends in the Adriatic seagrass communities of *Posidonia oceanica* (L.) Delile, *Cymodocea nodosa* (Ucria) Ascherson, *Zostera marina* L.: plant phenology, biomass partitioning, elemental composition and faunal features. In Faranda F., L. Guglielmo & G. C. Spezie (eds.), Mediterranean Ecosystems: Structures and Processes. Springer, Milano, 289–295.

ISTAT, 2000. Statistiche Metereologiche. Annuari 1986–2000. ISTAT, Roma.

Leriche, A., C. F. Boudouresque, B. Guillaume, P. Bonhomme & J. Denis, 2004. A one-century suite of seagrass bed maps: can we trust ancient maps? Estuarine, Coast and Shelf Science 59: 353–362.

Marbà, N., J. Cebrián, S. Enríquez & C. M. Duarte, 1994. Migration of large-scale subaqueous bedforms measured with seagrasses (*Cymodocea nodosa*) as tracers. Limnology and Oceanography 39(1): 126–133.

Morri, C., C. N. Bianchi, V. Damiani, A. Peirano, G. Romeo & L. Tunesi, 1988. L'ambiente marino tra Punta della Chiappa e Sestri Levante (Mar Ligure): profilo ecotipologico e proposta di carta bionomica. Bollettino dei Musei e degli Istituti Biologici dell'Università di Genova 52(Suppl.1986): 213–231.

Peirano, A. & C. N. Bianchi, 1997. Decline of the seagrass *Posidonia oceanica* in response to environmental disturbance: a simulation-like approach off Liguria (NW Mediterranean Sea). In Hawkins L. E. & S. Hutchinson (eds), The response of marine organisms to their environments. University of Southampton, UK, 87–95.

Pergent, G, C. Pergent-Martini & C. F. Boudouresque, 1995. Utilisation de l'herbier à *Posidonia oceanica* comme indicateur biologique de la qualité du milieu littoral en Méditerranée: état des connaissances. Mesogée 54: 3–27.

Relini, G., 2000. Nuovi contributi per la conservazione della biodiversità marina in Mediterraneo. Biologia Marina Mediterranea 7(3): 173–211.

Short, F. T., R. G. Coles & C. Pergent-Martini, 2001 Global seagrass distribution. In Short F. T. & R. G. Coles (eds), Global seagrass research methods. Elsevier Science, Amsterdam, 5–30.

Tunesi, L., M. E. Piccione & S. Agnesi, 2002. Progetto pilota di cartografia bionomica dell'ambiente marino costiero della Liguria. Proposta di un Sistema Informativo Geografico per la gestione di cartografie bionomiche e sedimentologiche. Quaderni ICRAM 2: 1–112.

Wabnitz C., M. Taylor, E. Green & T. Razak, 2003. From Ocean to Aquarium. UNEP-WCMC Biodiversity Series No 17, Cambridge, UK: 1–66.

Hydrobiologia (2007) 580:43–56
DOI 10.1007/s10750-006-0466-8

BIODIVERSITY IN ENCLOSED SEAS

Biodiversity evaluation of the macroalgal flora of the Gulf of Trieste (Northern Adriatic Sea) using taxonomic distinctness indices

Carlo Ceschia · Annalisa Falace · Richard Warwick

Abstract Recently a new index has been proposed for the evaluation of biodiversity: taxonomic distinctness. One of the positive features of this index is that it is neither sampling-effort nor sample-size dependent. Until now, its application has been limited to the assessment of zoobenthos and fish biodiversity.

The main objective of this paper was to test the applicability of this index to the macroalgal flora of the Gulf of Trieste (Northern Adriatic Sea). For this purpose the flora recently censused in this area was compared with a checklist of the entire region compiled from the literature. Two indices were mainly used for this study: average taxonomic distinctness based on presence/absence data (AvTD), and variation in taxonomic distinctness (VarTD). Their relationship with species richness was also assessed. The distinctness was compared with statistically significant limits estimated using randomisation tests made on the local master species list. On the same data set multivariate analysis based on a taxonomic similarity index was performed, and a 2nd stage MDS was used to compare results at four taxonomic levels.

The results confirmed that statistical over-threshold situations can be highlighted only by one index (AvTD or VarTD) or only by their relationship (AvTD vs VarTD) and not necessarily at the same time by both indices and their relationship. While the average distinctness (AvTD) did not show values significantly different from the expected ones, the variation (VarTD) in one site showed values significantly exceeding confidence limits. This situation has already been described for zoobenthic communities and explained as a consequence of a decrease in habitat diversity. In this case it might be the result of the reduced presence of hard substrata suitable for macroalgal colonisation.

The joint analysis of both parameters (AvTD vs VarTD) revealed one site exceeding the 95% confidence limit, which was not identified by analysing only one parameter at a time. This significant over-threshold pattern in the relationship of the distinctness indices could be explained by a relative increase of Rhodophyceae, attributable to the intensive grazing of the sea urchins at this site. The analysis of taxonomic distinctness indices at each site compared with the 95% probability funnels or ellipses derived from the regional species pool gave results consistent with the ones obtained

Guest editors: G. Relini & J. Ryland
Biodiversity in Enclosed Seas and Artificial Marine Habitats

C. Ceschia (✉) · A. Falace
Department of Biology, University of Trieste,
Via L. Giorgieri 10, Trieste 34127, Italy
e-mail: carlo.ceschia@tin.it

R. Warwick
Plymouth Marine Laboratory, Prospect Place,
West Hoe, Plymouth PL1 3DH, UK

using multivariate analysis. The results of this study suggest that the method may have more general validity.

Keywords Biodiversity · Taxonomic distinctness · Algal flora · Northern Adriatic Sea

Introduction

The influence of human disturbances may cause changes in the spatial and temporal distribution patterns of organisms, loss in species diversity and alteration of physical and biological habitat structure (Littler et al., 1983; D'Antonio, 1986; Brown et al., 1990; Airoldi et al., 1995; Airoldi & Virgilio, 1998; Gorostiaga et al., 1998; Walker & Kendrick, 1998; Munda, 2000; Benedetti-Cecchi et al., 2001).

A great number of ecological studies consider the number of species present in an area as the elementary descriptor of biodiversity (Convention on Biological Diversity (CBD), 1992; Heywood & Baste, 1995; Clayton, 1998; Huisman et al., 1998; Phillips, 1998). Drawing up check lists and establishing databases may represent an essential first step in the development of strategies for environmental management and conservation, and comparing present and past data constitutes a useful tool for the evaluation of ecosystem changes in the medium and long time (Stork & Samways, 1995; Stork et al., 1996; Huisman et al., 1998; Phillips, 1998; Giangrande, 2003; Martellos et al., 2004). During recent years the reconstruction of historical data has become of great interest in relation to changes and regressions of coastal biotic assemblages under the influence of natural and anthropogenic factors, as pointed out by several authors (Underwood, 1996; Munda, 2000; Falace, 2000; Giangrande, 2003). However, it is important to stress that variations in the number of recorded species in checklists produced at different times may often provide results that are difficult to interpret (Falace, 2000; Falace et al., 2005). The observed differences may depend not only on "natural factors" (e.g. diversity of habitats or environmental gradients) but also on sampling methods, specific systematic knowledge, nomenclature discrepancies from author to author, and systematic revisions over time. Moreover, as there are strong connections between the sampling scale and the processes which influence biodiversity, further difficulties connected with the comparison of species lists concern the definition of sampling areas (sampling stations that are not properly defined or insubstantial names to describe sampling sites), which make it difficult to compare biogeographical areas at different times (Falace, 2000; Falace et al., 2005).

To evaluate spatio-temporal biodiversity gradients, some studies have proposed the use of higher taxonomic ranks (Piazzi et al., 2002; Giangrande, 2003). For example, the biogeographical Rhodophyceae/Phaeophyceae index (Feldmann, 1937) has been proposed to assess the effects of environmental pollution on macroalgal communities (Giaccone, 1971; Drago et al., 1997), because of the general wider tolerance to disturbance of Rhodophyceae compared with the more sensitive Phaeophyceae. However, no unequivocal results have been obtained and this index proven not to be applicable everywhere (Verlaque, 1976; Falace, 2000; Falace et al., 2005). Functional groups or keystone species assessments have also been proposed to monitor biotic changes, but this subject is still debated and not yet completely accepted (Piazzi et al., 2002; Giangrande, 2003).

More recently taxonomic distinctness, that is a measure of the taxonomic structure of a community present in a specific site, has been proposed as a new biodiversity evaluation index (Warwick & Clarke, 1995, 2001).

Average Taxonomic Distinctness of a sampling site or region has been defined as the average taxonomic distance between any two randomly chosen species traced through a taxonomic hierarchy, or the average degree to which species in the assemblage are related to each other (Clarke & Warwick, 1998). It is therefore a measure of taxonomic spread rather than species richness. Disturbed biotic assemblages have been shown to comprise species that are closely related to one another, with low average taxonomic distinctness, while undisturbed communities include species belonging to a wide range of higher taxa. A second index is the Variation in Taxonomic

Distinctness, which is a measure of the variation in path lengths through the taxonomic tree. This index is usually high in disturbed situations as some taxa become over-represented and others under-represented. These two indices have been shown to be independent of species richness, and one of their most positive features is their lack of dependence on sampling-effort or sample-size (Clarke & Warwick, 1998). It is also possible to assess the significance of their departure from expectation by comparing them with the null hypothesis that the species present are random selections from the regional comprehensive species list using randomisation tests. The taxonomic distinctness index has been applied to a number of groups of organisms and environmental situations, for example nematodes (Clarke & Warwick, 1998, 1999), demersal fish (Hall & Greenstreet, 1998), reef corals and macrozoobenthos (Piepenburg et al., 1997; Somerfield et al., 1997; Mistri et al., 2000) but, even though it is not a new subject in terrestrial botanical studies (Dale et al., 1989), this statistical tool has not yet been applied to an algal flora.

Macroalgae are considered good descriptors of benthic communities and are widely utilised to monitor the coastal environments (Underwood & Peterson, 1988; Walker & Kendrick, 1998; Cormaci & Furnari, 1999; Falace, 2000; Piazzi et al., 2002). In several Mediterranean areas, increases in pollution and eutrophication have been accompanied by a qualitative and quantitative decline of macroalgal stands (Cormaci & Furnari, 1999; Benedetti-Cecchi et al., 2001). In the Adriatic Sea, during the last three decades, a significant floristic impoverishment of the most sensitive taxa occurred, as a result of anthropogenic disturbance (sewage, dredging, aquaculture, industrial and agricultural discharges) (Munda, 1991, 1993a, b, 2000; Sfriso et al., 1993; Cormaci & Furnari, 1999; Falace, 2000; Falace & Bressan, 2003).

The Gulf of Trieste is a semi-enclosed shallow area (maximum depth 25 m) of the Northern Adriatic Sea. The western part is characterised by shallower waters and sandy shores until Duino, where the coast becomes rocky with discontinuous pebbly shores and a little steeper bottom slope. The original features of the coast have been changed by anthropogenic activity, which becomes more evident approaching Trieste (Brambati & Catani, 1988) and from Miramare to Punta Sottile it is difficult to find the original littoral morphology. The freshwater input from the Isonzo and Timavo Rivers particularly affects the salinity from Sistiana to Miramare, while from Duino to Aurisina several underwater karst springs may result in local decreases of salinity (Mosetti, 1988). Rivers inputs also may cause persistent water turbidity and variation in the concentrations of inorganic nutrients (Burba et al., 1994). Organic detritus is not abundant but especially in the deeper layer and during summer time this component may cause a significant energetic input to the system. The Gulf of Trieste generally shows oligotrophic conditions offshore becoming mesotrophic or eutrophic coastward (Burba et al., 1994). Variations within the current systems are highly dependent on the Bora wind (ENE), which blows in an offshore direction (Mosetti, 1988). The water current inversion due to the Bora in the superficial layer represents the more effective mechanism for the water mass renewal in the Gulf of Trieste (Stravisi, 1988). Surface temperatures generally reach a maximum in August (23.5°C) and the minimum values are registered in February (7°C) (Mosetti, 1988). Close to the shore homothermy usually occurs from October to April while a thermocline is most evident in summer (Mosetti, 1988).

In recent years environmental stresses have profoundly changed the benthic algal vegetation in terms of floristic diversity and the dominant algal associations (Falace & Bressan, 2003). In particular recent researches conducted on algal colonisation on both natural and artificial substrata have highlighted a reduction of stand of Fucales that made the vegetation very uniform and dominated, from a physiognomic point of view, by turf-forming algae (Falace & Bressan, 1994, 1996, 1999a, b, 2003; Falace, 2000).

The most comprehensive work concerning the algal flora of the surroundings of Trieste (North Adriatic Sea) is by Pignatti and Giaccone (1967). Subsequent studies on the algal flora in this area are fragmentary or restricted to its poorer summer flora or to limited areas (Ghirardelli & Pignatti, 1968; Giaccone & Pignatti, 1972; Ghirardelli et al, 1973, 1974, 1975; Bressan & Godini, 1990;

Bressan et al., 1991; Bussani & Vucovic, 1992; Franzosini et al., 1983–1984; Franzosini & Bressan, 1988; Bressan et al., 2000).

A comprehensive study has been recently carried out (Falace, 2000; Falace & Bressan, 2003) aimed at evaluating the long-term floristic changes, by comparing the present flora with the one listed in the same area by Pignatti and Giaccone in 1967.

The aim of the present work was to test the applicability of the Taxonomic Distinctness indices to the macroalgal flora of the Gulf of Trieste (Northern Adriatic Sea), in order to verify the advantages of this method as a tool for the biodiversity evaluation and to establish conservation priorities.

Materials and methods

To test the method and to build up the floristic master list, data from Pignatti and Giaccone (1967) and Falace (2000) were employed. Pignatti and Giaccone (1967) reported the floristic data in a single comprehensive check list for the Gulf of Trieste, while Falace (2000) provided distinct lists for the nine sampling sites considered as representative of the entire studied area (Fig. 1). The congruence and continuity between the two studies, regarding the sampling methods and the choice of the sampling sites, were provided by the information exchange between the authors (Falace, 2000).

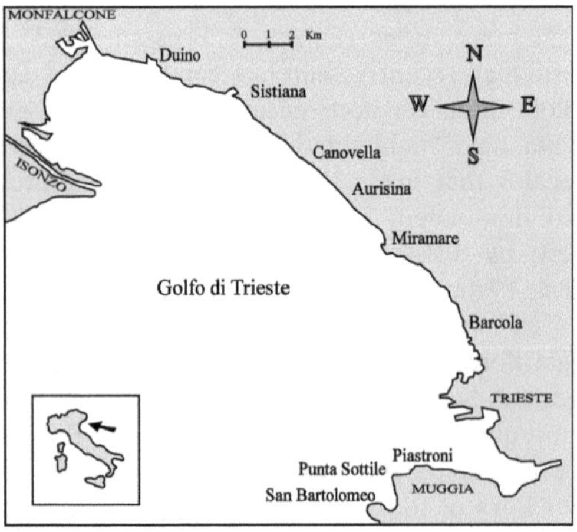

Fig. 1 Map showing locations of the sampling sites

Scientific names of the two lists were updated following Furnari et al. (2003) and the classification was arranged according to Silva et al. (1996).

The biodiversity in the nine sampling sites listed by Falace (2000) was assessed through:

- Evaluation of the diversity of each sampling site, by means of the Shannon diversity index and the quantitative Taxonomic Distinctness (Δ^*) on frequency data, and the Average Taxonomic Distinctness (AvTD) and its Variation (VarTD) on qualitative data, as proposed by Warwick & Clarke (1995). The need for data correction was evaluated before the calculation of qualitative Taxonomic Distinctness. The correlation (Pearson coefficient) between the results obtained with equal weighting between hierarchical levels (W: 11111) and weighting proportional to taxonomic richness, was verified. The assessment of Distinctness, using equal weighting between each hierarchical level resulted to be sufficient, and this method was employed in the following analysis.
- Expected distinctness tests to verify the significance of the departure of the taxonomic distinctness results from the expected values were performed (Warwick & Clarke, 1998; Clarke & Warwick, 1998; 2001b). The distinctness (AvTD, VarTD and their relationship) was compared with statistically significant 95% confidence limits (funnel and ellipse shaped) using randomisation tests based on the local master species list. Where statistically significant values resulted, the same analysis was repeated, compressing the two higher level ranks (W-2) or removing taxonomic groups, in order to evaluate their contribution to the values of the index (Clarke & Warwick, 1999).
- We also analysed the relationships between sites, by means of classifications performed with taxonomic similarity index (Izsak & Price, 2001), Bray-Curtis similarity measure (Bray & Curtis, 1957), Jaccard index (Jaccard, 1912) applied to species, genus, family and order qualitative data and monthly frequency data. The results of these classifications were then compared using MDS 2nd stage (Som-

erfield & Clarke, 1995) by applying the Spearman rank correlation. To evaluate grouping difference to the 6 similarity matrices the Pearson Chi-square coefficient was applied.Long-term biodiversity change evaluation was performed by comparing the AvTD and VarTD calculated on the Pignatti and Giaccone (1967) floristic data with the Falace (2000) data. The results were compared with statistically significant limits as before.

Results

The number of species (S), Shannon diversity (H′), Δ* and AvTD showed higher values at San Bartolomeo, Aurisina and Miramare (Table 1,

Table 1 Average and Variance of Taxonomic Distinctness referred to the sampling sites and calculated with equal weighting between taxonomic levels (*W* = 11111)

Sampling sites	S	AvTD	VarTD	Delta*	H′(loge)
DUINO	96	84.6	286.7	84.3	4.215
SISTINA	77	85.7	283.8	85.0	4.103
CANOVELLA	92	86.3	295.6	86.5	4.278
AURISINA	121	86.8	323.3	88.0	4.506
MIRAMARE	119	86.7	294.8	87.1	4.484
BARCOLA	102	85.2	361.0	84.5	4.312
PIASTRONI	105	84.7	360.8	86.6	4.389
PUNTA SOTTILE	90	84.5	391.0	86.2	4.284
SAN BARTOLOMEO	127	87.1	284.4	87.5	4.576

Fig. 2). The S, H′, Δ* and AvTD indices did not show any clear latitudinal trends since, excepting at San Bartolomeo, the highest values were observed in the middle of the latitudinal range (Miramare and Aurisina).

Comparison between the AvTD and the expected range of values based on randomised subsets of species from the master-list (Fig. 3) did not show any site outside the 95% confidence limits, while the VarTD at Punta Sottile (Fig. 4) had a higher value (391.02), exceeding the 95% confidence interval ($P = 0.025$). When compressing the two higher level ranks applying the W(-2) weighting the site values move within the confidence limits in the VarTD funnel.

The AvTD and VarTD of the sampling sites showed a weak negative correlation between them (Pearson correlation coefficient $r = -0.584$ significant at 1 tail with $P = 0.049$) and in accordance with the master list structural trend.

The analysis of the ellipse in a biplot of AvTD vs VarTD (Fig. 5) revealed that Duino was the only site outside the 95% confidence limit, related to the number of species present at this site (96 species). The removing of the Rhodophyceae from the master list and from the Duino data set (Fig. 6) changed the relationship between AvTD and VarTD of this site, now falling inside the 95% probability contour, and increased the correlation between AvTD and VarTD of the nine sites.

Classification performed using the taxonomic similarity (Fig. 7) defined three main groups of

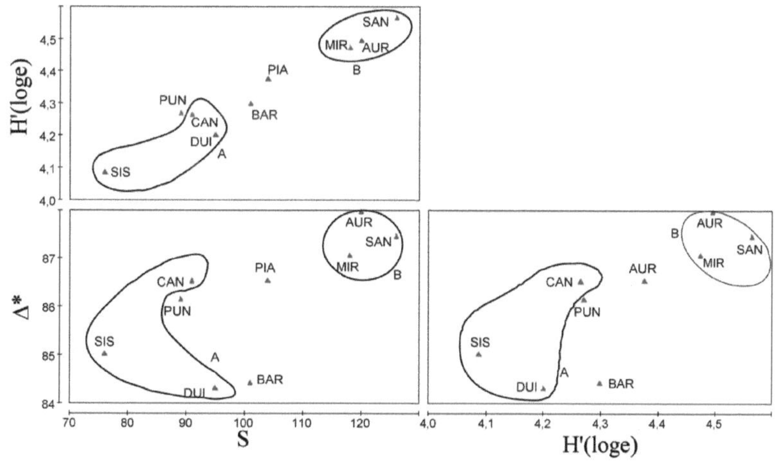

Fig. 2 Species richness (S), Diversity (H' Shannon index) and quantitative Taxonomic Distinctness (Δ*) plotted each one against the others. Grouping of sites in the northern part of Gulf (A) and of better ecological quality (B) are delineated

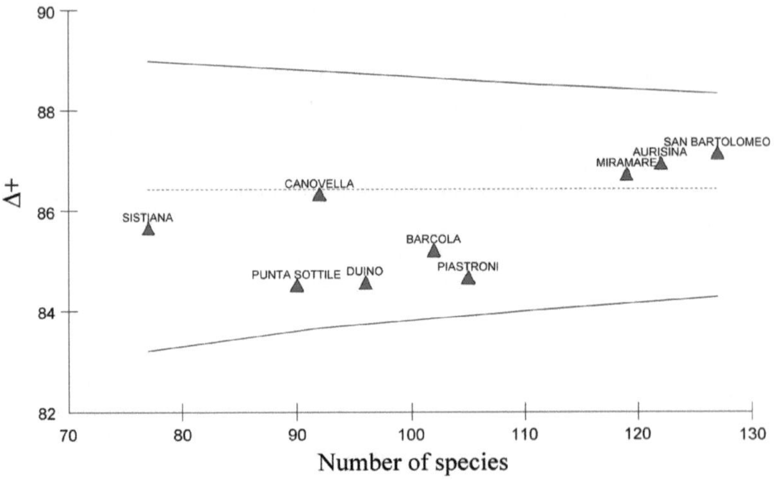

Fig. 3 Average taxonomic distinctness values Δ+ plotted against the observed number of species. Dashed line (simulated mean Δ+ value) and continuous lines (95% probability funnel) are calculated from 20000 random selections. We used equal weighting between taxonomic levels ($W = 11111$)

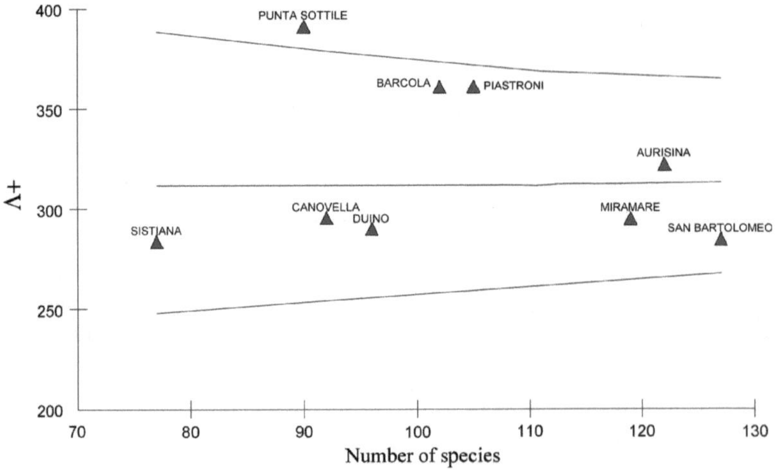

Fig. 4 Variance in taxonomic distinctness values Λ+ plotted against the observed number of species. Dashed line (simulated mean Λ+ value) and continuous lines (95% probability funnel) are calculated from 20000 random selections. We used equal weighting between taxonomic levels ($W = 11111$)

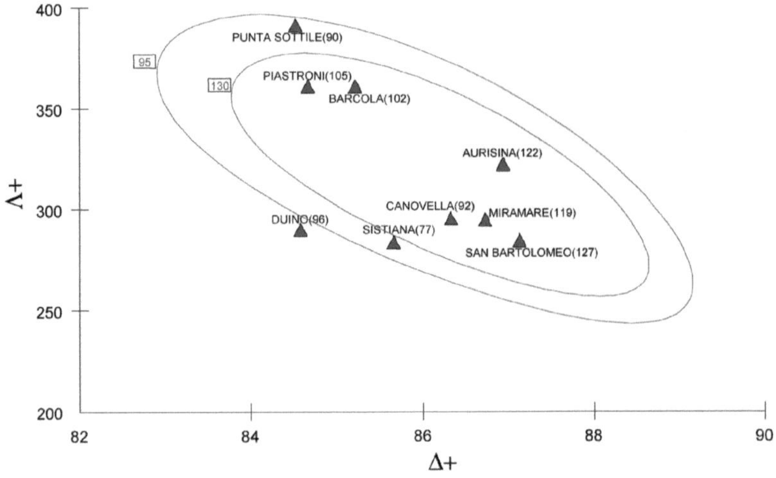

Fig. 5 Ellipse plot of 95% probability region for (AvTD, VarTD) pairs. The observed (Δ+, Λ+) values are superimposed on the plot. Only the Duino data set (96 species) departs significantly from expectation

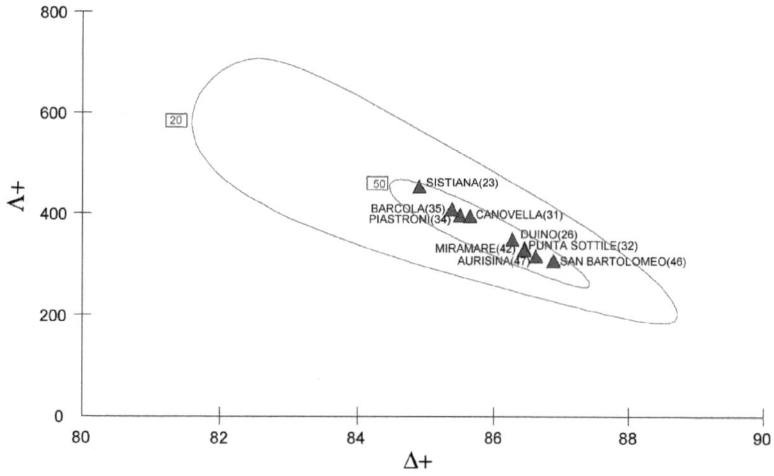

Fig. 6 Ellipse plot of 95% probability region for (AvTD, VarTD) pairs after removing the taxon Rhodophyceae. The observed (Δ+, Λ+) values are superimposed on the plot. Duino data set no longer depart significantly from expectation and there is an increase in correlation

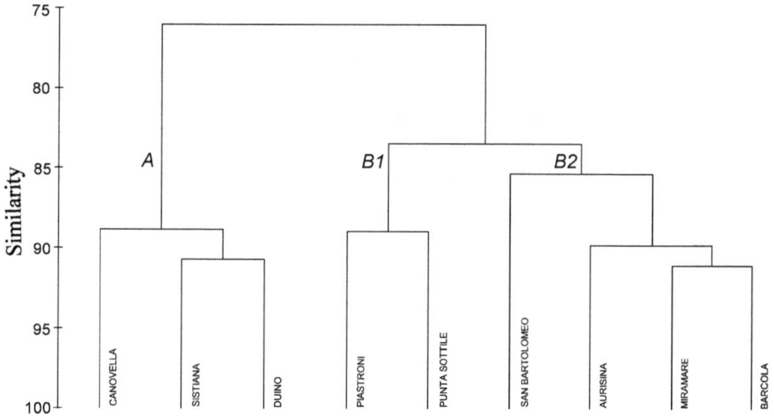

Fig. 7 Classification of the stations using taxonomic similarity (Izsak & Price, 2001) and complete linkage clustering

sites: A (Canovella, Sistiana and Duino), B1 (Piastroni, Punta Sottile), B2 (San Bartolomeo, Aurisina, Miramare, Barcola). Second stage MDS ordination (Fig. 8) indicated marked differences between patterns at the species and genus taxonomic levels compared with the family and order ones. The values of the Pearson Chi-square coefficient based on the crosstabs between classifications showed that taxonomic similarity (taxsim) does not give information different from that obtained at the species rank using the Bray-Curtis similarity on frequency data (spsim, $\chi^2 = 13.50$, d.f. = 4, $P < 0.05$) or the Jaccard index on binary data (spbin, $\chi^2 = 13.50$, d.f. = 4, $P < 0.05$).

The evaluation of long-term biodiversity changes showed a decreasing trend in AvTD, and an increase of VarTD. The flora of Pignatti and Giaccone (1967) resulted significantly different from the master list (Figs. 9 and 10) for both AvTD (AvTD = 87.752, $P = 0.001$) and VarTD (VarTD = 282.497, $P = 0.002$). Finally, also the ellipse plot of the relationships between the AvTD and VarTD confirmed these results. The Falace (2000) flora showed a lower Average Taxonomic Distinctness (85.644) and a higher Variation (343.256) but in this case neither AvTD and VarTD nor their relationship evidenced any significant difference from the expected values.

Discussion

According to previous observations (Falace, 2000; Falace & Bressan, 2003) all the diversity indices

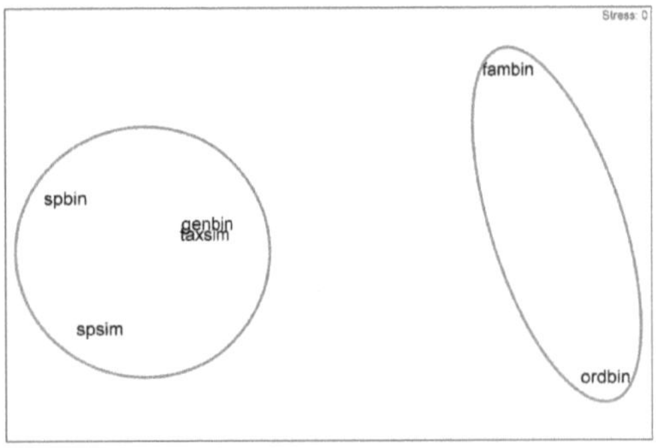

Fig. 8 Second-stage MDS based on 6 similarity matrices between sampling sites obtained using Jaccard index on binary data at different taxonomic levels (spbin, genbin, fambin, ordbin), using taxonomic similarity index on binary data (taxsim) and the Bray-Curtis similarity index on frequency data at species level (spsim)

tested agree in classifying San Bartolomeo, Aurisina and Miramare as higher diversity sites, while classifications and ordinations highlighted a strong similarity among sampling sites in the northern part of the Gulf (Canovella, Sistiana, Duino). A reduced algal cover, mainly represented by turf species, and a discontinuous vegetal colonisation characterise the latter (Falace, 2000). In particular the expected taxonomic distinctness test and the procedure involving removal of taxonomic groups at the species rank highlighted for Duino a significantly higher number of Rhodophyceae with a concomitant lower number of Phaeophyceae in comparison with the other sites. Moreover in this station brown algae also showed a lower number of families. The observed abundance of the echinoid *Paracentrotus lividus* (Lamarck) and its selective feeding behaviour (Verlaque & Nedelec, 1983; Verlaque, 1983, 1984; Pancucci et al., 1993; Falace & Bressan, 2002) have been recognized as the most probable cause for the Phaeophyceae reduction at this site. Furthermore in the northern part of the Gulf, lower down the sublittoral, bare slopes grazed by sea urchins with some remnants of crustose coralline algae, were usual.

The higher value of VarTD at Punta Sottile and the analysis of the results after the application of the weighting corrections, identify an algal assemblage with a taxonomic structure signifi-

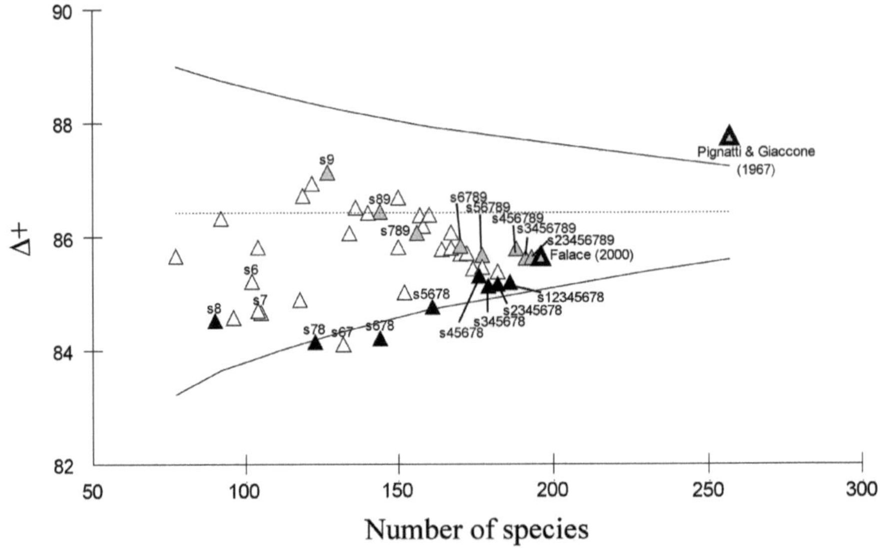

Fig. 9 Average taxonomic distinctness values Δ+ for the flora of the Gulf as described by Pignatti & Giaccone (1967) and by Falace (2000) plotted against the observed number of species. Dashed line (simulated mean Δ+ value) and continuous lines (95% probability funnel) are calculated from 20000 random selections from the regional species pool. We used equal weighting between taxonomic levels ($W = 11111$). Also cumulative data referred to adjacent stations are plotted. The stations are numbered progressively from Duino (1) to San Bartolomeo (9). Punta Sottile (8), the groups of stations containing it and San Bartolomeo (9) are highlighted, coloured black when San Bartolomeo is not included and grey in the other cases

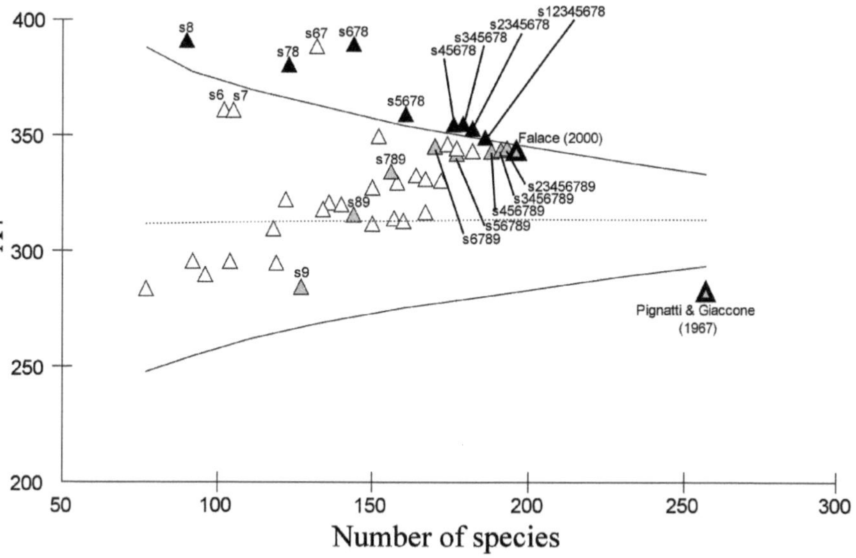

Fig. 10 Variation in taxonomic distinctness values Λ+ for the flora of the Gulf as described by Pignatti & Giaccone (1967) and by Falace (2000) plotted against the observed number of species. Dashed line (simulated mean Λ+ value) and continuous lines (95% probability funnel) are calculated from 20000 random selections from the regional species pool. We used equal weighting between taxonomic levels (W = 11111). Also cumulative data referred to adjacent stations are plotted. The stations are numbered progressively from Duino (1) to San Bartolomeo (9). Punta Sottile (8), the groups of stations containing it and San Bartolomeo (9) are highlighted, coloured black when San Bartolomeo is not included and grey in the other cases

cantly different from the one potentially present in the entire Gulf (as established by the master list). The higher unevenness in the cladistic structure at Punta Sottile was mainly found at the class and order taxonomic ranks. Similar results, reported by Clarke & Warwick (1999, 2001a) for the zoobenthic communities of the Isles of Scilly, were considered as a result of habitat diversity reduction. Some higher taxa are in fact associated with specific habitats, and the absence of such habitats may result in a more uneven distribution across the taxonomic spectrum compared to the regional average, which will encompass all habitat types. Distinctness parameters, in fact, are not only linked to disturbance by environmental pollution or anthropogenic impacts, but can also be strongly influenced by the edaphic features of the environment (Warwick & Clarke, 1995, 1998). At Punta Sottile the high VarTD values might be the result of a lower spatial heterogeneity, due to a reduced presence of hard substrata suitable for the macroalgal colonisation. Pebbles and coarse sand characterise in fact the substrata at Punta Sottile. Moreover during the sampling period this area was subjected to heavy coastline modifications, which further reduced suitable hard substrata, and made edaphic conditions worse (i.e. sediment, water transparency) (Falace, 2000). However, similar alteration was present in Piastroni without evident effects on the VarTD.

Data on the number of species, Shannon diversity and TD indices, were not indicative of a geographical trend from the north to south of the Gulf. Instead it may be supposed that the alteration of the coastline for human purposes and the anthropogenic pressures affect the algal communities. In fact, except at San Bartolomeo, the AvTD seems to show values that are positively correlated to the distance from the nearest highly urbanized area (Monfalcone harbour in the northeast part of the Gulf or Trieste in the south).

A perceived problem in assessing the significance of AvTD and VarTD departure from the null expectation, according to which both are referred to a random selection of the regional pool of species, is the definition of the regional pool of species itself. The aggregation of sites with different species compositions to obtain a comprehensive master list is unlikely to affect the AvTD or VarTD values. In fact the new added species are likely to spread themselves across the taxonomic tree without affecting the average values. In this study, by progressively adding up adjacent stations or geographically contiguous groups of them, the anomalously high value of VarTD at Punta Sottile is always evident, and is offset only by adding data from the San Bartolomeo site (Fig. 10). In some cases the combination of data from sites with AvTD and VarTD values that are not significantly different from the expected ones can result in a value that falls outside the statistical significance thresholds. This effect may be ephemeral and may disappear by simply adding another station (as happens for the group including Barcola and Piastroni); otherwise it may persist as is the case of Punta Sottile VarTD (Fig. 10).

The Taxonomic Distinctness indices of biodiversity are particularly appropriate for making such comparisons because of their lack of dependence on sampling effort (Warwick & Clarke, 2001). A complete census of all the species present in a wide region is clearly impossible and, in the case of taxonomic distinctness measures, unnecessary. This means that the evaluation of regional (gamma) diversity is possible by using a different number of samples and locations, or employing different sample sizes and survey techniques (Piepenburg et al., 1997; Hall & Greenstreet, 1998; Price et al., 1999; Rogers at al., 1999; Warwick & Light, 2002; Warwick & Turk, 2002).

The decreasing AvTD and increasing VarTD in the comparison of long term change indicate a significant loss in species diversity during the last three decades, in accordance with the effects of increased environmental changes (pollution, eutrophication, loss of habitats) reported by several authors (Munda, 1991, 1993a, b, 2000; Sfriso et al., 1993; Falace, 2000; Falace & Bressan, 2003). The floristic comparisons carried out after over 30 years (Falace, 2000; Falace & Bressan, 2003) pointed out a decrease in the number of species by 20% (from 258 to 207 species) compared to those collected in 1967. In particular a reduction of 28% of species amongst the Phaeophyceae (Ectocarpales and Fucales) and of 27% amongst the Chlorophyceae (Cladophorales) was recorded. The disappearance of a large number of epiphytic and sheltered Phaeophyta and Rhodophyta species was attributable to the decline of erect thallus algae and in particular of the larger Phaeophyceae on which they are often epiphytes. Furthermore, some species declined in quantity and were found only as single specimens in some sites or in low numbers.

One of the main features of the vegetation in the Gulf of Trieste was in fact prolific settlements of Fucales: *Fucus virsoides* J. Agardh in the eulittoral and of different *Cystoseira* species in the sublittoral (Pignatti and Giaccone, 1978). At present *Cystoseira barbata* (Stackhouse) C. Agardh and *Cystoseira compressa* (Esper) Gerloff & Nizamuddin are the only species belonging to this genus still present in the Gulf, though with reduced stands, probably surviving because of their wider tolerance to environmental stress (Falace, 2000).

The algal assemblages, at present characterised by the absence of well structured communities and vertical zonation, show a colonisation often scattered in small patches. Canopy species are often replaced by perennial turf-like mats of *Gelidium*, *Gelidiella* and *Pterosiphonia* species, especially in the lower eulittoral and subtidal zone (Falace, 2000; Falace & Bressan, 2003). The general disappearance of canopy algae in the Gulf is in accordance with a similar process observed on larger scale in the Adriatic or Mediterranean (Munda, 1993a, 2000; Cormaci & Furnari, 1999; Benedetti-Cecchi et al., 2001). A further phenomenon observed, which has been related to the increased turbidity (Falace, 2000; Falace & Bressan, 2003), is the upward migration of several species from the lower sublittoral to the eulittoral. During spring, in the northern part of the Gulf, from Duino to Miramare in the intertidal and upper subtidal zone some ephemeral species like *Porphyra leucosticta* Thuret, *Scytosiphon lomentaria* (Lyngbye) and *Ceramium ciliatum* (J. Ellis) Ducluzeau var. *ciliatum* may become dominant. Other species that may be dominant in the northernmost stations (Duino, Sistiana, Canovella) in the sublittoral and eulittoral zone are often linked to eutrophication (i.e. *Pterocladiella capillacea* (S.G. Gmelin) Santelices & Hommersand, *Ulva laetevirens* Areschoug, *Nitophyllum punctatum* (Stackhouse) Greville and *Codium fragile* (Suringar) Hariot subsp. *tomentosoides* (Goor) P.C. Silva) (Falace, 2000, Falace & Bressan, 2003).

Strangely, taxonomic distinctness tests classified the Pignatti and Giaccone census as significantly different from expectation based on the structure of the regional species pool, while describing as "normal" the recent flora. However, AvTD was above the 95% confidence intervals of the null distribution and VarTD below it, which is precisely the opposite of the situation that normally obtains in perturbed situations. These results are difficult to explain and most probably they are due to different systematic knowledge on specific groups of algae and to difficulties related to comparing floristic lists made by different authors. The results obtained from the comparison of the Taxonomic Distinctness between the old and the recent census are independent of the geographical spread of the respective sets of sampling sites because both describe in an equally representative way the floristic composition of the same area using similar methods. The relatively small geographical extension of the Gulf of Trieste, with its easily reachable coasts provide some confidence that the regional checklist is a comprehensive one, and that sampling of additional sites would not influence it. The potential impact on the results of the different number or location of sampling sites in this case can be assumed to be lower than the variability due to human factors or to the already described "author effect".

Another possible explanation is that, if the increase of AvTD and the decrease of VarTD have truly occurred, then one tail of the statistic population going out of significant limits does not represent a situation subject to disturbance but on the contrary a climax-like situation.

Conclusions

The analysis carried out and the Taxonomic Distinctness indices applied to the algal flora of the Gulf of Trieste have highlighted some situations of environmental stress, referred in particular to the northern area of the Gulf. On the contrary the algal colonisation in the southern sampling areas seemed not to display the effect of anthropogenic and ecological pressures.

Nevertheless, the Taxonomic Distinctness indices, like other indices, do not take into account the value and the ecological significance of each species, thus limiting their capability to define clear relationships between different impacts and their effects on the biotic component. Taxonomic distinctness based on frequency data (Δ^*) showed the same ranking of sites as Shannon Diversity (H′), and it is thus equally as sensitive as H′ as a disturbance index. However, AvTD and VarTD based on presence/absence data were quite insensitive as indicators of stress at sites that had been identified as disturbed by the quantitative indices, probably because all species are treated as equal and the rare or ephemeral species may cloud the picture.

In conclusion the Taxonomic Distinctness indices cannot substitute for other indices such as the number of species or Shannon diversity but, combined with them, may help in the interpretation of the information and provide an objective statistical threshold value for significance. In fact even if the Taxonomic Distinctness indices largely confirmed the trends in species diversity resulted by other indices, they also indicated that there are not simply differences in the number of species between locations, but also differences in the taxonomic spread, which are equally important.

Finally the expected distinctness test must be further investigated and developed also in order to distinguish "classes of environmental quality" for conservation purposes.

References

Airoldi, L., F. Rindi & F. Cinelli, 1995. Structure, seasonal dynamics and reproductive phenology of a filamentous turf assemblage on a sediment influenced, rocky subtidal shore. Botanica Marina 38: 227–237.

Airoldi, L. & M. Virgilio, 1998. Responses of turf-forming algae to spatial variations in the deposition of sediments. Marine Ecology Progress Series 165: 271–282.

Benedetti-Cecchi, L., F. Pannacciulli, F. Bulleri, P. S. Morchella, L. Airoldi, G. Relini & F. Cinelli, 2001. Predicting the consequences of anthropogenic disturbance: large-scale effects of loss of canopy algae on rocky shores. Marine Ecology Progress Series 214: 137–150.

Brambati, A. & G. Catani, 1988. Le coste e i fondali del Golfo di Trieste dall'Isonzo a Punta sottile: aspetti geologici, geomorfologici, sedimentologici e geotecnici. Hydrores Information 5(6): 13–28.

Bray, J. R. & J. T. Curtis, 1957. An ordination of the upland forest communities of Southern Wisconsin. Ecological Monographs 27: 325–349.

Bressan, G. & E. Godini, 1990. Alghe del Golfo di Trieste: guida allo studio. Atti Museo Civico Storia Naturale di Trieste 43: 1-201.

Bressan, G., L. Sergi & C. Welker, 1991. Variazioni della distribuzione batimetrica di macroalghe dell'infralitorale fotofilo nel Golfo di Trieste (Mare Adriatico). Bollettino Società Adriatica Scienze Naturali di Trieste 72: 107–126.

Bressan, G., F. Trebbi & L. Babbini, 2000. Variazione della distribuzione batimetrica di macrophytobenthos nel parco marino di Miramare (Golfo di Trieste) in rapporto a condizioni edafiche. Biologia Marina Mediterranea 7: 107–126.

Brown, V. B., S. A. Davies & R. N. Synnat, 1990. Long term monitoring of the effect of threated seawage effluent on the intertidale macroalgal community near Cape Schanck, Victoria, Australia. Botanica Marina 33: 85–98.

Burba, N., M. Cabrini, P. Del Negro, S. Fonda Umani & L. Dilani, 1994. Variazioni stagionali del rapporto N/P nel Golfo di Trieste. In Albertelli G., R. Cataneo-Vietti & M. Picazzo (eds). Atti X Congresso Associazione Italiana Oceanologia Limnologia, Alassio, 333–344.

Bussani, M. & A. Vucovic, 1992. Le alghe di Miramare. Hydrores Information 9: 1–48.

CBD, 1992. Convention on Biological Diversity. Text and Annexes, the Interim Secretariat for the CBD Geneva Executive Centre.

Clarke, K. R. & R. M. Warwick, 1998. A taxonomic distinctness index and its statistical properties. Journal of Applied Ecology 35: 523–531.

Clarke, K. R. & R. M. Warwick, 1999. The taxonomic distinctness measure of biodiversity: weighting of step lengths between hierarchical levels. Marine Ecology Progress Series 184: 21–29.

Clarke, K. R. & R. M. Warwick, 2001a. A further biodiversity index applicable to species lists: variation in taxonomic distinctness. Marine Ecology Progress Series 216: 265–278.

Clarke, K. R. & R. M. Warwick, 2001b. Changes in marine communities: An approach to statistical analysis and interpretation. 2nd edn. PRIMER-E Ltd., Plymouth Marine Laboratory, Plymouth, UK.

Clayton, M. N., 1998. Macroalgal Biodiversity: concepts and problems. Botanica Marina 41: 87–88.

Cormaci, M. & G. Furnari, 1999. Changes of the benthic algal flora of the Tremiti Islands (southern Adriatic), Italy. Hydrobiologia 398/399: 75–79.

D'Antonio, C. M., 1986. Role of sand in the domination of hard substrata by the intertidal alga *Rhodomela latrix*. Marine Ecology Progress Series 27: 263–273.

Dale, M., E. Feoli & P. Ganis, 1989. Incorporation of information from the taxonomic hierarchy in comparing vegetation types. Taxon 38: 216–227.

Drago, D., A. M. Mannino & S. Sortino, 1997. La vegetazione sommersa dei mari siciliani. L'epos, Palermo, 117 pp.

Falace, A., 2000. Variazioni fisionomiche spazio-temporali della vegetazione sommersa del Golfo di Trieste: analisi delle principali influenze ambientali. Ph.D. Thesis. University of Trieste.

Falace, A. & G. Bressan, 1994. Some observations on periphyton colonization of artificial substrata in the Gulf of Trieste (North Adriatic Sea). Bulletin Marine Sciences 55(2): 926–933.

Falace, A. & G. Bressan, 1996. Some observations of artificial structures situated in the proximity of underwater pipes off Lignano-Grado (North Adriatic). Southampton Oceanographic Centre: 313–318.

Falace, A. & G. Bressan, 1999a. Quantitative evaluation of algal community on an artificial reef in the Gulf of Trieste (Northern Adriatic Sea). Proceedings 7th International Conference on Artificial Reefs and Related Aquatic Habitats, Sanremo, Italy, 7–11 October, 173–178.

Falace, A. & G. Bressan, 1999b. Phytobenthic colonization on panels with different slope in the Gulf of Trieste (North Adriatic Sea). Proceedings 7th International Conference on Artificial Reefs and Related Aquatic Habitats, Sanremo, Italy, 7–11 October, 623–629.

Falace, A. & G. Bressan, 2002. A qualitative-quantitative analysis of the evolution of macroalgal colonisation on an artificial reef with antigrazing nets (Loano-Ligurian Sea). ICES Journal of Marine Science 59: 150–156.

Falace, A. & G. Bressan, 2003. Changes of algal flora in the Gulf of Trieste (Northern Adriatic Sea). Bocconea 16: 1033–1037.

Falace, A., A. Di Pascoli & G. Bressan, 2005. Valutazione della biodiversità nella Riserva Marina di Miramare (Nord Adriatico): macroalghe marine bentoniche. Biologia Marina Mediterranea 12(1): 88–98.

Feldmann, J., 1937. Recherches sur le vegetation marine de la Méditerranée. La côte des Alberes. Revue Algologique 10: 1–339.

Franzosini, C. & G. Bressan, 1988. Calibrazioni metodologiche nello studio del macrophytobenthos della Riserva- Parco Marino di Miramare (Trieste) Italy: 1. Rilievi senza prelievi. Atti Museo Civico Storia Naturale di Trieste 41: 143–159.

Franzosini, C., V. Verardo, L. A. Ghirardelli & G. Bressan, 1983–1984. La flora algale presso il Laboratorio di Biologia Marina di Aurisina-Filtri (Trieste – North Adriatic Sea): Macrophytobenthos. Nova Thalassia 6: 83–95.

Furnari, G., G. Giaccone, M. Cormaci, G. Alongi & D. Serio, 2003. Biodiversità marina delle coste italiane: catalogo del macrophytobenthos. Biologia Marina Mediterranea 10: 1–482.

Ghirardelli, E., G. Orel & G. Giaccone, 1973. L'inquinamento del Golfo di Trieste. Atti Museo Civico Scienze Naturali di Trieste 28: 431–450.

Ghirardelli, E., G. Orel & G. Giaccone, 1974. Evolution des peuplements benthiques du Golfe de Trieste. Revue International Océanographie. Méditerranée 35/36: 111–113.

Ghirardelli, E., G. Orel & G. Giaccone, 1975. Esperienze sullo scarico a mare a Trieste. Metodologie e ricerche per la valutazione degli effetti sul benthos. Ingegneria Ambientale 4: 414–418.
Ghirardelli, E. & S. Pignatti, 1968. Conséquences de la pollution sur les peuplements du "Vallone di Muggia" près de Trieste. Revue International Océanographie. Méditerranée 10: 111–112.
Giaccone, G., 1971. Significato biogeografico ed ecologico di specie algali delle coste italiane. Natura e Montagna 4: 40–45.
Giaccone, G. & S. Pignatti, 1972. Vegetazione algale costiera del Golfo di Trieste. Informatore Botanico Italiano 3: 188–189.
Giangrande, A., 2003. Biodiversity, conservation and the "taxonomic impediment". Acquatic Conservation 13: 451–459.
Gorostiaga, J. M., A. Santolaria, A. Secilla & I. Diez, 1998. Sublittoral benthic vegetation of the Easthern Basque coast (N-Spain): structure and environmental factors. Botanica Marina 41: 455–465.
Hall, S. J. & S. P. Greenstreet, 1998. Taxonomic distinctness and diversity measures: responses in marine fish communities. Marine Ecology Progress Series 166: 227–229.
Heywood, V. H. & R. T. Baste, 1995. Introducing Biodiversity. In Watson R. T & V. Heywood et al. (eds), Global Biodiversity Assessment. Cambridge University Press, Cambridge, 5–18.
Huisman, J. M., R. A. Cowan & T. J. Entwisle, 1998. Biodiversity of Australian Marine Macroalgae. A progress report. Botanica Marina 41: 89–93.
Izsak, C. & A. R. G. Price, 2001. Measuring ß-diversity using a taxonomic similarity index, and its relation to spatial scale. Marine Ecology Progress Series 215: 69–77.
Jaccard, P., 1912. The distribution of flora in the alpine zone. The New Phytologist 11: 37–50.
Littler, M. M., D. R. Martz & D. S. Littler, 1983. Effects of recurrent sand deposition on rocky intertidal organisms: importance of substrate heterogeneity in a fluctuating environment. Marine Ecology Progress Series 11: 129–139.
Martellos, S., G. Alongi, A. Falace, M. Cormaci, G. Giaccone & P. L. Nimis, 2004. Biodiversità on-line: una chiave interattiva delle alghe dell'Alto Adriatico e della Sicilia Orientale. Oral communication 35th SIBM-Genoa, 19–20 July.
Mistri, M., V. U. Ceccherelli & R. Rossi, 2000. Taxonomic distinctness and diversity measures: responses in lagoonal macrobenthic communities. Italian Journal of Zoology 67: 297–301.
Mosetti, F., 1988. Condizioni idrologiche della costiera Triestina. Hydrores Information 5(6): 29–38.
Munda, I., 1991. Algal resources in polluted sites of the Northern Adriatic (vicinity of Piran). Acta Adriatica 32: 682–704.
Munda, I., 1993a. Changes and degradation of seaweed stands in the Northern Adriatic. Hydrobiologia 260/261: 239–253.
Munda, I., 1993b. Impact of pollution on benthic marine algae in the Northern Adriatic. International Journal Environmental Studies 43: 185–199.
Munda, I., 2000. Long-term marine floristic changes around Rovinj (Istrian coast, North Adriatic) estimated on the basis of historical data from Paul Kuckuck's field diaries from the end of the 19th century. Nova Hedwigia 71: 1–36.
Pancucci, M. A., P. Panayotidis & A. Zenetos, 1993. Morphological changes in sea urchins populations as a response to environmental stress. In Aldrich J. C. (ed.), Quantified Phenotipic Responses in Morphology and Physiology. JAPAGA, Ashford, 247–257.
Phillips, J. A., 1998. Marine conservation initiatives in Australia: their relevances to the conservation of macroalgae. Botanica marina 41: 95–103.
Piazzi L., G. Pardi, D. Balata, E. Cecchi & F. Cinelli, 2002. Seasonal dynamics of a subtidal North-Western Mediterranean Macroalgal Community in relation to depht and substrate inclination. Botanica Marina 45: 243–252.
Piepenburg, D., J. Voss & J. Gutt, 1997. Assemblages of sea stars (Echinodermata: Asteroidea) and brittle stars (Echinodermata: Ophiuroidea) in the Weddel Sea (Antarctica) and off Northeast Greend (Arctic): a comparison of diversity and abundance. Polar Biology 17: 305–322.
Pignatti, A. & G. Giaccone, 1967. Studi sulla produttività primaria del fitobentos nel Golfo di Trieste - I: Flora sommersa del Golfo di Trieste. Nova Thalassia 3: 1–17.
Price, A. R. G., M. J. Keeling & C. J. O'Callaghan, 1999. Ocean-scale patterns of 'biodiversity' of Atlantic asteroids determined from taxonomic distinctness and other measures. Biological Journal of the Linnean Society 66: 187–203.
Rogers, S. I., K. R. Clarke & J. D. Reynolds, 1999. The taxonomic distinctness of coastal bottom-dwelling fish communities of the North-east Atlantic. Journal of Animal Ecology 68: 769–782.
Sfriso, A., A. Marcomini, B. Pavoni & A. A. Orio, 1993. Species composition, biomass, and net primary production in shallow coastal waters: the Venice Lagoon. Bioresource Technology 44: 235–250.
Silva, P. C., P. W. Basson & R. L. Moe, 1996. Catalogue of the Benthic Marine Algae of the Indian Ocean. University of California. Publications in Botany 29: 1–1259.
Somerfield, P. J. & K. R. Clarke, 1995. Taxonomic levels, in marine community studies, revisited. Marine Ecology Progress Series 127: 113–119.
Somerfield, P. J., F. Olsgard & M. R. Carr, 1997. A further examination of two new taxonomic distinctness measures. Marine Ecology Progress Series 154: 303–306.
Stork N. E. & M. J. Samways, 1995. Inventorying and monitoring. In Watson R. T. & V. Heywood et al. (eds), Global Biodiversity Assessment. Cambridge University Press, Cambridge: 453–543.
Stork, N. E., M. J. Samways & C. Eeley, 1996. Inventorying and monitoring biodiversity. Tree 11: 34–40.
Stravisi, F., 1988. Caratteristiche oceanografiche del Golfo di Trieste. Hydrores 5(6): 39–45.

Underwood, A. J., 1996. Detection, interpretation, prediction and management of environmental disturbances; some roles for experimental marine ecology. Journal of Experimental Marine Biology and Ecology 200: 1–27.

Underwood, A. J. & C. H. Peterson, 1988. Towards an ecological framework for investigating pollution. Marine Ecology Progress Series 46: 227–234.

Verlaque, M., 1976. Etude de l'impact du rejet thermique de Martigues-Ponteau sur le macrophytobenthos. Tethys 1: 19–46.

Verlaque, M., 1983. Alimentation des juveniles de *Paracentrotus lividus* (Lamarck). Préférences alimentaire de l'espéce et impact sur le phytobenthos de substrat rocheux de Corse (Méditerranée, France). Symbioses 15: 223–224.

Verlaque, M., 1984. Biologie des juveniles de l'oursin herbivore *Paracentrotus lividus* (Lamarck): séléctivité du broutage et impact de l'espèce sur le communautes algales de substrat rocheux de Corse (Méditerranée, France). Botanica Marina 26: 401–424.

Verlaque, M. & H. Nedelec, 1983. Biologie de *Paracentrotus lividus* (Lamarck) sur substrat rocheux de Corse (Méditerranée, France): alimentation des adultes. Vie et Milieu 33: 191–201.

Walker, D. I. & G. A. Kendrick, 1998. Threats to macroalgal diversity: marine habitat destruction and fragmentation, pollution and introduces species. Botanica Marina 41: 105–112.

Warwick, R. M. & K. R. Clarke, 1995. New 'biodiversity' measures reveal a decrease in taxonomic distinctness with increasing stress. Marine Ecology Progress Series 129: 301–305.

Warwick, R. M. & K. R. Clarke, 1998. Taxonomic distinctness and environmental assessment. Journal of Applied Ecology 35: 532–543.

Warwick, R. M. & K. R. Clarke, 2001. Practical measures of marine biodiversity based on relatedness of species. Oceanography and Marine Biology. Annual Review 39: 207–231.

Warwick, R. M. & J. Light, 2002. Death assemblages of molluscs on St. Martin's Flats, Isles of Scilly: a surrogate for regional biodiversity? Biodiversity and Conservation 11: 99–112.

Warwick, R. M. & S. M. Turk, 2002. Predicting climate change effects on marine biodiversity: comparison of recent and fossil molluscan death assemblages. Journal of the Marine Biological Association of the United Kingdom 82: 847–850.

Hydrobiologia (2007) 580:57–75
DOI 10.1007/s10750-006-0465-9

BIODIVERSITY IN ENCLOSED SEAS

Biodiversity of settled material in a sediment trap in the Gulf of Trieste (northern Adriatic Sea)

Tamara Cibic · Oriana Blasutto · Serena Fonda Umani

Abstract Phytoplankton succession and sinking rates were studied from January to December 2003 at a coastal station in the Gulf of Trieste (northern Adriatic Sea), 200 m offshore, in a relatively undisturbed area. A conical sediment trap, moored at 15 m depth (water depth 17 m), was used. The hypothesis if the presence of benthic and epiphytic diatoms can lead to an overestimation of the vertical fluxes was tested. To evaluate primary and secondary sedimentation contributions, planktonic, benthic and epiphytic diatoms were distinguished. Benthic species abundance varied throughout the year and it was related to resuspension that strongly influenced sinking rates. All over the year, diatoms were the prevailing class in the trap material accounting for 75.32% of the settled cells, while flagellates represented 24.11%. Dinophyceae and resting cells constituted minor components, accounting for 0.43% and 0.14%, respectively. The gross sedimentation rates ranged from 0.006×10^8 cell m^{-2} d^{-1} in the second week of May to 6.30×10^8 cell m^{-2}d^{-1} in the third week of January with a mean annual value of $1.09 \pm 1.43 \times 10^8$ cell m^{-2} d^{-1}. To the primary sedimentation rate *Pseudo-nitzschia seriata* of the group *"Nitzschia seriata complex"* contributed for 49.77% followed by *Chaetoceros* spp. (23.88%). The major contributor to the secondary sedimentation rate was the diatom *Paralia sulcata*, accounting for 24.76%. Epiphytic diatoms contributed for 11.19% and 12.27% on annual average gross abundance and biomass, respectively, reaching even 72.04% of gross abundance and 56.06% of gross biomass in the second week of August. The correlation between temperature and the logarithm of the epiphytic biomass was statistically significant, with $r = 0.66$ and $P < 0.001$. Both in the cluster analysis and in the PCA four main groups were formed, where benthic and epiphytic species were separately gathered. Planktonic, benthic and epiphytic forms accounted for 50.78%, 36.95% and 12.27%, respectively, calculated on the annual average biomass. Therefore, vertical fluxes can be overestimated of 50% or more if benthic and epiphytic species are not rejected.

Guest editors: G. Relini & J. Ryland
Biodiversity in Enclosed Seas and Artificial Marine Habitats

T. Cibic (✉)
Biological Oceanography Department, National Institute of Oceanography and Experimental Geophysics, Via A. Piccard, 54, 34010 Trieste, Italy
e-mail: tcibic@ogs.trieste.it

O. Blasutto
Regional Environmental Protection Agency – FVG, Piazza Collalto, 15, 33057 Palmanova, Udine, Italy

S. Fonda Umani
Department of Biology, University of Trieste, Via A. Valerio, 28/A, 34127 Trieste, Italy

Keywords Sediment trap · Biodiversity · Benthic diatoms · Phytoplankton · Gulf of Trieste

Introduction

Phytoplankton abundance and biomass variability, community structure and succession, including suspended benthic microalgae, are generally controlled by several interacting factors that change during the annual cycle of a pelagic ecosystem (Margalef, 1978; Heiskanen, 1998). The dominant life forms of algal cells are selected by the water column's physical structure and by nutrient availability. In shallow coastal ecosystems temperature and light control phytoplankton succession, while water column mixing is one of the most important factors affecting organic matter sedimentation (Margalef, 1978).

Sediment traps are widely used in oceanic and coastal environments to measure the vertical flux of particulate material in the water column. These measurements are commonly used to estimate carbon loss rates from the euphotic surface layer. Sediment traps can be also used to determine the biodiversity of settled material (Heiskanen, 1995). Several researchers have investigated the downward flux of the settling material by mooring or drifting sediment traps in the northern Adriatic Sea (Giordani & Frignani, 1988; Puškarić et al., 1992; Matteucci & Frascari, 1997; Miquel et al., 1999; Giani et al., 2001) and in the Gulf of Trieste (Faganeli, 1989; Posedel & Faganeli, 1991; Faganeli et al., 1995). In the latter environment the nature and the contribution of living particles to trap collections have received relatively little attention (Wassmann et al., 1998).

The relevant presence of benthic diatoms as temporary members of the phytoplankton is a well known phenomenon. In the water column of coastal environments, also epiphytic and epilithic diatoms, resuspended by tidal currents and waves, significantly contribute to the total phytoplankton abundance and biomass (de Jonge & van Beusekom, 1992; Lucas et al., 2001 and references therein).

In this study phytoplankton biodiversity in a sediment trap was investigated in order to test the hypothesis if the presence of benthic and epiphytic diatoms can lead to an overestimation of the vertical flux of particulate material. In sublittoral areas when sediment traps are used for carbon, hydrogen and nitrogen fluxes determinations, the resuspension events and the proliferation of epiphytic flora should be taken into consideration. An overestimation of the organic material fluxes is likely to occur if benthic and epiphytic forms are not rejected. For this reason, in this paper, planktonic, benthic and epiphytic species were separately considered and their contribution to the gross sedimentation rate was estimated.

Study site

The Gulf of Trieste, located at the north-western end of the Adriatic Sea, is a shallow embayment of about 600 km^2 and a coastline of about 100 km. It is almost completely surrounded by land except to the southwest, where it is limited by an imaginary line connecting Punta Tagliamento in Italy with Punta Salvore in Slovenia and it is isolated from the rest of the Adriatic by a sill from Grado to the Salvore peninsula (Ogorelec et al., 1991); 10% of its area is <10 m and maximum depth is about 25 m. Average salinities range from 33 psu to 38 psu at the surface and from 36 psu to 38.5 psu at the bottom. Annual temperatures fluctuate from 8°C to ≥24°C at the surface and from 8°C to ≥20°C at the bottom. Tidal amplitude is about 1.5 m, which is the highest in the Mediterranean Sea (Cardin & Celio, 1997). Water enters the Gulf from the southeast and circulation at the surface is predominantly from southeast to northwest. Sedimentation is controlled mainly by river input rather than by marine currents (Brambati & Catani, 1988).

Main natural factors influencing the characteristics of the composition, evolution and persistence of marine life in the Gulf of Trieste, are winds and stratification of the water column. The strongest wind, called Bora, comes from the northeast and breaks the otherwise strong stratification of the gulf waters. Vertical mixing can be markedly enhanced by wave motion induced by strong Bora gales that can produce significant resuspension phenomena of the sediment (Celio et al., 2002).

Materials and methods

Salinity and temperature were measured on vertical profiles using a Multiparameter Probe OCEAN SEVEN 316 IDRONAUT or an SBE 19 plus SEACAT PROFILER twice a month at a station (T21) 50 m nearby the sediment trap.

The sediment trap (Model Technicap PPS4/3, with a collecting area of 0.05 m^2) was moored at a coastal site of the Gulf of Trieste, 200 m offshore (45°42′ 1.15″ N, 13°42′55.52″ E), within the Marine Reserve of Miramare, sheltered from boats, fishing and swimmers, representing a relatively undisturbed area. The trap was positioned at 15 m (2 m above the bottom), where the bottom depth is approximately 17 m (Fig. 1a, b). The mooring system is made up of a heavy platform which is connected to a mooring buoy by a line and a little ballast, positioned at a fixed distance from the sediment trap. The main advantage of this mooring system is that the water column above the trap mouth is kept clear, without any lines constituting an attachment area for epiphytic flora and fauna.

Sampling was carried out from 7th January to 15th December 2003 (the sediment trap was not available from 25th January to 22nd April). During the third week of August the trap was

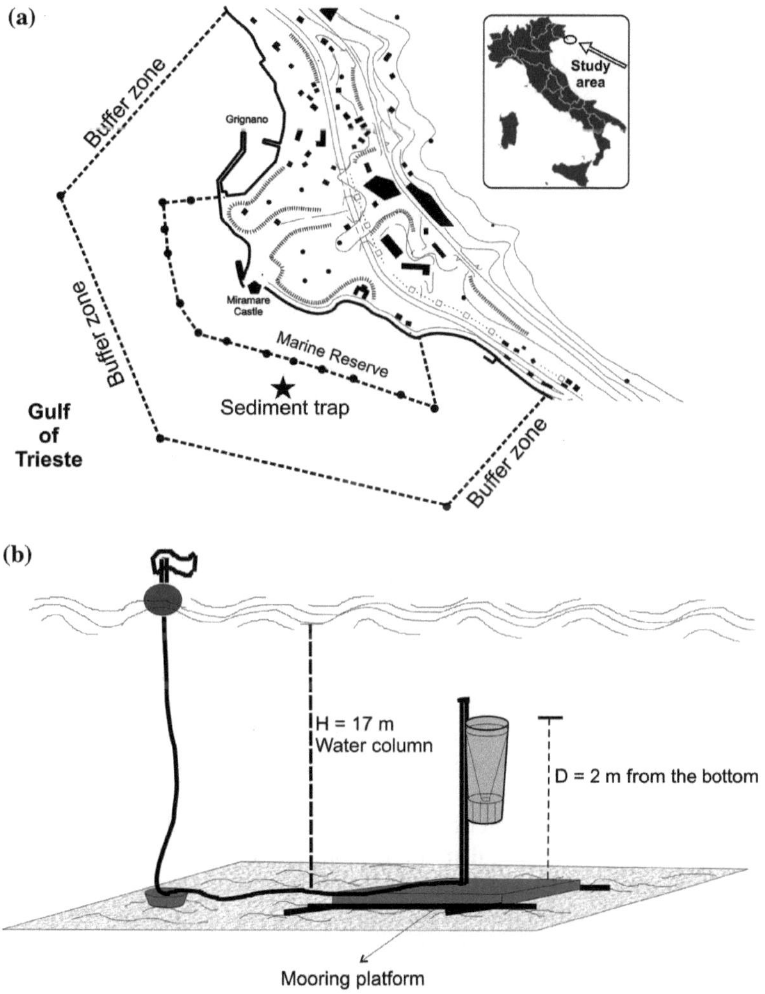

Fig. 1 Location of the sediment trap (**a**) and its schematic representation (**b**)

landed and subjected to cleaning. Flora and fauna were scraped off the inner wall of the trap which was lowered again. The maintenance was performed within a day. Seven sample bottles (260 ml) were filled with filtered seawater (0.22 μm Nucleopore filter) with in situ salinity. Neutralized formalin (pH 8.2–8.3) was added to yield a 4% final concentration to prevent decomposition and microbial growth. The bottles were screwed into a rotary disk perpendicular to the funnel; when the disk rotated, one sample bottle moved forward and the next bottle replaced it at a set time previously programmed. Each bottle was set for 6 days, so the temporal sequence used for the sampling was based on weeks composed of 6 days, e.g. the first week of January went from the 1st to the 6th, the second one from the 7th to the 12th, and so on. The collected bottles were sealed and stored at 4°C until analysis. The recovered trap samples were kept homogenous by agitation and transferred into 6 falcon tubes containing 50 ml. Among these 6 falcon tubes, 3 ones, having the same weight, were chosen for 3 replicates, manually stirred, diluted from 1:2 to 1:50 when necessary and transferred into a counting Lund chamber. For the qualitative and quantitative determination of phytoplankton flux the viable cells at the time of fixing were counted under a Leitz inverted light microscope (Utermöhl, 1958) using a 32× objective. Cell counts of phytoplankton assemblages were carried out to the genus and, when possible, to the species level using floras of Van Heurck (1899), Germain (1981), Dexing et al. (1985), Ricard (1987) and Tomas (1997).

The sedimentation rate was calculated from the number of viable cells considering the area of the sediment trap, the volume of each bottle and the time of exposure. Sedimentation rates were converted into relative abundance (RA, %). Microalgal biovolume was calculated to assess the RA as biomass (carbon content) of occurring specimens varying in shape and size. A set of geometric shapes and mathematical equations was used for biovolume calculations from microscopically measured linear dimensions that included the entire range of microalgal shapes (Hillebrand et al., 1999). The carbon content of microalgal cells was calculated from the transformation of cell volume to plasma volume including an estimate of the vacuole volume, and the calculation of cell carbon was based on the plasma volume multiplied by a factor of 0.11 (Strathmann, 1967). Finally, relative biomass values were calculated (RB, %).

To evaluate primary and secondary sedimentation contributions, we distinguished planktonic, benthic and epiphytic diatoms according to the ecology of each species. Both benthic and epiphytic species contributed to secondary sedimentation rates. Epiphytic species develop inside the sediment trap. Benthic diatoms sink in the trap due to resuspension phenomena. Planktonic diatoms can have origin from direct sedimentation but can also be deposited and subsequently resuspended together with the benthic species. Microscopical analysis of the phytobenthos performed during the year of study revealed that on average only 9% of the species in the sediment were planktonic (data not published). Moreover, planktonic diatoms observed in the sediment are usually in appalling condition and only species with a hardy frustule, e.g. *Pseudo-nitzschia seriata*, can be recognized. For these reasons we considered that the contribution of planktonic species to secondary sedimentation was negligible.

Gross sedimentation rate was obtained considering primary and secondary sedimentation contributions. All rates were estimated both as abundance (ABU, cell m^{-2} d^{-1}) and biomass (BIOM, mg C m^{-2} d^{-1}).

Univariate diversity analysis was performed using PRIMER–5 software (Clark & Warwick, 2001), considering richness (Margalef, d), equitability (Pielou, J′), diversity (Shannon, H′) and dominance (Simpson, λ) (Shannon & Weaver, 1949; Simpson, 1949; Pielou, 1966; Margalef, 1986).

For cluster analysis based on species and principal component analysis (PCA) the number of species was reduced performing a ranking of the variables which were ordered following their decreasing specific variance. The first 23 species, which corresponded to a cumulative variance of 99.99%, were used (Orloci, 1978). Cluster analysis and PCA were carried out using MATEDIT software (Burba et al., 1992). For cluster analysis based on species the correlation coefficient was

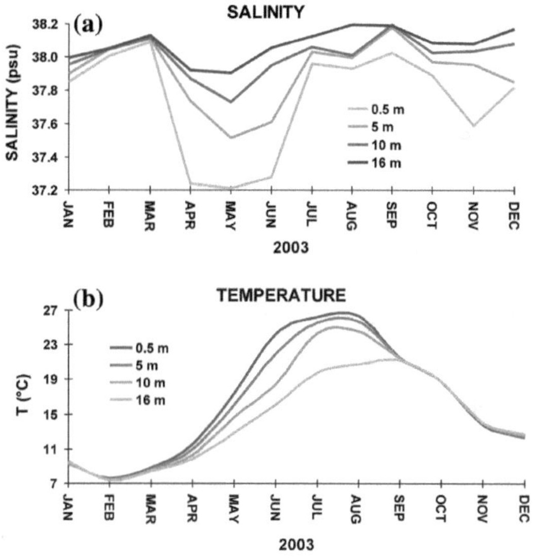

Fig. 2 Salinity (**a**) and temperature (**b**) recorded at 4 depths, 50 m nearby the sediment trap

employed to transform the data matrix and the complete linkage was applied. For cluster analysis based on samplings the absolute distance and the complete linkage were applied. PCA was based on r algorithm (correlation coefficient). Simultaneous ordination of species and samplings was obtained using first and second autovectors and principal components.

Results

Thermohaline parameters are shown in Fig. 2a, b. Two distinct periods were recognized: a spring–summer period of stratified water masses and an autumn–winter period with a well-mixed water column. The stability of the water column was stronger from April to September. The lowest temperature was measured in February (7.4°C at the bottom). The highest temperature was observed in August at the surface layer

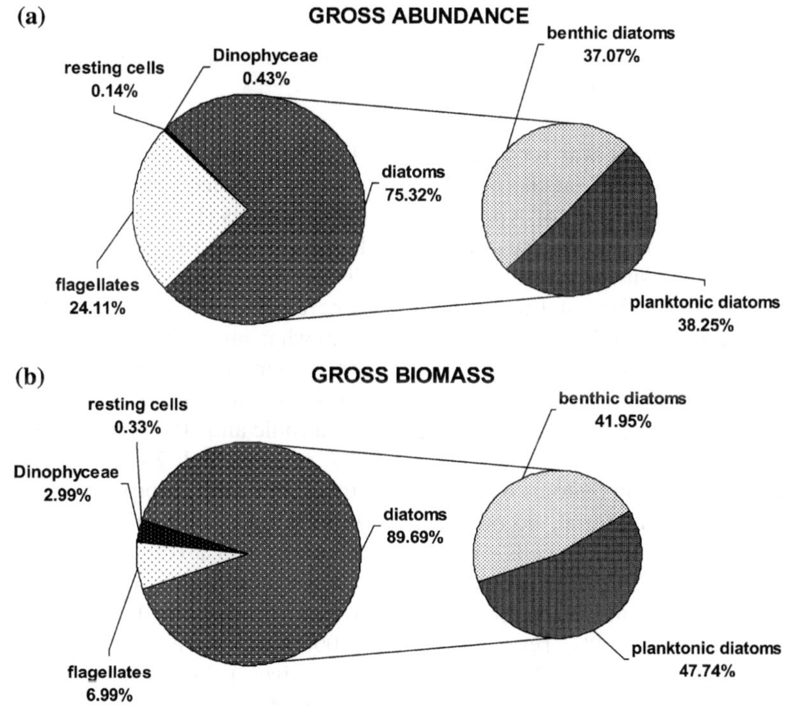

Fig. 3 Gross sedimentation rate expressed as abundance (**a**) and biomass (**b**)

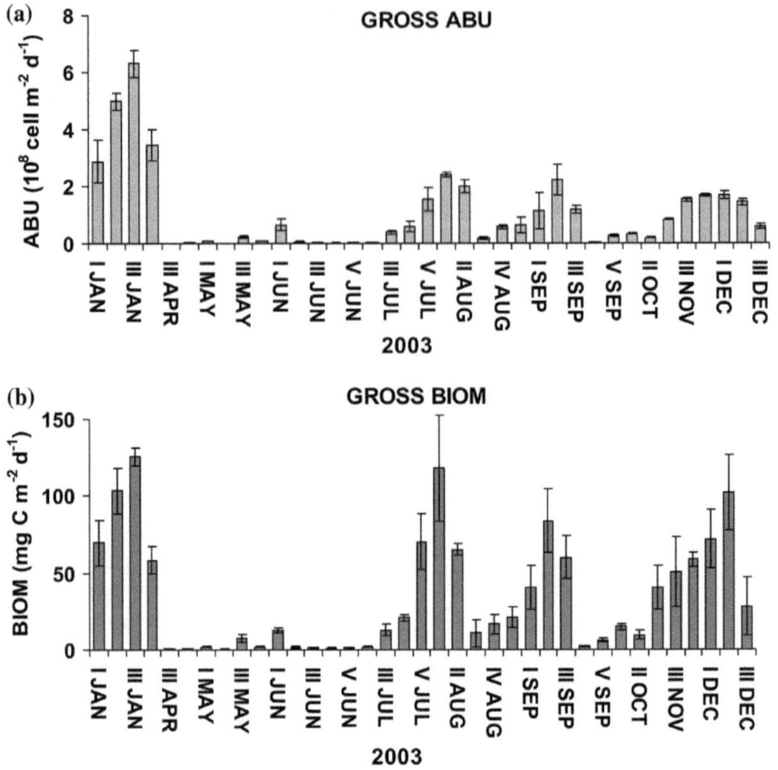

Fig. 4 Gross abundance (**a**) and gross biomass (**b**) variations during the year of study. Data are presented as mean values of three replicates with standard deviations

(26.3°C), while at the bottom it was recorded in September (21.5°C). During 2003, at the bottom layer, salinity was >38.00 psu except for April and May. The absolute minimum (37.21 psu) and a relative minimum (37.59 psu) of salinity were recorded in May and November at the surface layer due to a moderate input of the Isonzo River. The highest average values of salinity through the water column were reached in March and September (38.11 ± 0.02 psu and 38.14 ± 0.08 psu, respectively).

Throughout the year of study diatoms were the prevailing taxa in the trap material, accounting for 75.32% of the total settled cells, while flagellates represented 24.11%. Dinophyceae and resting cells were minor components accounting for 0.43% and 0.14%, respectively (Fig. 3a). Diatoms represented 89.69% of the total biomass, while flagellates and Dinophyceae accounted for 6.99% and 2.99%, respectively (Fig. 3b). Considering different characteristics of diatoms like ecology, physiology, shape and size of the species (Hendey, 1976; Dexing et al., 1985; Round et al., 1992; Hoagland et al., 1993) we discriminated between benthic and planktonic genera. This distinction was, at times, difficult and somewhat an arbitrary task (Philibert & Prairie, 2002). Among all the 133 Bacillariophyceae taxa found in the trap, we identified 68 benthic, 29 planktonic and 10 epiphytic species (for a total of 107 species and 26 genera). Benthic diatoms (together with epiphytic ones) represented 49.22% of the total diatoms and planktonic diatoms 50.78% (Fig. 3a).

Minimum values of gross ABU and BIOM were recorded in the second week of May (0.006×10^8 cell m^{-2} d^{-1} and 0.38 mg C m^{-2} d^{-1}, respectively). Maxima for both parameters were observed in the third week of January (6.30×10^8 cell m^{-2} d^{-1} and 125.37 mg C m^{-2} d^{-1}).

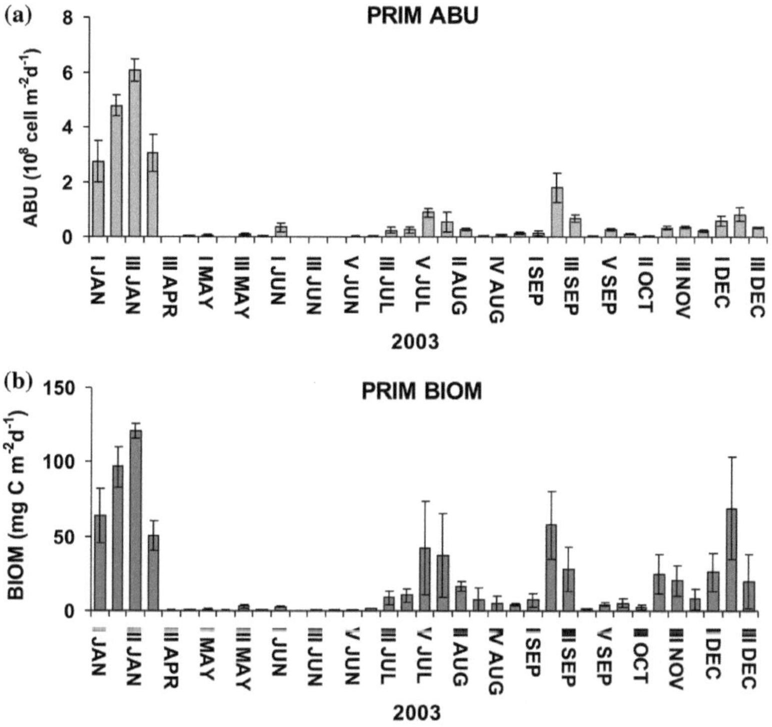

Fig. 5 Primary abundance (**a**) and primary biomass (**b**) variations during the year of study. Data are presented as mean values of three replicates with standard deviations

Gross mean annual ABU and BIOM were $1.09 \pm 1.43 \times 10^8$ cell m^{-2} d^{-1} and 35.00 ± 37.52 mg C m^{-2} d^{-1} (Fig. 4a, b).

Planktonic Bacillariophyceae together with flagellates, Dinophyceae and resting cells were responsible for primary sedimentation. Planktonic diatoms were the major component of the settled material heavily weighing on primary sedimentation rate. The maximum ABU and BIOM of primary sedimentation were recorded in the third week of January (6.07×10^8 cell m^{-2} d^{-1} and 120.75 mg C m^{-2} d^{-1}, respectively). On the contrary, the minimum BIOM was observed in the second week of June (0.20 mg C m^{-2} d^{-1}), while the lowest ABU was recorded one month before (0.004×10^8 cell m^{-2} d^{-1}) (Fig. 5a, b).

The highest secondary sedimentation rates, both for ABU and BIOM, were recorded during the first week of August (1.89×10^8 cell m^{-2} d^{-1} and 80.46 mg C m^{-2} d^{-1}, respectively). The lowest secondary sedimentation rates for ABU and BIOM were recorded during the second week of May (0.002×10^8 cell m^{-2} d^{-1} and 0.034 mg C m^{-2} d^{-1}, respectively) (Fig. 6a, b).

The most abundant species of diatoms, which accounted for 75.32% of the total settled cells (89.69% as biomass), found in the trap throughout the year (Table 1) were: *Pseudo-nitzschia seriata* which accounted for 49.77% of primary ABU and *Chaetoceros* spp. (23.88%), while to primary BIOM *Pseudo-nitzschia seriata* accounted for 31.14%, *Thalassionema frauenfeldii* for 13.92%, followed by *Chaetoceros* spp. and *Rhizosolenia* spp. with 8.38% and 7.18%, respectively (Fig. 7a, b). Secondary ABU was determined by *Paralia sulcata* (24.76%), *Navicula mollis* and *N. corymbosa* together (12.48%), *Navicula* spp.1 and *Diatoma* cfr. *vulgare* (10.74% and 10.11%, respectively). On the other hand, *Diatoma* cfr. *vulgare, Paralia sulcata, Synedra* spp. and *Plagiotropis* cfr. *gaussii* together reached 38.03% of the total secondary carbon flux (Fig. 8a, b).

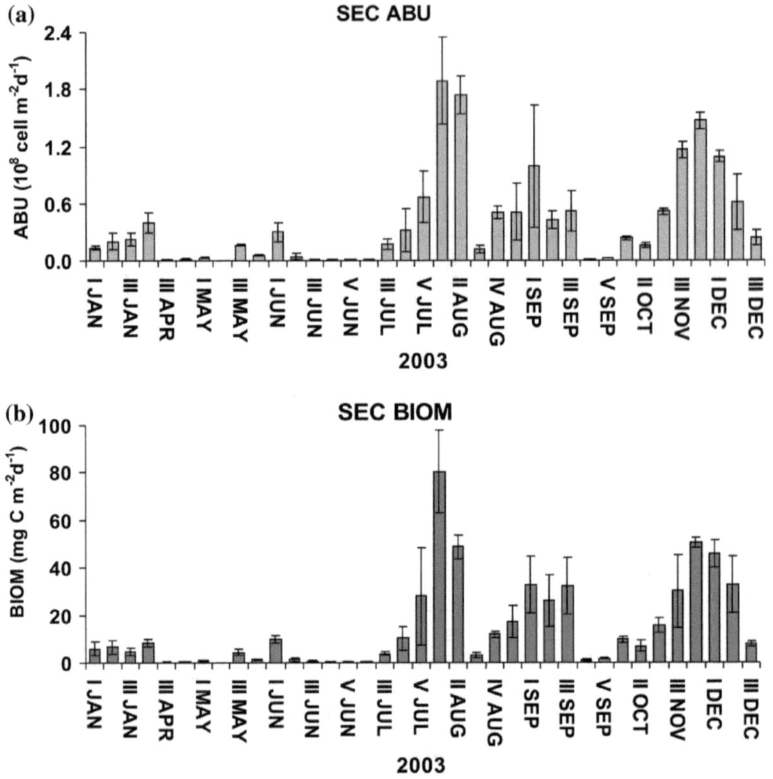

Fig. 6 Secondary abundance (**a**) and secondary biomass (**b**) variations during the year of study. Data are presented as mean values of three replicates with standard deviations

In Fig. 9a, b we can observe the incidence of epiphytic species in the summer period. To include them in secondary ABU and BIOM considerably affected temporal trends of settling rates. A drop in the epiphytic ABU was observed in the third week of August, when the maintenance of the trap was performed.

In Fig. 10 the linear regression between temperature and the logarithm of the epiphytic BIOM is presented. The correlation was statistically significant, with $r = 0.66$, $n = 25$, $P < 0.001$.

The planktonic species *Pseudo-nitzschia seriata* and *Chaetoceros* spp., the epiphytic *Diatoma* cfr. *vulgare* and the benthic *Paralia sulcata* reached high values of ABU or BIOM in certain periods of the year, heavily weighing on sedimentation rates. *Pseudo-nitzschia seriata* was found in the trap only in January while *Chaetoceros* spp. settled in the trap from the fifth week of June to the end of the year, with the highest ABU in the second week of September. *Diatoma* cfr. *vulgare* was observed especially in July and August; *Paralia sulcata* was present over the entire year of study with the highest ABU in the fourth week of November (Fig. 11).

Using univariate diversity analysis we found the maximum value of λ (0.81) in conjunction with the minimum value of H′ (0.59), J′ (0.19), d (0.91) and the minimum number of species (S = 21) during the third week of January. On the contrary, the lowest value of λ (0.08) was observed in the third week of August, in conjunction with the highest value of J′ (0.81). The maximum value of H′ (3.00) was calculated in the fifth week of August, while the maximum value of d was recorded in the fourth week of July (2.15) with the highest S (46) (Table 2).

Table 1 Floristic list. *Navicula* spp.1 < 20 μm, 20 μm < *Navicula* spp.2 < 40 μm and *Navicula* spp.3 > 40 μm

CHRYSOPHYTA
Chrysophyceae
Dictyocha fibula Ehrenberg
Dictyocha speculum Ehrenberg
Dictyocha spp.
Bacillariophyceae
**Achnanthes* spp.
**Amphiprora* spp.
**Amphora arenaria* Donkin
**Amphora coffaeformis* Kützing
**Amphora hyalina* Kützing
**Amphora* cfr. *lineolata* Ehrenberg
**Amphora ostrearia* Brébisson
**Amphora ovalis* Kützing
**Amphora rhombica* Kitton
**Amphora* spp.
Asteromphalus spp.
Asterionellopsis glacialis Round
°*Auricula insecta* Grunow
°*Auricula* spp.
Azpetia nodulifera Schmidt
Azpetia spp.
°*Bacillaria paxillifera* Hendey
Bacteriastrum delicatulum Cleve
Bacteriastrum spp.
**Campylodiscus decorus* var. *pinnatus* Peragallo
Cerataulina pelagica Hendey
Cerataulina spp.
Chaetoceros decipiens Cleve
Chaetoceros teres Cleve
Chaetoceros spp.
**Climacosphenia moniligera* Ehrenberg
**Cocconeis* spp.
Coscinodiscus spp.
Cyclotella cfr. *comta* Kützing
Cyclotella glomerata Bachmann
Cyclotella spp.
**Cylindrotheca closterium* Lewin & Reimann
**Cymbella* spp.
°*Diatoma* cfr. *vulgare Bory*
Diatoma spp.
**Diploneis bomboides* Cleve
**Diploneis bombus* Ehrenberg
**Diploneis ovalis* Cleve
**Diploneis smithii* Cleve
**Diploneis* spp.
Ditylum brightwellii Grunow (vide Van Heurck)
**Epithemia* spp.
**Eunotia* cfr. *lunaris* Grunow
Fragilaria cfr. *crotonensis* Kitton
Fragilaria spp.
Fragilariopsis spp.
**Grammatophora marina* Kützing
Guinardia delicatula Hasle
Guinardia flaccida Peragallo
Guinardia striata Hasle

Table 1 continued

CHRYSOPHYTA
Guinardia cfr. *tubiformis* Hasle
Guinardia spp.
**Gyrosigma acuminatum* Rabenhorst
**Gyrosigma attenuatum* Rabenhorst
**Gyrosigma balticum* Rabenhorst
**Gyrosigma fasciola* Griffith et Henfrey
**Gyrosigma macrum* Griffith et Henfrey
**Gyrosigma spencerii* Griffith et Henfrey
**Gyrosigma* spp.
Hemiaulus hauckii Grunow (vide Van Heurck)
Leptocylindrus danicus Cleve
**Licmophora flabellata* Agardh
**Licmophora gracilis* Grunow
**Licmophora* spp.
Lioloma cfr. *pacificum* Hasle
Lioloma spp.
**Melosira* spp.
**Navicula* cfr. *alpina* Ralfs
°*Navicula corymbosa* Cleve
**Navicula directa* W. Smith
**Navicula* cfr. *divergens* Ralfs
**Navicula* cfr. *longa* Ralfs
**Navicula* cfr. *liber* W. Smith
°*Navicula mollis* Cleve
**Navicula* spp.1
**Navicula* spp.2
**Navicula* spp.3
**Nitzschia acicularis* W. Smith
**Nitzschia acuminata* Grunow
**Nitzschia angularis* W. Smith
°*Nitzschia dissipata* Grunow
**Nitzschia fasciculata* Grunow
°*Nitzschia fruticosa* Hustedt
Nitzschia longissima Ralfs
**Nitzschia lorenziana* Grunow
**Nitzschia lorenziana* var. *densestriata* A. Schmidt
**Nitzschia obtusa* var. *nana* Grunow
**Nitzschia panduriformis* Gregory
**Nitzschia punctata* Grunow
**Nitzschia sigma* Smith
**Nitzschia sigma* var. *intercedens* Grunow
**Nitzschia sigma* var. *sigmatella* Grunow
**Nitzschia sigmoidea* Smith
°*Nitzschia subtubicola* Germain (nova specie)
**Nitzschia tryblionella* Hantzsch
**Nitzschia vermicularis* Grunow
**Nitzschia* spp.
**Paralia sulcata* Cleve
**Pinnularia cardinalis* W. Smith
**Pinnularia viridis* Ehrenberg
**Pinnularia* spp.
**Plagiotropis* cfr. *gaussii* Paddock
**Pleurosigma aestuarii* W. Smith
**Pleurosigma angulatum* W. Smith
**Pleurosigma elongatum* W. Smith
**Pleurosigma formosum* W. Smith
**Pleurosigma minutum* Grunow

Table 1 continued

CHRYSOPHYTA
**Pleurosigma normanni* Ralfs
**Pleurosigma* spp.
**Podosira* spp.
Proboscia alata Sundström
Proboscia alata indica Sundström
Pseudo-nitzschia cfr. *seriata* Hasle
Pseudo-nitzschia spp.
Rhizosolenia imbricata Brightwell
Rhizosolenia spp.
**Rhopalodia gibba* O. Muller
**Rhopalodia* spp.
°*Striatella unipunctata* Agardh
**Surirella* spp.
°*Synedra* cfr. *gallionii* Ehrenberg
**Synedra* spp.
°*Tabellaria fenestrata* Kützing
°*Tabellaria flocculosa* Kützing
Thalassionema frauenfeldii Hallegraeff
Thalassionema nitzschioides Mereschkowsky
Thalassionema spp.
Thalassiosira eccentrica Cleve
Thalassiosira spp.
**Toxarium hennedyanum* Grunow
**Toxarium undulatum* Bailey
**Tropidoneis lepidoptera* Cleve
**Tropidoneis longa* Cleve
**Tropidoneis* spp.
DYNOPHYTA
Ceratium furca Claparède & Lachmann
Ceratium spp.
Gonyaulax spp.
Gymnodinium spp.
cfr. *Gymnodinium* spp.
Prorocentrum spp.
Prorocentrum micans Ehrenberg
EUGLENOPHYTA
Eutreptia spp.
RESTING SPORES AND CYSTS
Chaetoceros resting spores spp.
Pyrocystis lunula Schütt
Dinoflagellate cysts spp.

(*) = benthic species, (°) = epiphytic species

The result of the hierarchical clustering based on species is represented in Fig. 12. The dendrogram revealed a clear distribution of epiphytic and planktonic species within the two main groups A and B.

In Fig. 13 the cluster analysis on the basis of samplings revealed three main groups. The group A gathered all samplings performed in January. In the group B the majority of samplings were positioned. Going from the left to the right, samplings in this group showed an increase in abundance. In fact, samplings in the subgroup E, characterized by the highest abundances, were positioned on the far right of the cluster. Finally, the group C was composed of the two samplings, performed in the first and second week of August, when the highest ABU of epiphytic species was observed.

Figure 14 shows an ordination plot that accounts for 37.84% of the total variance. The principal component 1 (PC1) axis explained 25.90% of the total variance, while the PC2 axis explained 11.94% of the remaining variance. Samplings and species could be seasonally gathered in four groups on the basis of the temperature at the bottom. In the group I the majority of samplings carried out from June to the beginning of October were located, when high temperature at the bottom was recorded. In the group II samplings performed in January were gathered, when very low temperature was observed. Samplings from April to June were located in the group III. In this period the warming of the water column began. Finally, November and December were positioned in the group IV, when the temperature of the water body started to lower.

Discussion

During the year of study Bacillariophyceae was the prevailing class found in the trap material, followed by a smaller percentage of flagellates and Dinophyceae. Since the carbon content of flagellates is lower than that of diatoms and the presence of Dinophyceae was only occasional, almost all the total settled carbon was due to diatoms.

The abundances of vertical fluxes found during our study were on average lower than those reported by other authors (Passow, 1991; Heiskanen & Kononen, 1994) for traps located at depths ranging from 10 m to 60 m in the Baltic Sea, while they were slightly higher than those obtained from traps moored at 50 m depths in the Northeastern Pacific Ocean (Silver & Gowing, 1991). The qualitative and quantitative analyses of the phytoplankton are usually performed on water samples collected from different depths

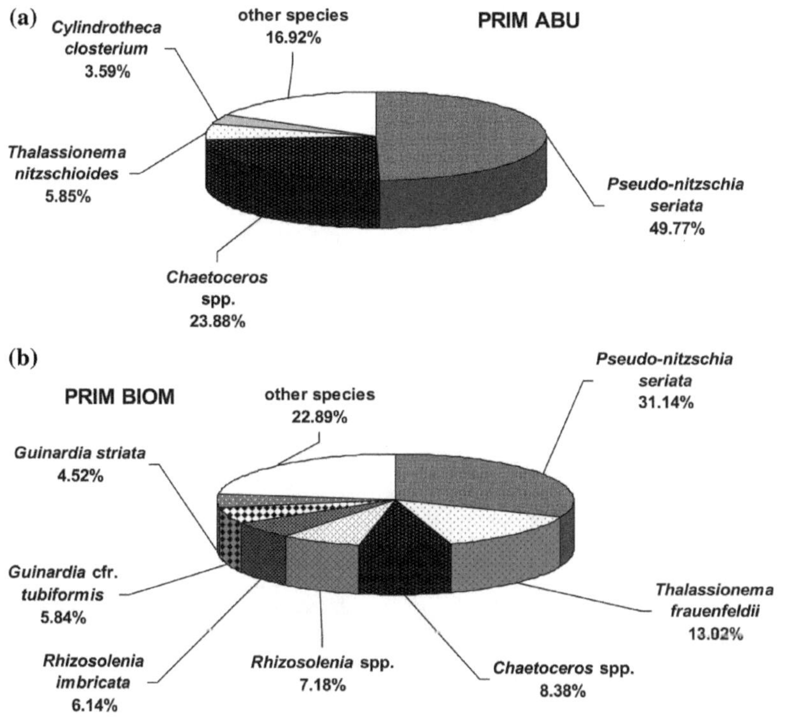

Fig. 7 Dominant species responsible for the primary sedimentation expressed as abundance (**a**) and biomass (**b**)

along the water column by means of bottles or other types of holders (Ribera d'Alcalà & Saggiomo, 1990; Cabrini et al., 2000; Comisso et al., 2003). From such point samples it is not possible to gather information about sedimentation processes over an extended period of time. For this reason, our data cannot be numerically compared with these authors' results.

Gross sedimentation rates, expressed both as ABU and BIOM, showed a high annual variability. Three periods with different sedimentation rates were observed. The first period, in January, was characterized by a high sedimentation rate caused by *Pseudo-nitzschia seriata*, which reached relatively high abundances in the water column (data not published). In the second period, from April to the first week of July, we observed a very scarce sedimentation, probably due to both the scarce presence of phytoplankton in the water column and the lack of resuspension from the bottom. In the third period, from the third week of July to the end of the year, the quantity of material found in the trap was wavering. This variability was more evident for gross BIOM than for gross ABU. During the second part of the year three relative peaks of gross sedimentation were observed: one in the first week of August, when a high abundance of *Diatoma* cfr. *vulgare* was recorded; a second one in the second week of September, which was caused by a high abundance of *Chaetoceros* spp.; the third one, limited to BIOM, was observed in the second week of December. The lack of the ABU peak in this week was due to the presence of species with high biomass but relatively low abundance. These were *Thalassionema frauenfeldii*, whose RA accounted for only 11.71%, while its relative biomass (RB) accounted for 46.67%, *Plagiotropis* cfr. *gaussii* with RA = 2.22% and RB = 9.93% and other large size species like *Lioloma pacificum*, *Rhizosolenia* spp., and *Surirella* spp.

In the Gulf of Trieste the observed seasonal cycle of primary sedimentation did not follow the general pattern described for other coastal areas

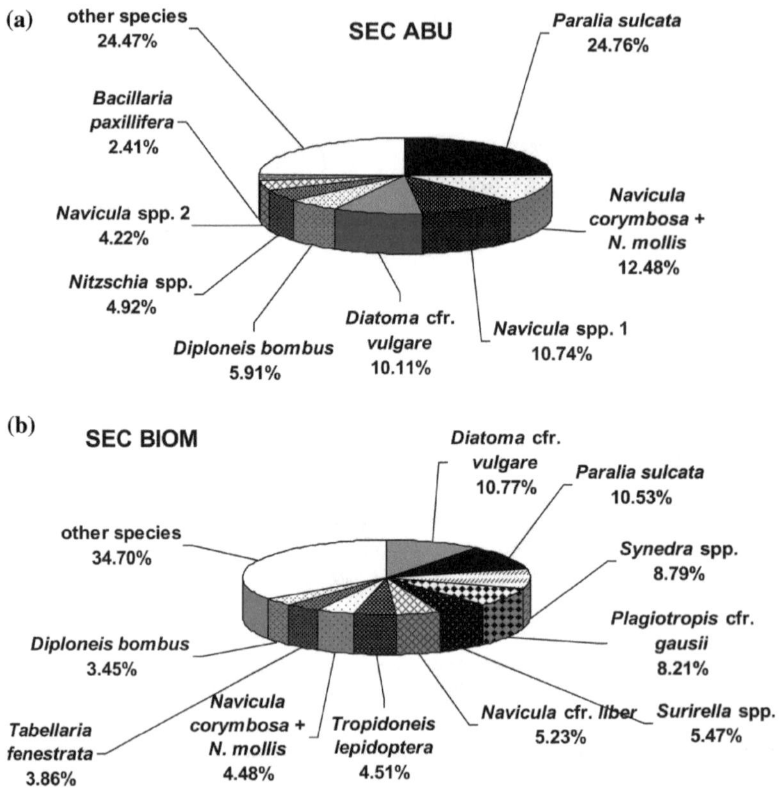

Fig. 8 Dominant species responsible for the secondary sedimentation expressed as abundance (**a**) and biomass (**b**)

in temperate and boreal zones, where the primary sedimentation rates are particularly high in spring (Lutter et al., 1989; Wassman, 1991; Heiskanen & Kononen, 1994; Tallberg & Heiskanen, 1998). The primary sedimentation rates, observed during the year of study, were probably low since during 2003 the phytoplankton in the water column of the Gulf of Trieste reached its lowest abundance in the last eighteen years (Fonda Umani et al., 2004).

Observing the primary sedimentation pattern we can better notice the divergence between ABU and BIOM in the second week of December, confirming that planktonic species were the major responsible for the higher BIOM peak. In the second and third week of September *Chaetoceros* spp. sank in the trap from the water column where it was present in relatively high abundance (data not published). Also in the second and third week of December *Chaetoceros* spp. was present in the trap material, but it was not observed throughout the water column. Moreover, while the *Chaetoceros* spp. chains found in the trap were usually not longer than 2–3 cells, with scarce chlorophyll inside almost disintegrated frustules, in December the cells belonging to this genus were particularly viable, photosynthetically active and had formed long chains. We infer that these cells were derived from resting spores laying on the sediment and settled into the trap because they were dragged by bottom currents.

The trend of benthic diatoms found in the trap depended on the quantity of resuspended sediment from the bottom. This, in turn, depended on the strength of the wind rather than on the abundance of the microphytobenthic community. In the Gulf of Trieste the quantity and species composition of the microphytobenthos follows a seasonal pattern. Microphytobenthic abundances

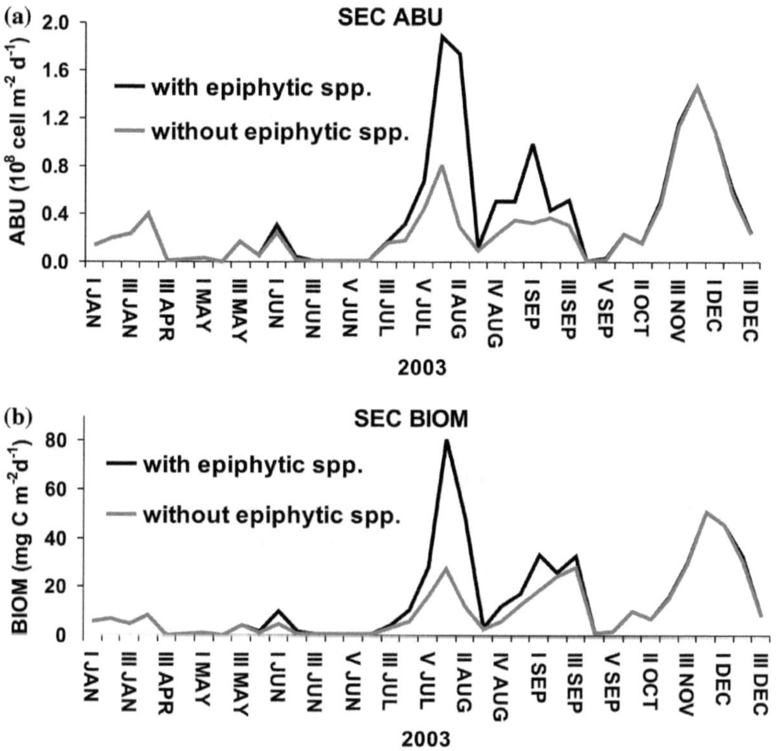

Fig. 9 Secondary abundance (**a**) and secondary biomass (**b**) trends with and without epiphytic diatoms

are lower in winter and higher in summer, never reaching phytoplanktonic abundances observed during blooming periods in the water column (Blasutto et al., 2003).

Secondary sedimentation patterns can be divided into two periods: the first, from the first week of January to the first week of July characterized by low rates and the second one, from the third week of July to the end of the year, typified by high sedimentation rates. In autumn and winter resuspension events, caused by strong winds (www.dts.units.it/OM/mens_TS/WD0312.gif), were responsible for the observed high sedimentation rates.

In the third and fourth week of January a decrease in secondary BIOM was recorded in correspondence to an increase in secondary ABU ascribable to a high RA of flagellates that have a low carbon content and therefore a low RB (RA = 45.00% and 58.14%, RB = 20.41% and 30.97%, respectively). A similar situation was recorded in the fourth week of August, where the high secondary ABU could be attributed to the presence of small sized species (*Navicula mollis*, *N. corymbosa*, *Navicula* spp.1 and *Navicula* spp.2). In the Gulf of Trieste the genus *Navicula* is well represented with high species variability. The dimensional range is wide and species identification is possible only by means of scanning electron microscopy (SEM). Using the light microscope species identification was possible only in few cases; therefore, it was decided to arbitrarily group specimens of this genus into three size classes: *N.* spp.1 up to 20 μm, *N.* spp.2 from 20 μm to 40 μm and *N.* spp.3 larger than 40 μm (Blasutto et al., 2003). During the third week of September high BIOM and low ABU were observed probably because of resuspension due to large sized benthic species like *Pleurosigma formosum* (RA = 0.75%, RB = 7.40%), *Navicula* cfr. *divergens* (RA = 1.00%, RB = 5.08%) and *Navicula* cfr. *liber* (RA = 0.50%, RB = 4.19%).

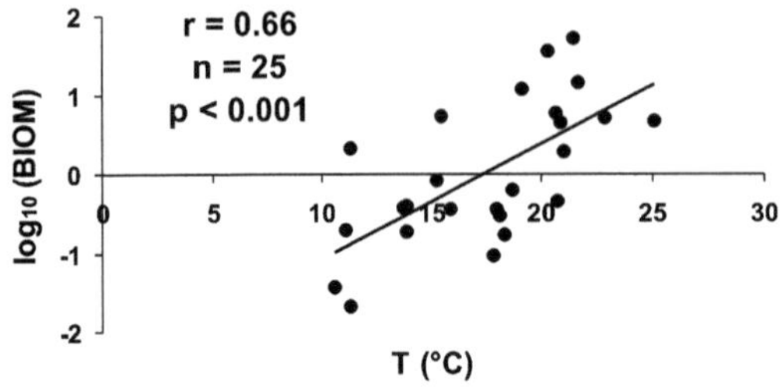

Fig. 10 Linear regression between temperature and the logarithm of epiphytic BIOM. For the calculation of the logarithm, epiphytic BIOM values equal to zero were not considered

In summer the highest secondary ABU and BIOM were due to the presence of epiphytic species in the trap samples. High temperature during the summer supported the development of those species on the inner trap wall, where they adhere finding a favorable microhabitat and producing a great amount of tubes. In order to observe the biodiversity of the settled material using the sediment trap we took into consideration also the epiphytic species. Their presence and abundance were not strictly bound to the water column or the sediment communities but only to the structure of the trap. Therefore, including epiphytic species in the assessment of sedimentation rates can result in an overestimation of the data.

The presence of epiphytic species in the sediment trap is a problem related with the closeness to the coast. These species occasionally sink in the trap, dragged by local currents and waves from the shoreline, where they live attached to macroalgae or rocky substrata. In the trap samples we also found other species, such as *Auricula* spp., *Bacillaria paxillifera* and *Synedra* spp., whose

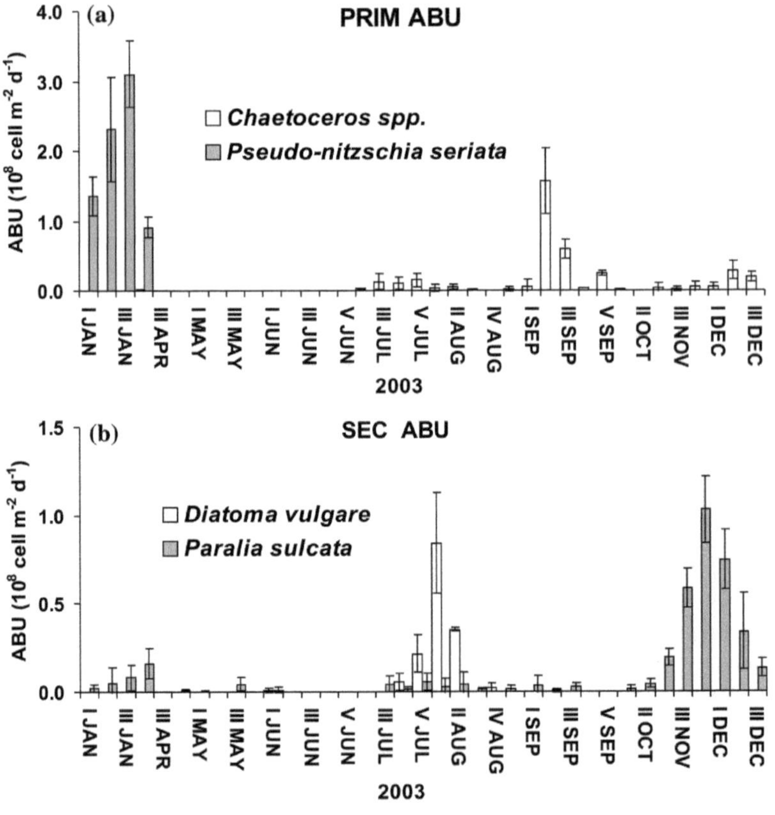

Fig. 11 The most abundant species of the primary (**a**) and secondary (**b**) sedimentation. Data are presented as mean values of three replicates with standard deviations

Table 2 Richness (d), equitability (J′), diversity (H′) and dominance (λ) indices

	Margalef	Pielou	Shannon	Simpson
Sample	d	J′	H′	λ
I Jan	1.13	0.24	0.76	0.76
II Jan	0.92	0.24	0.72	0.76
III Jan	0.91	0.19	0.59	0.81
IV Jan	1.00	0.50	1.54	0.43
III Apr	2.04	0.71	2.54	0.14
IV Apr	1.70	0.60	2.07	0.26
I May	1.32	0.52	1.67	0.38
II May	1.81	0.62	2.09	0.21
III May	1.82	0.73	2.62	0.12
IV May	2.14	0.74	2.75	0.11
I Jun	1.74	0.67	2.40	0.16
II Jun	2.07	0.69	2.51	0.15
III Jun	1.93	0.67	2.37	0.18
IV Jun	1.78	0.74	2.54	0.14
V Jun	1.86	0.74	2.58	0.13
I Jul	1.49	0.46	1.53	0.48
III Jul	1.52	0.67	2.30	0.18
IV Jul	1.54	0.76	2.65	0.10
V Jul	2.15	0.76	2.92	0.08
I Aug	1.95	0.70	2.64	0.16
II Aug	1.54	0.56	1.99	0.27
III Aug	2.07	0.81	2.99	0.08
IV Aug	1.99	0.68	2.54	0.15
V Aug	2.09	0.79	3.00	0.08
I Sep	1.78	0.58	2.03	0.32
II Sep	1.95	0.39	1.48	0.51
III Sep	2.06	0.58	2.19	0.28
IV Sep	1.97	0.40	1.42	0.54
V Sep	1.39	0.25	0.84	0.74
I Oct	1.95	0.80	2.92	0.09
II Oct	1.80	0.76	2.69	0.12
II Nov	1.47	0.73	2.50	0.13
III Nov	1.47	0.62	2.15	0.23
IV Nov	1.27	0.47	1.57	0.45
I Dec	1.32	0.59	1.98	0.27
II Dec	1.52	0.70	2.43	0.14
III Dec	1.68	0.63	2.23	0.19

presence was due to the sheltered microhabitat established inside the trap, where they could proliferate undisturbed. In the trap they reached high abundances, while in the water column they were rare or totally absent. Consequently, to evaluate secondary sedimentation rate we considered also *Striatella unipunctata*, *Diatoma* cfr. *vulgare*, *Nitzschia fruticosa* and species producing tubes, such as *Navicula mollis*, *N. corymbosa*, *Nitzschia subtubicola* and *N. dissipata*. *Striatella unipunctata* is a typically epiphytic species capable of producing mucous-polysaccharide stalks

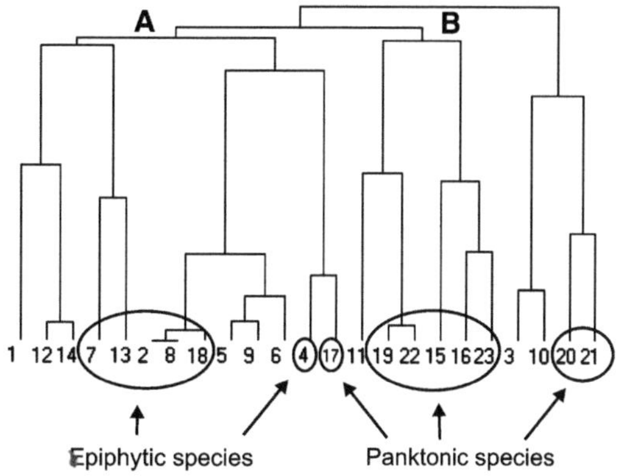

Fig. 12 Cluster analysis based on species. The first 23 species, which corresponded to a cumulative variance of 99.99%, were used for the hierarchical clustering. 1 = *Bacillaria paxillifera*, 12 = *Pleurosigma* spp., 14 = *Chaetoceros* spp., 7 = *Nitzschia dissipata*, 13 = *Tabellaria flocculosa*, 2 = *Diatoma* cfr. *vulgare*, 8 = *Nitzschia fruticosa*, 18 = *Guinardia* cfr. *tubiformis*, 5 = *Navicula* spp. 1, 9 = *Nitzschia* spp., 6 = *Navicula* spp. 2, 4 = *Navicula mollis*, 17 = *Guinardia delicatula*, 11 = *Pleurosigma elongatum*, 19 = *Pseudo-nitzschia seriata*, 22 = *Thalassiosira eccentrica*, 15 = *Cyclotella glomerata*, 16 = *Cylindrotheca closterium*, 23 = Centrales undet., 3 = *Diploneis bombus*, 10 = *Paralia sulcata*, 20 = *Thalassionema frauenfeldii*, 21 = *Thalassionema nitzschioides*

that are unidirectional structures by which it adheres to various substrata. *Diatoma* cfr. *vulgare, Auricula* spp. and *Synedra* spp. adhere to other cells or to a substratum with small globular apical pads (Round et al., 1992). In our samples also *Nitzschia fruticosa* was occasionally found, forming starry colonies or adhering to *Synedra* spp. Among species producing tubes, *Navicula corymbosa* was observed with a characteristic singular chain disposition of the cells embedded in a mucilage tube, while the cells of *Navicula mollis, Nitzschia subtubicola* and *N. dissipata* were organized in more complex and irregular multiple chains (Dexing et al., 1985). Another species worth mentioning is *Bacillaria paxillifera*, which forms colonies whose characteristic movement makes the cells slide forward while adhering to one another for the full length of the raphe (Hendey, 1976). This diatom was often found in the same periods during which epiphytic species were observed, presumably swept up from the sediments.

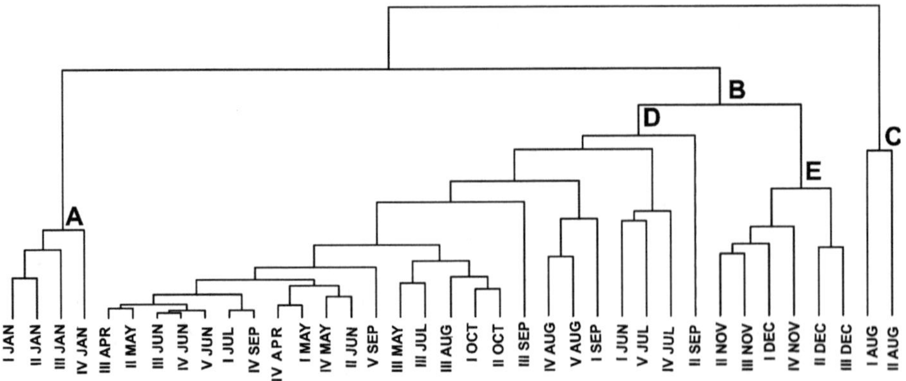

Fig. 13 Cluster analysis on the basis of samplings. Absolute distance and complete linkage were applied

Epiphytic species weighed for 11.19% and 12.27% on annual average gross ABU and BIOM, respectively. These diatoms proliferated significantly in summer reaching even 72.04% of gross ABU and 56.06% of gross BIOM in the second week of August. The correlation between temperature and the epiphytic BIOM was statistically significant. In July and August the epiphytic microalgae reached the highest BIOM, whereas in winter they were almost totally absent.

Among planktonic diatoms *Pseudo-nitzschia seriata* and *Chaetoceros* spp. dominated primary ABU. While *Pseudo-nitzschia seriata* was found in the settled material only in January, reaching high ABU, *Chaetoceros* spp. was present almost all the year, with several relative maxima and the highest ABU in the second week of September. Similarly, *Paralia sulcata* was the prevalent

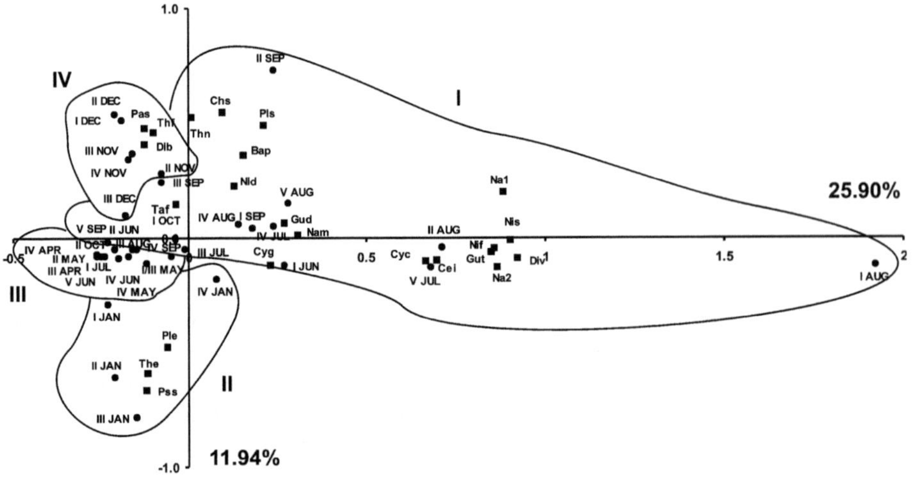

Fig. 14 PCA, based on r algorithm, of species (■) and samplings (•) obtained using the first and the second autovectors. *Bacillaria paxillifera* = Bap, *Pleurosigma* spp. = Pls, *Chaetoceros* spp. = Chs, *Nitzschia dissipata* = Nid, *Tabellaria flocculosa* = Taf, *Diatoma* cfr. *vulgare* = Div, *Nitzschia fruticosa* = Nif, *Guinardia* cfr. *tubiformis* = Gut, *Navicula* spp. 1 = Na1, *Nitzschia* spp. = Nis, *Navicula* spp. 2 = Na2, *Navicula mollis* = Nam, *Guinardia delicatula* = Gud, *Pleurosigma elongatum* = Ple, *Pseudo-nitzschia seriata* = Pss, *Thalassiosira eccentrica* = The, *Cyclotella glomerata* = Cyg, *Cylindrotheca closterium* = Cyc, Centrales undet. = Ceu, *Diploneis bombus* = Dib, *Paralia sulcata* = Pas, *Thalassionema frauenfeldii* = Thf, *Thalassionema nitzschioides* = Thn

species both for secondary ABU and BIOM. This benthic species is a non-motile centric diatom, living loosely linked with the sediments (Werner, 1977; Round, 1985). In our samples *Paralia sulcata* was the prevalent species both for secondary ABU and BIOM. Moreover, it was also very frequent, since it was found in 27 out of 37 weekly samples, reaching the highest ABU in the fourth week of November. On the contrary, *Diatoma* cfr. *vulgare*, an epiphytic species, whose presence was due to the particular microhabitat inside the trap, was observed only in July and August. Nonetheless, its contribution to secondary sedimentation, both as ABU and BIOM, was noticeable.

Among the 23 species used to perform the cluster analysis and the PCA, an equal number of planktonic, benthic and epiphytic forms was observed, with 9, 8 and 6 species, respectively. Within the 23 species, which corresponded to a cumulative variance of 99.99%, there was not the prevalence of a single living form, but the three forms were numerically similar.

In the group I of the PCA all the epiphytic species were gathered. The position of the sampling performed during the first week of August was due to the highest ABU of epiphytic species found in the trap material over the study period. Therefore this sampling differed from the others performed during the summer period. The group III was characterized by the absence of species. In fact, from April to the end of June the paucity of phytoplankton was observed. In this period also secondary ABU was very low and epiphytic diatoms did not proliferate in the sediment trap. *Pseudo-nitzschia seriata* was positioned in the group II, were the samplings, characterized by very low temperature, were located. This is in agreement with Bernardi Aubry et al. (2004) who consider the taxa *Pseudo-nitzschia* a tipical diatom of cold waters. In the group IV two typically benthic species were found, *Paralia sulcata* and *Diploneis bombus*. Their position confirmed the relevance of resuspension phenomena during the last two months of 2003.

Samplings in the cluster analysis were positioned on the basis of the composition of the community and the abundances of the species. Although both groups A and C were characterized by high abundances, they were very far from each other due to the presence of different species. In the first and second week of August epiphytic diatoms reached the highest abundances, while in January they were totally absent. Another difference between the two groups was the dominance of *Pseudo-nitzschia seriata* in January. The position of the groups A and C in the cluster was observed also in the PCA. The group A in the cluster corresponded to group II in the PCA. The first week of August, both in the cluster and in the PCA, was positioned far from all the other samplings. The subgroup E of the dendrogram gathered samplings performed in November and December when benthic species were prevalent in the trap material. In the PCA the same samplings were gathered in the group IV together with *Paralia sulcata* and *Diploneis bombus*.

In conclusion, the use of the sediment trap let us to collect data regarding resuspension phenomena and secondary sedimentation mostly due to benthic species. Besides, the establishment of a microhabitat inside the trap and the development of several epiphytic species led to an increase in biodiversity. The importance of epiphytic diatoms becomes relevant especially when the sediment trap is positioned within the euphotic layer, where the light is still sufficient to allow their proliferation. On the other hand, benthic diatoms become relevant if the sediment trap is located near the bottom, where sediment resuspension can cause the settling of benthic forms into the trap. Our study revealed that among all the settled Bacillariophyceae, planktonic, benthic and epiphytic forms accounted for 50.78%, 36.95% and 12.27%, respectively, calculated on the annual average biomass. Particularly when the sediment trap is used to estimate carbon, hydrogen and nitrogen fluxes from the water column it is important to consider the presence of all these three life forms. Therefore, vertical fluxes can be overestimated of 50% or more if epiphytic and benthic species are not rejected.

Acknowledgements This study was carried out as a part of the European Community INTERREG III Italy—Slovenia project. We wish to thank Dr. M. Giani for the technical support and the staff of the Marine Reserve of Miramare for their logistic support and for the maintenance of the sediment trap. We are also grateful to Dr. N. Burba for helping us with the statistical analyses. We wish to express our appreciation to anonymous reviewers for their constructive criticisms of the manuscript.

References

Bernardi Aubry, F., A. Berton, M. Bastianini, G. Socal & F. Acri, 2004. Phytoplankton succession in a coastal area of the NW Adriatic, over a 10-year sampling period (1990–1999). Continental Shelf Research 24: 97–115.

Blasutto, O., C. De Vittor, E. Sdrigotti, F. De Prà & S. Fonda Umani, 2003. The use of biochemical parameters to study microphytobenthic community in the Gulf of Trieste (northern Adriatic Sea). Biologia Marina Mediterranea 10: 319–326.

Brambati, A. & G. Catani, 1988. Le coste e i fondali del Golfo di Trieste dall'Isonzo a Punta Sottile: aspetti geologici, geomorfologici, sedimentologici e geotecnici. Hydrores Information 5: 13–28.

Burba, N., E. Feoli, M. Malaroda & V. Zuccarello, 1992. Un sistema informativo per la vegetazione. Software per l'archiviazione della vegetazione italiana e per l'elaborazione di tabelle. Manuale di utilizzo dei programmi. CETA Book 2, Gorizia.

Cabrini, M., S. Cok & I. Pecchiar, 2000. Phytoplankton structure and dynamic in the coastal zone of the Gulf of Trieste. Biologia Marina Mediterranea 7: 850–853.

Cardin, V. & M. Celio, 1997. Cluster Analysis as a statistical method for identification of the water bodies present in the Gulf of Trieste (northern Adriatic Sea). Bollettino di Geofisica Teorica ed Applicata 38: 119–135.

Celio, M., C. Comici & A. Bussani, 2002. Thermohaline Anomalies in the spring and early summer of 2000 in the Gulf of Trieste. P.S.Z.N. Marine Ecology 23: 101–110.

Clark, K. R. & R. M. Warwick, 2001. Change in Marine Communities: An Approach to Statistical Analysis and Interpretation. PRIMER-E, Plymouth U.K.

Comisso, S., I. Pecchiar & M. Cabrini, 2003. Biodiversity evolution of microphytoplankton community of the Gulf of Trieste (January 2001–January 2002). Biologia Marina Mediterranea 10: 987–989.

de Jonge, V. N. & J. E. E. van Beusekom, 1992. Contribution of resuspended microphytobenthos to total phytoplankton in the Ems Estuary and its possible role for grazers. Netherlands Journal of Sea Research 30: 91–105.

Dexing, J., L. Junmin, C. Zhaodi & L. Shicheng, 1985. Marine Benthic Diatoms in China. China Ocean Press, Beijing.

Faganeli, J., 1989. Sedimentation of particulate nitrogen and amino acids in shallow coastal waters (Gulf of Trieste, northern Adriatic). Marine Chemistry 26: 67–80.

Faganeli, J., N. Kovac, H. Leskovsek & J. Pezdic, 1995. Sources and fluxes of particulate organic matter in shallow coastal waters characterised by summer macroaggregate formation. Biogeochemistry 29: 71–88.

Fonda Umani, S., A. Beran, S. Parlato, D. Virgilio, A. Zollet, A. De Olazabal, B. Lazzarini & M. Cabrini, 2004. *Noctiluca scintillans* MACARTNEY in the northern Adriatic Sea: long-term dynamics, relationships with temperature and eutrophycation, and role in the food web. Journal of Plankton Research 26: 545–561.

Germain, H., 1981. Flore des Diatomées eaux douces et saumâtres. Société Nouvelle des Éditions Boubée, Paris.

Giani, M., A. Boldrin, G. Matteucci, F. Frascari, M. Gismondi & S. Rabitti, 2001. Downward fluxes of particulate carbon, nitrogen and phosphorus in the north-western Adriatic Sea. The Science of the Total Environment 266: 125–134.

Giordani, P. & M. Frignani, 1988. A sediment trap experiment in NW Adriatic coastal waters. Proceedings of the VIII Italian Oceanology and Limnology Congress, 445–456.

Heiskanen, A. -S., 1995. Contamination of sediment trap fluxes by vertically migrating phototrophic microorganisms in the coastal Baltic Sea. Marine Ecology Progress Series 122: 45–58.

Heiskanen, A. -S., 1998. Factors governing sedimentation and pelagic nutrient cycles in northern Baltic Sea. Monographs of the Boreal Environment Research No. 8.

Heiskanen, A. -S. & K. Kononen, 1994. Sedimentation of vernal and late summer phytoplankton communities in the coastal Baltic Sea. Archiv für Hydrobiologie 131: 175–198.

Hendey, N. I., 1976. An Introductory Account of the Smaller Algae of British Coastal Waters. Otto Koeltz Science Publishers, West Germany.

Hillebrand, H., C. -D. Dürselen, D. Kirschtel, U. Pollingher & T. Zohary, 1999. Biovolume calculation for pelagic and benthic microalgae. Jounal of Phycology 35: 403–424.

Hoagland, K. D., J. R. Rosowski, M. R. Gretz & S. C. Roemer, 1993. Diatom extracellular polymeric substances: function, fine structure, chemistry and physiology. Jounal of Phycology 29: 537–566.

Lucas, C. H., C. Banham & P. M. Holligan, 2001. Benthic-pelagic exchange of microalgae at a tidal flat. 2. Taxonomic analysis. Marine Ecology Progress Series 212: 39–52.

Lutter, S., J. P. Taasen, C. C. E. Hopkins & V. Smetacek, 1989. Phytoplankton dynamics and sedimentation processes during spring and summer in Balsfjord, Northern Norway. Polar Biology 10: 113–124.

Margalef, R., 1978. Life-forms of phytoplankton as survival alternatives in an unstable environment. Oceanologica Acta 1: 493–509.

Margalef, R., 1986. Ecologia. Omega, Barcellona.

Matteucci, G. & F. Frascari, 1997. Fluxes of suspended materials in the north Adriatic Sea (Po prodelta area). Water, Air, and Soil Pollution 99: 557–572.

Miquel, J. C., S. W. Fowler, T. Hamilton, J. Heilmann, J. La Rosa & M. Carrol, 1999. Carbon fluxes and export in the northern and central Adriatic Sea measured with drifting sediment traps. In Hopkins,

T. S., A. Artegiani, G. Cauwet, D. Degobbis & A. Malej (eds) Ecosystem Research No. 32. The Adriatic Sea (EUR 18834). European Commission, 1999: 509–524.

Ogorelec, B., M. Misic & J. Faganeli, 1991. Marine geology of the Gulf of Trieste (northern Adriatic): sedimentological aspects. Marine Geology 99: 79–92.

Orloci, L., 1978. Multivariate Analysis in Vegetation Research. Junk, Den Haag, Boston.

Passow, U., 1991. Species-specific sedimentation and sinking velocities of diatoms. Marine Biology 108: 449–455.

Pielou, E. C., 1966. Shannon's formula as a measure of species diversity: its use and misuse. The American Naturalist 118: 463–465.

Philibert, A. & Y. T. Prairie, 2002. Is the introduction of benthic species necessary for open-water chemical reconstruction in diatom-based transfer functions? Canadian Journal of Fisheries and Aquatic Sciences 59: 938–951.

Posedel, N. & J. Faganeli, 1991. Nature and sedimentation of suspended particulate matter during density stratification in shallow coastal waters (Gulf of Trieste, northern Adriatic). Marine Ecology Progress Series 77: 135–145.

Puškarić, S., S. W. Fowler & J. C. Miquel, 1992. Temporal changes in particulate flux in the northern Adriatic Sea. Estuarine, Coastal and Shelf Science 35: 267–287.

Ribera d'Alcalà, M. & V. Saggiomo, 1990. Campionamento. In Innamorati, M., I. Ferrari, D. Marino & M. Ribera d'Alcalà (eds), Metodi nell'ecologia del plancton marino. Nova Thalassia, vol. 11. Edizioni Lindt, Trieste: 5–8.

Ricard, M., 1987. Atlas du Phytoplancton Marin Volume II: Diatomophycées. Éditions du Centre National de la Recherche Scientifique, Paris.

Round, F. E., 1985. The Ecology of Algae. Cambridge University Press, Cambridge.

Round, F. E., R. M. Crawford & D. G. Mann, 1992. The Diatoms. Cambridge University Press, Cambridge.

Silver, M. W. & M. M. Gowing, 1991. The "particle" flux: origins and biological components. Progress in Oceanography 26: 75–113.

Simpson, E. H., 1949. Measurement of diversity. Nature 163: 688.

Shannon, C. E. & W. Weaver, 1949. The Mathematical Theory of Communication. Illinois Press, Urbana, Illinois.

Strathmann, R. R., 1967. Estimating the organic carbon content of phytoplankton from cell volume or plasma volume. Limnology and Oceanography 12: 411–418.

Tallberg, P. & Heiskanen A. -S., 1998. Species-specific phytoplankton sedimentation in relation to primary production along an inshore-offshore gradient in the Baltic Sea. Journal of Plankton Research 20: 2055–2070.

Tomas, C. R., 1997. Identifying Marine Phytoplankton. Academic Press, San Diego.

Van Heurck, H., 1899. Traité des Diatomées. Édité aux Frais de L'Auteur, Anvers.

Utermöhl, H., 1958. Zur Vervollkommnung der quantitativen Phytoplankton-Methodik. Mitteilungen. Internationale Vereiningung für Theoretische und Angewandte Limnologie 9: 1–38.

Wassmann, P., 1991. Dynamics of primary production and sedimentation in shallow Fjords and polls of western Norway. Oceanography and Marine Biology. An Annual Review 29: 87–154.

Wassmann, P., S. Fonda Umani, J. E. Ypma, M. Reigstad, S. Cok, C. Salvi & G. Cauwet, 1998. Suspended biomass and vertical flux in the Gulf of Trieste, Adriatic Sea: Two summer scenarios. In Hopkins, T. S., A. Artegiani, G. Cauwet, D. Degobbis & A. Malej (eds) Ecosystems Research No. 32. The Adriatic Sea (EUR 18834). European Commission, 1999, 537–544.

Werner, D., 1977. The Biology of Diatoms. Blackwell Scientific Publications, Oxford, U.K.

Hydrobiologia (2007) 580:77–84
DOI 10.1007/s10750-006-0464-x

Phylogeography of the sea urchin *Paracentrotus lividus* (Lamarck) (Echinodermata:Echinoidea): first insights from the South Tyrrhenian Sea

V. Iuri · F. P. Patti · G. Procaccini

Abstract The sea urchin *Paracentrotus lividus* (Lamarck) (Echinodermata: Echinoidea) is an Atlanto-Mediterranean species abundant in the littoral zone, where it occurs in the sublittoral down to 20 m. The aim of our work is to investigate the genetic patterns of *P. lividus* along the South Tyrrhenian coasts. Five specimens were collected in six localities, from the Gulf of Naples and the Cilento coast. The nuclear rDNA ITS2 spacer and the two mitochondrial genes 16S and COI were used for the analysis. The three markers utilised did not show any structure among populations from the Gulf of Naples. All populations appear to be polyphyletic, with Cilento samples more differentiated from the others. This suggests the existence of phylogeographic structure at larger geographic scale. Absence of genetic structure has been interpreted taking into consideration theoretical dispersal of the planktotrophic larvae, which can survive for 4–8 weeks before settlement, marine current patterns and persistence of the species in the area.

Guest editors: G. Relini & J. Ryland
Biodiversity in Enclosed Seas and Artificial Marine Habitats

V. Iuri · F. P. Patti (✉) · G. Procaccini
Laboratorio di Ecologia del Benthos, Stazione Zoologica "A.Dohrn", P.ta S. Pietro, 80077 Ischia, Napoli, Italy
e-mail: fpatti@szn.it

V. Iuri
e-mail: iuri@szn.it

Keywords Paracentrotus lividus · mtDNA · ITS2 · Larval dispersal · Marine currents · Phylogeography

Introduction

Factors affecting population structure vary widely between habitats. Among terrestrial animals and plants, habitat fragmentation, topography, soil types, watershed characteristics, and a host of other factors are known to affect species distribution and gene flow within species ranges (Palumbi et al., 1997). Some of these factors can be sometimes stable over long periods, and allow the built up of substantial genetic differentiation (Avise, 1992). Genetic discontinuities are also present where gene flow is interrupted by clear geographic boundaries (Avise et al., 1979; Patton & Smith, 1989). Genetic boundaries among populations of marine species are often more difficult to discern, and the physical and biological factors that determine gene flow patterns are poorly understood (Palumbi et al., 1997).

Several phylogeography studies have been performed utilizing molecular markers and in particular mtDNA regions (e.g. dragonflies, Artiss, 2004; urchins, Palumbi et al., 1997). Some marine invertebrates display little differentiation in mtDNA markers over vast areas. Examples include lobsters in the genera *Jasus* (Ovenden et al., 1992) and *Panulirus* (Silberman et al., 1994), and several shallow-water species of sea urchin in the genera *Heliocidaris* and *Strongylocentrotus* (Palumbi & Wilson, 1990; Palumbi & Kessing, 1991; McMillan et al., 1992). In the sea urchin species *Echinothrix diadema*, both mtDNA and allozyme markers suggest recent and massive trans-Pacific gene flow, perhaps through periodic larval transport during El Niño events across a broad expanse of open ocean (Lessios et al., 1998).

Genetic drift due to habitat selective pressure enhances population differentiation. Stable and unstable sites, for example, are characterised by different communities. In the sea urchin *Paracentrotus lividus* (Lamarck), unstable sites characterised by episodic pressure of selective forces, experience high mortalities of young individuals; stable communities rely instead on annual settlements and feature a lower and more homogeneous mortality which allows the development of well-structured populations (Turon et al., 1995).

Paracentrotus lividus is an Atlanto-Mediterranean species abundant in the littoral zone, where it occurs in the sublittoral down to 20 m (Turon et al., 1995). It is characteristic of the rocky biotic community and abounds in one of the most important biocenoses of the Mediterranean Sea, the *Posidonia oceanica* (Delile) meadows. At the moment there is no knowledge about genetic diversity and connectivity of populations along the coastline at different geographic distances. Massive urbanization along almost the whole Mediterranean coastline has created extensive gaps in the distribution of natural populations, theoretically enhancing the possibility of genetic structure and population differentiation.

Development to the adult stage in many marine organisms involves a long and complicated feeding larval stage. Such planktotrophic larvae are highly adapted to spend weeks hunting plantonic organisms before settlement (McMillan et al., 1992). This long-living phase is thought to be important in establishing the geographic ranges of many marine animals (McMillan et al., 1992). *P. lividus* has a planktotrophic larva called *pluteus*, which can survive in the plankton for 20–40 days (Pedrotti, 1993). Marine currents drive circulation of *plutei* between distant localities.

Phylogeography of benthic species in the Mediterranean basin is still poorly investigated. Even less is known about particular areas, such as the Gulf of Naples. Here, the SE circulation of open sea currents can isolate inner waters creating a slow cyclonic gyre which leaves the coastal waters in a still slower motion. Open sea currents flowing towards the NW can enter the bay, and provide a renewal of inner waters (De Maio et al., 1985).

The aim of this work is to investigate population subdivision and genetic structure in the sea urchin *P. lividus* at regional scale. For this purpose we used three different DNA regions, two mitochondrial, and one nuclear, with different theoretical mutations rates.

Our prediction is that active larval dispersal ensures sufficient gene flow among populations of the Gulf of Naples, including the Island of Ischia, allowing panmixia in the studied area. Our results will also allow the identification of the most suitable genetic markers for inferring phylogeographic relationships in *P. lividus*, at different geographical scales.

Materials and methods

Sample collection, DNA extraction and sequencing

Specimens were collected in the following localities: Lacco Ameno, Castello Aragonese and La Nave (Ischia Island); Gajola and Castel dell'Ovo (Gulf of Naples); Pioppi (Cilento coast) (Fig. 1). Five specimens were randomly sampled in each site by scuba diving (Table 1).

Genomic DNA was isolated from EtOH 70% fixed tube feet, with a standard Proteinase K protocol and RNase incubation overnight, involving three different organic washes: the first based on Phenol, the second on Phenol/Chloroform/Isoamylic Alcohol, and the third on Chloroform/Isoamylic Alcohol. DNA pellet was precipitated with 2-Propanol, washed with EtOH 70% and resuspended in ddH_2O.

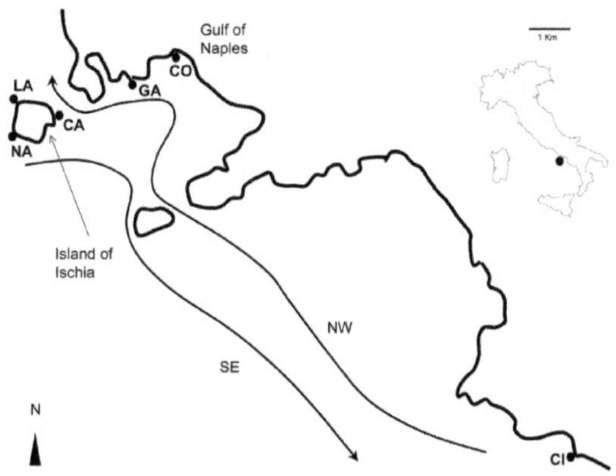

Fig. 1 Sampling sites. Dominant marine currents are also indicated; the main direction along the Tyrrhenian coast is the NW. LA (Lacco Ameno), CA (Castello Aragonese), NA (La Nave), GA (Gajola), CO (Castel dell'Ovo), CI (Pioppi, Cilento)

PCR (polymerase chain reaction) amplifications were performed using the following primers: for nuclear rDNA ITS2 region the forward primer ITS3 (White et al., 1990), and the reverse specific primer D1R (Patti et al., 2001); for 16S, the universal primers drawn on sea urchins sequences 16Sar (forward) and 16Sbr (reverse) (Palumbi et al., 1991); for COI (Cytochrome Oxidase I), the specific primers COI748f (forward; GGA TTT GGA ATG ATT TCA CAC GTA) and COIPLr (reverse; CGG TAG AAG GTG TTT CGT CAA).

Forty amplification cycles were performed for each marker, with the following profiles: 94°C for 1 minute, 44°C for 1 min and 72°C for 2 min, for the ITS2; 94°C for 30 sec, 48°C for 40 sec and 72°C for 1 min, for the COI; 94°C for 30 sec, 52°C for 30 sec and 72°C for 1 min, for the 16S.

Sequence reactions were obtained with the BigDye Terminator Cycle Sequencing technology (Applied Biosystems), purified in automation using a robotic station "Biomek FX" (Beckman Coulter) and performed on an Automated Capillary Electrophoresis Sequencer "3730 DNA Analyzer" (Applied Biosystems). Sequences data were aligned using Clustal W (Thompson et al., 1994) and alignments were adjusted in BioEdit 4.8.5 (Hall, 1999) computer software.

Phylogenetic analysis

Phylogenetic and molecular evolutionary analyses were conducted using PAUP*4.0 (Swofford, 2003).

Three different analyses have been performed for each marker: Maximum Likelihood (ML), choosing randomly between the 56 evolutionary models, Maximum Parsimony (MP) and Neighbor Joining (NJ) Kimura 2 parameters; gaps have been considered as fifth character. Each tree has been tested with 1000 bootstrap replicates and has been drawn and analysed with TreeView 1.5.2 (Page, 1996).

The same analyses have been applied to the three consensus sequences obtained, one for each geographical area, to verify the markers resolution level.

Network analyses were performed with the Median Joining Networks 4.1 (Bandelt et al., 1999). This program constructs networks from recombination-free population data without resolving ties. The software combines features of Kruskal's algorithm for finding minimum spanning trees by favouring short connections,

Table 1 Number of samples analysed for each population

	# samples			Distance					
	16S	COI	ITS2	Lacco Ameno	Castello Aragonese	La Nave	Gajola	Castel dell'Ovo	Pioppi
Lacco Ameno (5)	5	5	4	0	3.25	4.75	14	17.5	72
Castello Aragonese (5)	5	5	4		0	7	11	14.5	68
La Nave (6)	5	6	5			0	17	20.5	72
Gajola (5)	5	5	4				0	3.75	65
Castel dell'Ovo (5)	5	5	5					0	66.5
Pioppi (5)	3	5	2						0

In the second part of the table distances in nautical miles between sites are given. In brackets are the number of individuals collected for each sampling site

and Farris's maximum-parsimony (MP) heuristic algorithm, adding sequentially new vertices called median vectors (Bandelt et al., 1999).

Alternative phylogeographical hypothesis have been tested with the non parametric test of Templeton (1983), from PAUP*4.0 Maximum Parsimony, generating a tree in which we constrained the two populations of Gajola and Castel dell'Ovo together, to evaluate the possibility of a separation between the Gulf of Naples and the Island of Ischia. Nucleotide diversity (Pi) and the haplotype frequencies were calculated for all molecular markers using the DNASp 5.53 (Rozas & Rozas, 1999).

Results

Alignments considered in the analyses consisted of 509bp for the 16S gene, 646bp for the COI region and 286bp for the ITS2. Sequence polymorphism analysis (Pi) indicated that 16S is the most conserved marker whereas ITS2 is the most variable one. The number of polymorphic sites is 8% in ITS2, 5.4% in COI and 1.6% in 16S. Five sites are informative for Parsimony for 16S (1%), eight for ITS2 (2.8%), and eighteen for COI (2.8%) (Table 2).

In general there was not clear separation either among all populations or between the Gulf of Naples and the Island of Ischia for the three molecular markers utilized, which show a different level of resolution. Only ML trees are shown (Fig. 2).

With the COI marker all the analyses give almost identical trees. The two main distinct COI clades group together individuals from both Ischia and the Gulf of Naples populations. In one of them (clade A), the three Ischia populations (CA, LA and NA) cluster together with one individual from Gajola (GA43) and one individual from Castel dell'Ovo (CO72). In the second one (clade B) individuals from the two Gulf of Naples populations group together with individuals from La Nave (NA, NAB-C-D). Only the clade A is present in the other mt marker (16S). All the individuals present in the 16S clade A (with all the three analyses) are also present in the same clade identified by COI, but not vice versa (Fig. 2).

Table 2 Descriptive parameters in the three molecular markers utilized

Markers/data	$N_{(bp)}$	PS	Pis	Pi
16S (28)	509	8	5	0.00365
COI (31)	646	35	18	0.01140
ITS2 (24)	286	23	8	0.01361

Number of sequences included in the alignment is indicated in parenthesis. $N_{(bp)}$: number of sites; PS: polymorphic sites; Pis: Parsimony informative sites; Pi: nucleotide diversity

Higher differences, instead, are present between the mitochondrial and the nuclear markers. In the ML, ITS2 tree, in fact, only two individuals, characteristic of the mt clade A (GA43 and NAE), group together although with low bootstrap support. In ITS2, differences are present among trees obtained with the three methods, although never coherent with any geographical pattern.

MP trees have been also tested with the non parametric test of Templeton (1983) confirming our hypothesis of panmixia along the Tyrrhenian coast for all the trees obtained. The probability calculated on the extra steps is always lower than the significant P value of 0.005.

With the Median Joining method and a new population from Pioppi (Cilento coast), that is about 200 km distant from the Gulf of Naples area, it has been possible to clarify better the power of resolution of the three molecular markers. Individuals from Pioppi never cluster in the same group, with any of the markers utilized. In the 16S network they stick out from the main groups on individual long branches. This is less pronounced in the COI and completely absent in the ITS2 networks (Fig. 3). ML, MP and NJ analyses applied on consensus sequences support a clear separation between the Cilento sites and the area of the Gulf of Naples/Island of Ischia (data not shown). The separation clearly results from the analysis of genetic distance values. Pairwise values of Ischia Vs Cilento and the Gulf of Naples Vs Cilento are 0.0718 and 0.0699 respectively, whereas distance is equal to 0 between Ischia and the Gulf of Naples.

The network analysis has been conducted on all samples with the two mitochondrial markers,

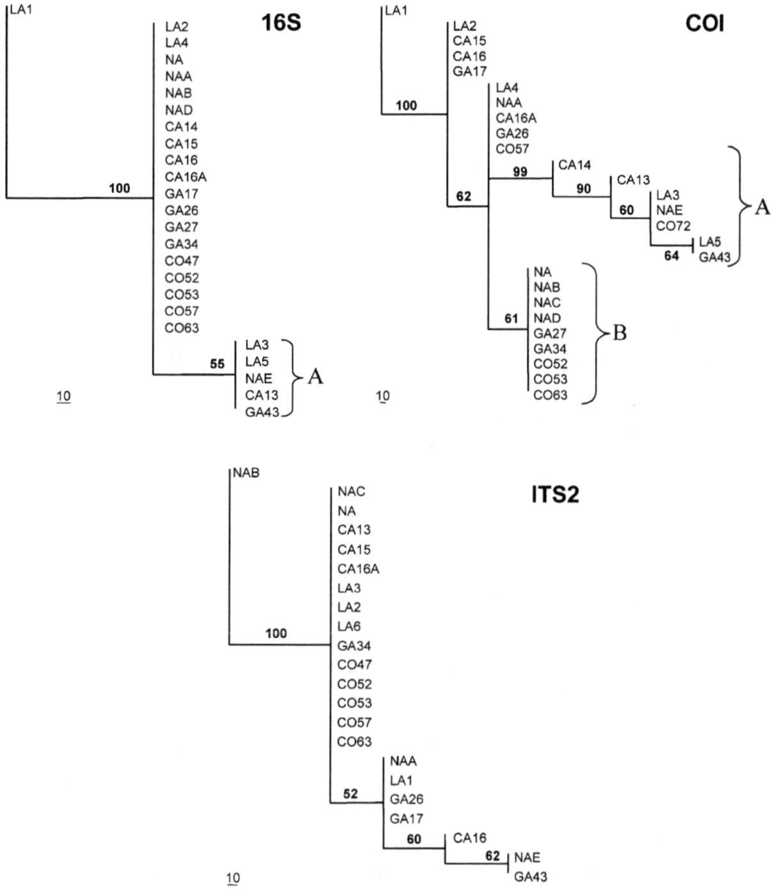

Fig. 2 Maximum Likelihood (ML) analyses for all the three molecular markers utilized (1,000 bootstrap replicates) in the Gulf of Naples. For population identification, see Fig. 1

COI and 16S, grouped together, for a total of almost 970bp (Fig. 4). Three main haplotypes have been detected, two of which are shared between Ischia and Gulf of Naples populations (O-P), while a single one is recovered only in Ischia (Q) (Table 3). Many individual-specific haplotypes have also been found.

Discussion

Phylogeographic hypothesis

The results of our analysis suggest either that *Paracentrotus lividus* is panmictic at the geographical scale investigated or that genetic structure is below the power of resolution of the molecular markers utilized.

COI resulted to be the more efficient of the three markers utilized. The same marker has also been utilized in a recent study on the disjunction between Mediterranean and Atlantic populations of *Paracentrotus lividus* (Duran et al., 2004). Also at this geographical level the species presented an almost complete panmixia, with dominant haplotypes scattered all over the sampling area.

In marine organisms, two main factors strongly influence phylogeographic patterns at different geographical scales: larval dispersal and action of marine currents.

P. lividus has feeding larvae that probably spend weeks or months in the plankton before metamorphosis (Uehara & Shingaki, 1984), hence having high dispersal potential. Long distance larval dispersal is characteristic of several marine organisms with planktotrophic larvae. High

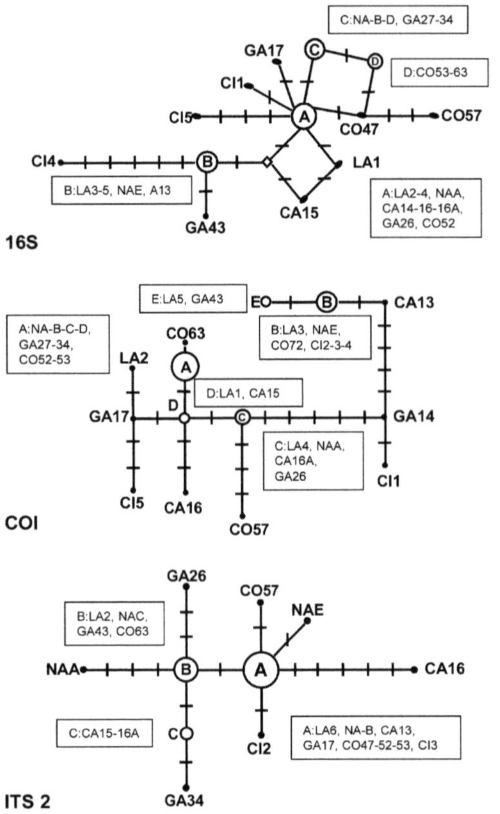

Fig. 3 Median-Joining Networks for the three molecular markers. Circles identify haplogroups; rhombi indicate median vectors. The area of each circle is proportional to the number of individuals; number of ticks corresponds to number of steps between each group and/or individuals. For population identification, see Fig. 1

dispersal in our area is confirmed by the discovery of identical mtDNA sequences in individuals collected in distinct localities. Identical mtDNA

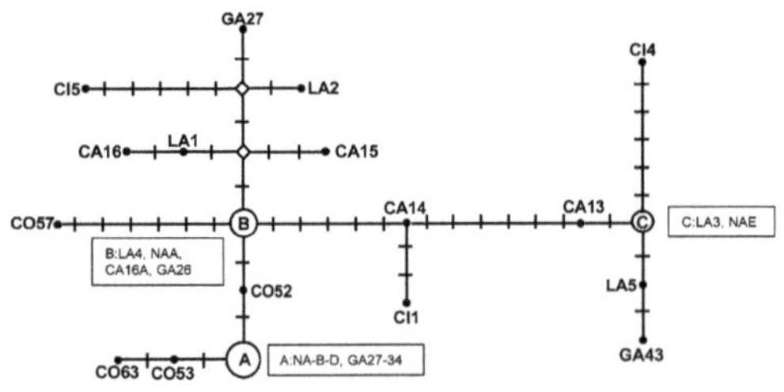

Fig. 4 Median-Joining Networks for the two mitochondrial markers grouped together. Circles identify haplogroups, rhombi indicate median vectors. For population identification, see Fig. 1

sequences in geographically distant localities have also been found in other marine species with planktotrophic larvae (Avise, 1994; Palumbi, 1994). Similar results, indicating high dispersal and potential panmixia, at different spatial scales, were obtained by Palumbi (1996) on the sea urchin *Strongylocentrotus purpuratus* (Stimpson) sampled along the Pacific coast of North America.

Even if the phylogeographic structure of *P. lividus* in the Gulf of Naples is not resolved, patterns of gene flow are recognizable. This is the case in the three populations of Castel dell'Ovo, Gajola and La Nave, where gene flow could be mediated by current patterns linking these sampling sites more tightly than others.

The observed genetic patchwork could be identified as a shallow gene tree model with sympatric lineages (Category IV, Avise, 2000). This pattern is defined as characteristic of high-gene-flow species of modest or small effective size "whose populations have not been sundered by long-term biogeographic barriers" (Avise, 2000). This tree category entails broad sympatry of lineages with presumably recent evolutionary connections, such as recent gene flow through natural dispersal (Avise, 2000).

In general, local heterogeneity in mtDNA is typically shallow in the sense of involving closely related haplotypes (Avise, 2000). What is surprising in our analysis is the high level of sequence diversity among haplotypes, with some of them separated by up to 15 steps in the COI mtDNA region here utilized (see, for example LA and CI samples in Fig. 3). The high level of haplotype diversity encountered here can be related with the demographic expansion of *P. lividus* in the late Pleistocene (Duran et al., 2004), as for many marine organisms with a high level of haplotype diversity. It has been hypothesized that *P. lividus* expanded in the Mediterranean after the last glacial maximum (about 18,000 years ago), from donor populations with enormous population size. This should have favoured both the existence of many rare haplotypes (Watterson, 1984) and an excess of rare mutations (Rogers & Harpending, 1992).

The high variability detected among the Pioppi samples (CI), could also be related to the physical and geographic characteristics of sampled areas.

Table 3 Main haplotypes for each molecular marker, and the two mitochondrial ones grouped together, for each population. Haplotypes O-P-Q correspond to haplogroups A-B-C of Fig. 4, respectively

	16S				COI				ITS2				16S/COI		
Populations/haplotypes	A	B	C	D	E	F	G	H	I	L	M	N	O	P	Q
LA	2	0	2	0	0	1	1	1	1	2	2	0	0	1	1
NA	1	3	1	0	4	1	1	0	0	2	1	0	3	1	1
CA	3	0	1	0	0	0	1	1	0	1	0	2	0	1	0
GA	1	2	0	0	2	0	1	0	1	1	1	0	2	1	0
CO	1	0	0	2	4	1	0	0	0	3	1	0	0	0	0
CI	0	0	0	0	0	3	0	0	0	1	0	0	0	0	0
Total	8	5	4	2	10	6	4	2	2	10	5	2	5	4	2

Haplotypes belonging to a single individual have not been shown

The Gulf of Naples together with the Island of Ischia could be considered a semi-enclosed habitat. The lack of structure within our dataset suggests that marine currents freely disperse larvae over the whole geographic area considered.

Geographic patterns of mtDNA haplotypes are heterogeneous, even if one haplotype, belonging only to the Island of Ischia, and a host of rare haplotypes belonging only to a specific area, have been identified.

Further analysis conducted by means of more sensitive population genetic markers (e.g. microsatellites and SNPs), should be performed in order to confirm the pattern of population structure.

Acknowledgements The authors wish to thank the following for supplying specimens of *Paracentrotus lividus*: B. Iacono, A. Soria, the Fishing Service and the Molecular Biology Service of the Stazione Zoologica "A. Dohrn".

References

Artis, T., 2004. Phylogeography of a facultatively migratory dragonfly, Libellula quadrimaculata (Odonata: Apnisoptera). Hydrobiologia 515: 225–234.

Avise, J. C., 1992. Molecular population structure and biogeographic history of a regional fauna: a case history with lessons for conservation and biology. Oikos 63: 62–76.

Avise, J. C., 1994. Molecular markers, Natural History, and Evolution. Chapman and Hall, New York: 511 pp.

Avise, J. C., 2000. Phylogeography. The History and Formation of Species. Harvard University Press, 447 pp.

Avise, J. C., C. Giblin, J. Larem, J. C. Patton & R. Lansman, 1979. Mitochondrial DNA Clones and Matriarchal Phylogeny within and among Geographic Population of the Pocket Gopher, *Geomys pinetis.* Proceedings of the National Academy of Sciences USA 76: 6694–6698.

Bandelt, H. J., P. Forster & A. Röhl, 1999. Median-Joining Networks for Inferring Intraspecific Phylogenies. Molecular Biology and Evolution 16: 37–48.

De Maio, A., M. Moretti, E. Sansone, G. Spezie & M. Vultaggio, 1985. Outline of Marine Currents in the Bay of Naples and Some Considerations on Pollutant Transport. Il Nuovo Cimento 8: 955–969.

Duran, S., C. Palacìn, M. A. Becerro, X. Turon & G. Giribet, 2004. Genetic diversità and population structure of the commercially harvested sea urchin *Paracentrotus lividus* (Echinodermata, Echinoidea). Molecular Ecology 13: 3317–3328.

Hall, T.A., 1999. BioEdit: a user-friendly biological sequence alignment editor and analysis program for Windows 95/98/NT. Nucleic Acids Symposium Series 41: 95–98.

Lessios, H. A., B. D. Kessing & D. R. Robinson, 1998. Massive gene flow across the world's most potent marine biogeographic barrier. Proceedings Royal Society of London B 265: 583–588.

McMillan, W. O., R. A. Raff & S. R. Palumbi, 1992. Population genetic consequences of developmental evolution in sea urchin (genus *Heliocidaris*). Evolution 46: 1299–1312.

Ovenden, J. R., D. J. Brasher & R. W. G. White, 1992. Mitochondrial DNA analyses of the rock lobster *Jasus edwardsii* supports an apparent absence of population subdivision throughout Australasia. Marine Biology 112: 319–326.

Page, R. D. M., 1996. TREEVIEW: an application to display phylogenetic trees on personal computers. Computer Applications in the Biosciences 12: 357–358.

Palumbi, S. R. & A. C. Wilson, 1990. Mitochondrial DNA diversity in the sea urchin Strongylocentrotus purpuratus and S. droebahiensis. Evolution 44: 403–415.

Palumbi, S. R. & B. D. Kessing, 1991. Population biology of the Trans-Artic exchange: MtDNA sequence sim-

ilarity btween Pacific and Atlantic sea urchin. Evolution 45: 1790–1805.
Palumbi, S. R., 1994. Reproductive isolation, genetic divergence, and speciation in the sea. Annual Review of Ecology and Systematics 25: 547–572.
Palumbi, S. R., 1996. What can molecular genetics contribute to marine biogeography? An urchin's tale. Journal of Experimental Marine Biology and Ecology 203: 75–92.
Palumbi, S. R., A. Martin, S. Romano, W. O. McMillan & L. Stice, G. Grabowski, 1991. The simple fool's guide to PCR. Ver. 2. Available from author, University of Hawaii, Honolulu.
Palumbi, S. R., G. Grabowsky, T. Duda, L .Geyer & N. Tachino, 1997. Speciation and population genetic structure in tropical pacific sea urchins. Evolution 51: 1506–1517.
Patti, F. P. & M. C. Gambi, 2001. Phylogeography of invasive polychaete *Sabella spallanzani* (Sabellidae) based on nuclear sequence of internal spacer (ITS2) of nuclear rDNA. Molecular Ecology Progress Series 215: 169–177.
Patton, J. L. & M. Smith, 1989. Population structure and the genetic and morphologic divergence among pocked gopher species (Genus *Thomomys*) pp. 284–304 *in* D.Otte and J. Endler (eds), Speciation and its consequences. Sinauer, Sunderland, MA.
Pedrotti, M. L, 1993. Spatial and temporal distribution and recruitment of echinoderm larvae in the Ligurian Sea. Journal of the Marine Biology Association of the UK 73: 513–530.
Rogers, A. R. & H. Harpending, 1992. Population growth makes waves in the distribution of pairwise genetic differences. Molecular Biology and Evolution 9: 552–569.
Rozas, J. & R. Rozas, 1999. DNA SP version 3: an integrated program for molecular population genetics and molecular evolution analysis. Bioinformatics 15: 174–175.
Silberman, J. D., S. K., Sarver & P. J. Walsh, 1994. Mitochondrial DNA variation and population structure in the spiny lobster *Panulirus argus*. Marine Biology 120: 601–608.
Swofford, D. L., 2003. PAUP*. Phylogenetic Analysis Using Parsimony (*and other methods). Version 4. Sinauer Associates, Sunderland, MA.
Templeton, A. R., 1983. Phylogenetic inference from restriction endonuclease cleavage site maps with particular reference to the evolution of humans and the apes. Evolution 37: 221–244.
Thompson, J. D., D. G. Higgins, T. & J. Gibson, 1994. CLUSTAL W: improving the sensitivity of progressive multiple sequence alignment through sequence weighting, position specific gap penalties and weight matrix choice. Nucleic Acids Research, submitted, June 1994.
Turon, X., G. Giribet, S. López & C. Palacin, 1995. Growth and population structure of *Paracentrotus lividus* (Echinodermata: Echinoidea) in two contrasting habitats. Marine Ecology Progress Series 122: 193–204.
Uehara, T., M. Shingaki & K. Taira, 1984. Taxonomic studies in the sea urchin, genus *Echinometra*, from Okinawa and Hawaii. Zoological Science 3: 1114.
Watterson, G. A., 1984. Allele frequencies after a bottleneck. Theoretical Population Biology 26: 387–407.
White, T. J., T. Bruns, S. Lee & J. Taylor, 1990. Amplification and direct sequencing of fungal ribosomal RNA genes for phylogenetics. PCR Protocols: A Guide and Application. Part Three. Genetics and Evolution. Academic Press, Inc.

Hydrobiologia (2007) 580:85–96
DOI 10.1007/s10750-006-0463-y

Community structure of the macroinfauna inhabiting tidal flats characterized by the presence of different species of burrowing bivalves in Southern Chile

E. Jaramillo · H. Contreras · C. Duarte

Abstract Several species of bivalves coexist at the lower intertidal of large tidal flats located in the enclosed or inland coast of the northern area of the Nord-Patagonic archipelagos on the Chilean coast (ca. 40–42°S): *Tagelus dombeii* (Lamarck), *Mulinia edulis* (King & Broderip), *Venus antiqua* King & Broderip, *Semele solida* (Gray), *Gari solida* (Gray) and *Diplodonta insconspicua* Philippi. To explore possible spatial variation in the community structure of the macroinfauna inhabiting sediments with different assemblages of these bivalves, seasonal sampling was carried out during 2003–2004 at two tidal flats of that area. Higher species richness and specimen densities of the macroinfauna occurred in sediments with the higher densities of bivalves, especially in sediments where the deep burrower *T. dombeii* reaches its greatest abundances. Our results suggest that, apart from presence of bivalves, the burrowing depth of these organisms is also important in promoting the abundance of macroinfauna. Our results are in contrast with earlier conceptualizations for community organization of the soft bottom macroinfauna inhabiting intertidal flats, related to biological interactions occurring among different phyletic groups, such as that arguing that suspension feeding bivalves (such as *T. dombeii* and *V. antiqua*) will negatively affect the recruitment of species with planktonic larvae, by filtering them before they become established in the substrate. Thus, it is concluded that beneficial effects of bivalve bioturbation overcome that negative effects on the macroinfauna, although detrimental effects may well occur at bivalve densities higher than those studied here.

Keywords Macroinfauna · Bivalves · Tidal flats · Southern Chile

Guest Editors: G. Relini & J. Ryland
Biodiversity in Enclosed Seas and Artificial Marine Habitats

Electronic Supplementary Material Supplementary material is available for this article at http://dx.doi.org/10.1007/s10750-006-0463-y and accessible for authorised users

E. Jaramillo (✉) · H. Contreras · C. Duarte
Instituto de Zoología, Universidad Austral de Chile, Valdivia, Chile
e-mail: ejaramil@uach.cl

Introduction

Bivalves usually dominate the biomass of infaunal communities in sedimentary habitats such as tidal flats (Peterson, 1977; Legendre et al., 1997). Their key role in the community ecology of the soft bottom macroinfauna has been widely investigated (see Peterson, 1977 and Dame, 1996 for references). The burrowing activity of bivalves in fact affects the vertical distribution and stability of sediments (cf. Rhoads & Young, 1970; Nowell et al., 1981; Hall, 1994), the nutrient fluxes from the sediment to the water column (Vaughn & Hakenkamp, 2001; Kanaya et al., 2005; Michaud et al., 2006) and the oxygen availability (Michaud et al., 2005), all factors that condition macroinfauna dynamics and abundances (Gutiérrez et al., 2000). Moreover, suspension and deposit feeding bivalves produce faeces and pseudo-faeces that increase organic matter content of sediments facilitating penetration of the macroinfauna into the sediment column (Gutiérrez et al., 2000; Vaughn & Hakenkamp, 2001). All the above suggest that bivalves promote species richness and population abundance of the infaunal organisms living in tidal flats (cf. Reise, 1983), a pattern also found in sedimentary bottoms of freshwater streams (Vaughn & Spooner, 2006).

Several species of bivalves coexist in the lower intertidal zone of large tidal flats (in the range of 20–30 ha each), located at the enclosed or inland coast (i.e. not exposed to the breaking waves of the Pacific Ocean) of the northern area of the Nord-Patagonic archipelagos along the Chilean coast (ca. 40–42°S). The most common species are the razor clam *Tagelus dombeii* (Lamarck), the clams *Mulinia edulis* (King & Broderip), *Venus antiqua* King & Broderip, *Semele solida* (Gray), *Gari solida* (Gray) and *Diplodonta insconspicua* Philippi (e.g. Lardies et al., 2001; Stead et al., 2002). Even though they are similar in shell length (up to ~50–60 mm), the burial depths of *T. dombeii, M. edulis, V. antiqua, S. solida* and *G. solida* are quite different: that of the first species (a deep burrower) may go down to nearly 30 cm, that of *M. edulis* (a mid-depth burrower) is 5–10 cm, while that of the other three clams (near-surface burrowers) are restricted to the upper 4–5 cm of the sediment (similar to that of the small clam *D. insconspicua* which has an adult size close to 30 mm).

In this study we aimed to explore possible spatial variation in the community structure of the macroinfauna (those retained by a 500 μm sieve) inhabiting sediments with different assemblages of bivalves. We hypothesize that, if bivalves inhabiting tidal flats of the Nord-Patagonic archipelagos do actually promote macroinfauna, species richness and population abundances should be richer and higher in sediments with higher abundances of bivalves. To test this hypothesis we compared the community structure of the intertidal macroinfauna in three paired scenarios: (1) sediments dominated in number by the deep burrower *T. dombeii* versus sediments with lower abundances of this species but with other near surface burrowers (*V. antiqua, S. solida, G. solida* and *D. insconspicua*), (2) sediments dominated in number by the mid-depth burrower *M. edulis* which co-inhabits with *T. dombeii* and *V. antiqua* (both in lower abundances) versus nearby sediments without bivalves, and (3) sediments dominated in number by the near surface burrower *V. antiqua* which co-inhabits with *M. edulis* versus nearby sediments without bivalves.

Materials and methods

Study sites

The tidal flats chosen to represent each of the above three paired scenarios were those of Pelluco for scenario 1 (41°29′ S, 72°54′ W) and Compu (42°52′ S, 73°54′ W) (Fig. 1) for scenarios 2 and 3. While the first experiences tidal ranges close to 6 m during spring tides, at Compu the range is only about 3 m (Viviani, 1979). Samples were collected at Pelluco during spring tides of February, June, August, November and December 2003 and March 2004, while those at Compu were collected during spring tides of February, July, August, October and December 2003 and March 2004. Site 1 of Pelluco (dominated by *T. dombeii*) was located at the low tide level, while site 2 (with lower density of bivalves) was nearly 20 m upshore. Site 1 of Compu (dominated by *M. edulis*) was located at the low tide level. Sites 2 (dominated by *V. antiqua*) and 3 (with no bivalves at all) were nearly 15 m apart each other and about 20 m upshore from site 1.

Collection of samples

Sediment samples were collected at each site from the centre of five randomly located parcels (1 × 0.5 m) using plastic cylinders (7.5 cm in diameter) pushed to a depth of 15 cm into the

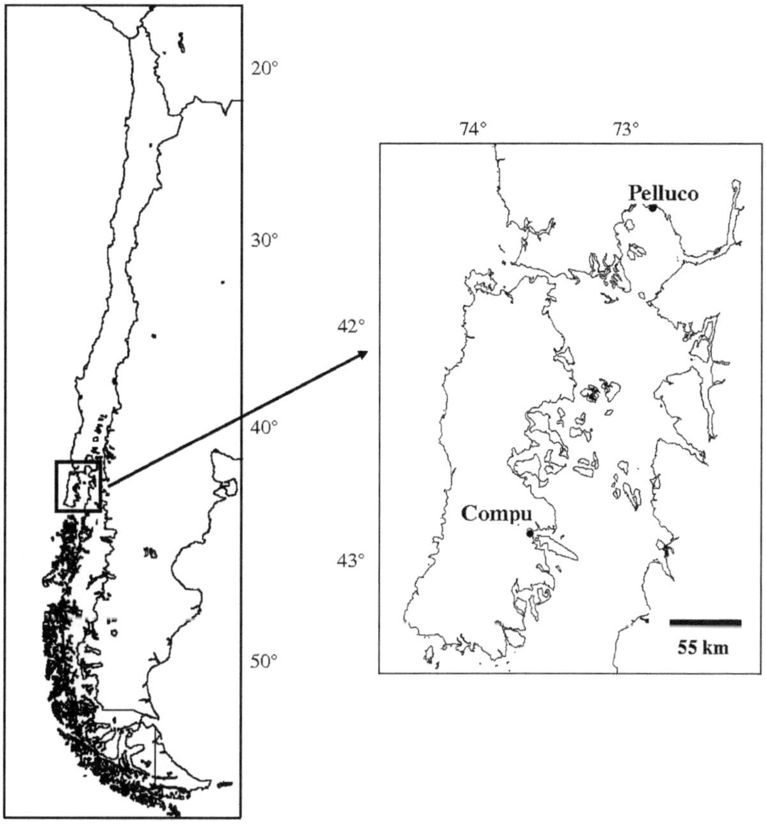

Fig. 1 Location of the tidal flats studied on the inland coast of the Nord-Patagonic archipelagos of the Chilean coast

substrate. A subsample of sediments for textural and granulometric analyses of sediments was collected with a plastic cylinder 2.5 cm in diameter pushed 5 cm deep into the sediments collected within the larger cylinder. These samples were frozen (–20°C) until further analysis (see below). The remaining sediments were passed through a sieve with a 500 μm mesh and the residue was preserved in 10% formaldehyde until laboratory sorting for faunal analyses. After removing the sand sample, the sediment of each parcel was excavated to a depth of about 30 cm to collect all visible bivalves.

Laboratory and data analyses

Samples for textural and granulometric analysis were thawed to examine spatial variability in the percentages and mean grain size of sand particles (63–2000 microns). Percentages of sand were calculated according to Anderson et al. (1981) while mean grain size of sands was determined with a settling tube (Emery, 1938) and the moments computational method (Seward-Thompson & Hails, 1973). Total organic matter (TOM) was estimated after calculations of weight differences between samples incinerated at 550°C for 6 h and previously dried at 60°C for 24 h. Volume of bivalves at each sampling date was estimated using regression equations resulting from preliminary analyses (unpublished data) aimed to relate volume displacement and shell length of individual bivalves across a full range of body sizes (i.e. shell lengths). Two-way ANOVA with sites and months as factors was performed to compare the temporal variability of sediment characteristics, species richness and macroinfaunal abundances at both flats (Sokal &

Rohlf, 1995). The bivalves collected by hand were excluded from the estimates of species richness and specimen's densities. The *a posteriori* test, Tukey's Honest-Significant-Difference or HSD (Sokal & Rohlf, 1995) was used for the data of Compu to compare sites when the two-way ANOVA indicated significant differences for them. Biological relationships between the sites sampled at each flat were assessed using cluster analyses and non-metric multidimensional scaling (MDS). Both analyses were based upon a similarity matrix calculated with the Bray Curtis similarity coefficient after double root transformation of abundance data as run by the PRIMER program (Plymouth Routines in Multivariate Ecological Research) (Carr, 1997). The usefulness of the MDS analyses (i.e. display of relationships between sites) was evaluated with the stress statistics: values <0.1 indicates that the depiction of relationships is good, while if stress values are >0.2 the depiction is poor (Clarke, 1993).

Results

The sediments

Sand particles were represented by similar percentages at Pelluco and Compu: 86–97% and 84–96%, respectively (Fig. 2). Mean grain sizes of sand varied 429–600 μm at Pelluco and 328–442 μm at Compu; thus, coarser sands occurred at Pelluco (Fig. 2). The percentage of total organic matter (TOM) were 0.36–2.28% and 0.70–3.78% at Pelluco and Compu, respectively (Fig. 2). Two-way ANOVA showed significant variability of percentages and mean grain size of sands and TOM at both flats for sites, months and interactions (sites × months) for the majority of analyses (Table 1). Percentages of sand were significantly higher at site 2 of Pelluco than at site 1 of the same flat (results of ANOVA, Table 1). The same variable had a significantly higher value at site 3 of Compu (without bivalves) than at the sites 1 and 2, which did not differ among themselves in

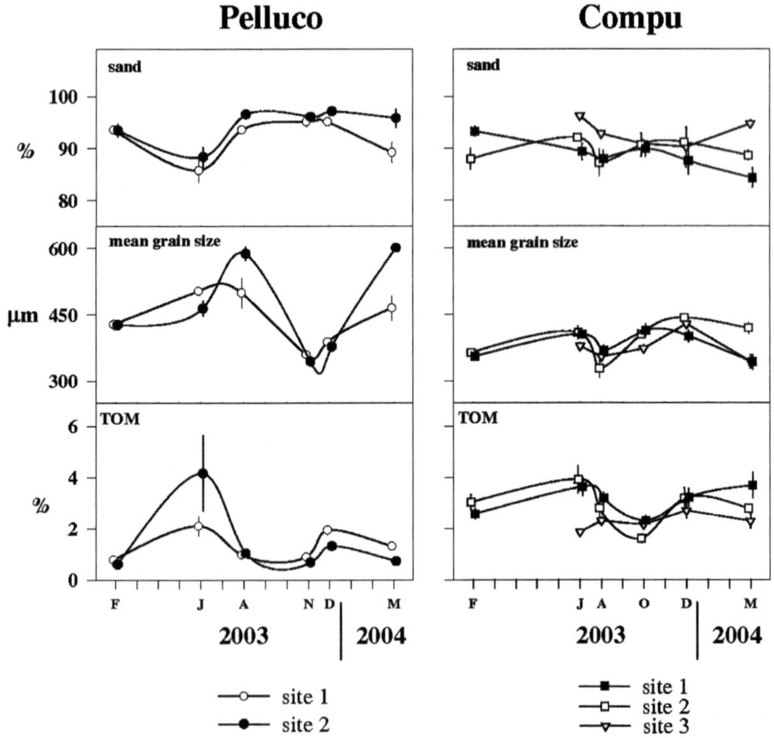

Fig. 2 Temporal variability in the percentages of sand, mean grain size of sands and total organic matter (TOM) at the sites studied at each flat. The values are means ±1 standard error

Table 1 Summary of the two way ANOVA carried out to test for differences in sedimentological characteristics at the sites studied. Results of Tukey's HSD are also given for the tidal flat of Compu (three sites)

	Flat	Source of variation	*F* ratio	*P* value	Results of ANOVA (Pelluco) and Tukey's HSD test (Compu)
% sand	Pelluco	Sites	18.00	<0.01	site 1 < site 2
		Months	16.40	<0.01	
		Interactions	1.75	0.16	
	Compu	Sites	14.59	<0.01	site 1 = site 2 < site 3
		Months	3.99	<0.01	
		Interactions	3.37	<0.01	
Mean grain size	Pelluco	Sites	9.86	<0.01	site 1 < site 2
		Months	59.59	<0.01	
		Interactions	11.34	<0.01	
	Compu	Sites	6.10	<0.01	site 1 = site 2; site 1 = 3; site 2 > site 3
		Months	18.22	<0.01	
		Interactions	5.04	<0.01	
TOM	Pelluco	Sites	0.11	0.73	site 1 = site 2
		Months	12.48	<0.01	
		Interactions	3.10	0.031	
	Compu	Sites	13.24	<0.01	site 1 = site 2 > site3
		Months	7.12	<0.01	
		Interactions	3.15	<0.01	

percentages of sands (results of Tukey's HSD, Table 1). Values of mean grain size of sands showed a similar trend to that shown by the between-site variability of percentages of sand at each flat (Table 1). Values of TOM did not differ significantly among sites at Pelluco, while results of the Tukey's HSD test showed significantly higher concentration of TOM at sites 1 and 2 of Compu which did not differ among themselves (Table 1).

The bivalve assemblage

Five species of clams were found in the sediments of site 1 at Pelluco: *T. dombeii*, *V. antiqua*, *S. solida*, *G. solida* and *D. insconspicua* (Fig. 3). *Tagelus dombeii* was the most abundant species with densities as high as 44 and 49 individuals per 0.5 m^2 during December 2003 and March 2004, respectively. During most sampling dates, *T. dombeii* represented more than 80% of the total abundance of bivalves at this site. The total volume occupied by the individuals of this bivalve also reached the highest values (up to nearly 275 cc during December 2003 and March 2004); during all sampling months except November 2003, volumes represented more than 80% of the total volume of bivalves at this site (Fig. 3). *Tagelus dombeii*, *V. antiqua* and *D. insconspicua* were also found at the sediments of site 2 of Pelluco. With the exception of December 2003, the population numbers of these species were usually lower than three individuals per 0.5 m^2. The volume occupied by *T. dombeii*, *V. antiqua* and *D. insconspicua* were also low at this site (<10 cc) (Fig. 3).

Three species of clams were found at site 1 in Compu: *M. edulis*, *V. antiqua* and *T. dombeii* (Fig. 3). *Mulinia edulis* was the dominant species, either in numbers or volume (up to 52 individuals per 0.5 m^2 and 1262 cc respectively, during February 2003). Population abundances and volume of *M. edulis* species varied around 58–90% and 75–96% of the total abundance and volume of total bivalves, respectively. Two species of clams were collected at the sediments of site 2 of Compu: *V. antiqua* and *M. edulis* (Fig. 3). The dominant species was *V. antiqua* with numbers as high as nearly 25 individuals per 0.5 m^2 during February 2003; thus, its representation varied around 73–91% of the total abundance. During all sampling months (with the exception of July 2003), the total volume occupied by *V. antiqua* at this site (up to nearly 348 cc during February 2003) exceeded 70% of the total volume of bivalves (Fig. 3).

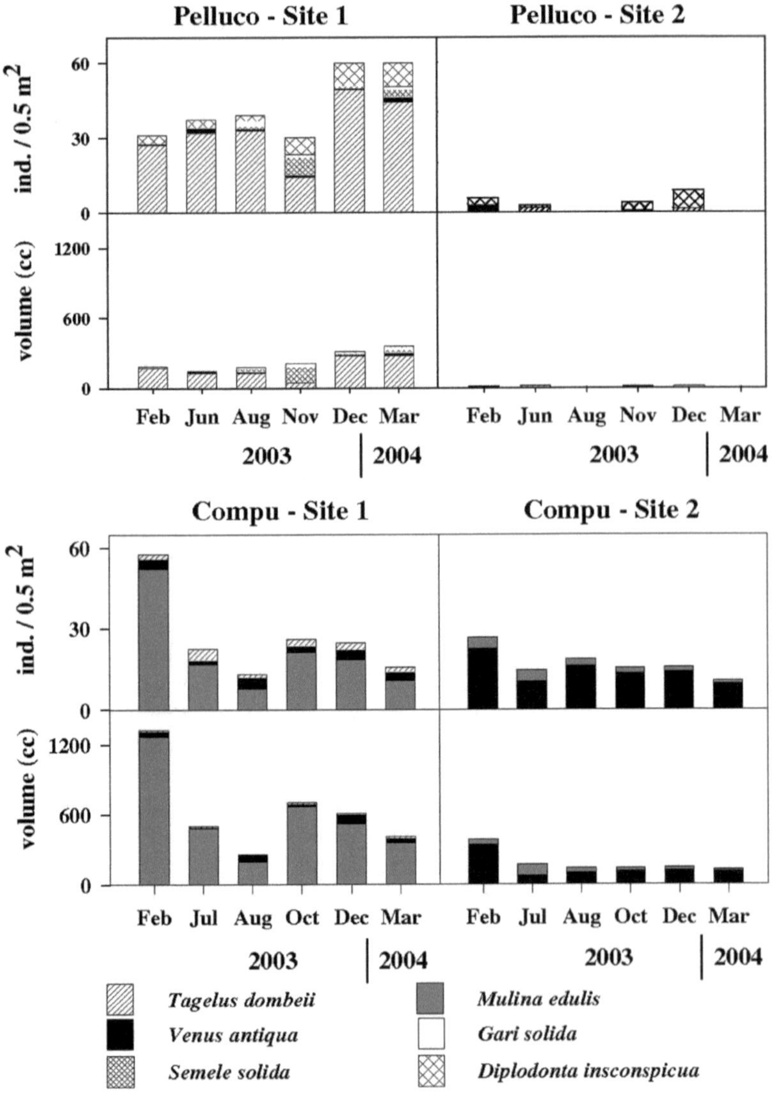

Fig. 3 Temporal variability in population abundances and volume occupied by bivalves at the tidal flats of Pelluco and Compu

The macroinfaunal assemblage

The bivalves collected by hand are excluded from the following estimates of species richness and specimen densities. Overall, a total of 55 and 40 species were collected at sites 1 and 2 of Pelluco (see Electronic Supplementary Material); both sites had a taxonomic similarity of 0.67 (Jaccard Index = JI), sharing 38 species. The number of species collected at sites 1, 2 and 3 of Compu were 39, 27 and 22 (see Electronic Supplementary Material). The taxonomic similarities and species shared for these sites were: 1–2: JI = 0.61, 25 species shared, 1–3: JI = 0.42, 18 species shared and 2–3: JI = 0.40, 14 species shared. Polychaete worms were the most represented group with 28 and 21 species at sites 1 and 2 of Pelluco and 24, 15 and 12 species at sites 1, 2 and 3 of Compu (see Electronic Supplementary Material). During most of the sampling dates, the dominant species at Pelluco and Compu was the archiannelid *Polygordius* sp. This worm reached 33.1 and

25.2% of the overall mean abundance at sites 1 and 2 of Pelluco and 63.6, 48.6 and 75.3% of the overall mean abundance at sites 1, 2 and 3 of Compu (see Electronic Supplementary Material).

The highest mean number of macroinfaunal species and abundances were found at site 1 of Pelluco dominated by *T. dombeii*: 16 and 27 species and 124 and 346 individuals per 0.004 m^2, respectively (Fig. 4). That figures varied between 9 and 13 species and 25 and 91 individuals per 0.004 m^2 at site 2 of the same flat. The following figures were estimated for Compu: 5 and 11 species and 18 and 95 individuals per 0.004 m^2 for site 1 (dominated by *M. edulis*), 2 and 9 species and 14 and 103 individuals per 0.004 m^2 for site 2 (dominated by *V. antiqua*), and 2 and 5 species and 3 and 53 individuals per 0.004 m^2 for site 3 (without bivalves) (Fig. 4). The results of a two-way ANOVA showed significant variability of number of species and abundances of the total macroinfauna at both flats for sites, months and interactions (sites × months) for all the analyses (Table 2). Number of species was significantly higher at site 1 of Pelluco as compared with site 2 of the same flat (results of ANOVA, Table 2). The same variable showed the highest value at site 1 of Compu where number of species decreased significantly from site 1 to site 2 and to site 3 (results of Tukey's HSD test, Table 2). The results of ANOVA and Tukey's HSD test carried out to compare between-site abundances of the total macroinfauna at each flat showed significant higher values at those sites dominated or inhabited by bivalves (site 1 of Pelluco and sites 1 and 2 of Compu, respectively) (Table 2).

Figure 5 shows the results of cluster analyses. For Pelluco, most of the sampling dates of site 1 (dominated by *T. dombeii*) linked together to a similarity value close to 58%; on the other hand, the sampling dates of site 2 formed two groups, one of them more similar to sampling dates of site 1 than to the other dates in the same site. The cluster for sampling dates in Compu shows a gradual linking of sampling dates for sites 1 and 2 and a rather clear separation for sampling dates of site 3. The macroinfaunal relationships between sampling stations resulting from MDS analyses are depicted in the plots shown in Fig. 6. The values of the stress statistic (<0.20) indicate that the depiction of the relationship for each tidal flat is good enough to provide some conclusions (Clarke, 1993). The MDS plot for Pelluco show that in general, samples of sites 1 and 2 have a clear separation along the *x* axis.

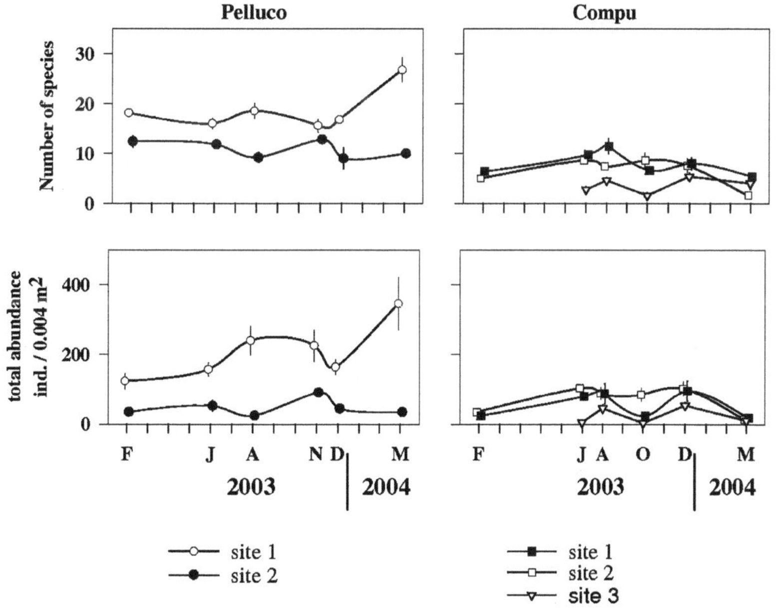

Fig. 4 Temporal variability in number of species and specimen abundances of the total macroinfauna at the sites studied at each flat. The values are means ±1 standard error

Table 2 Summary of the two way ANOVA carried out to test for differences in number of species and specimen abundances of the total macroinfauna at the sites studied. Results of Tukey's HSD are also given for the tidal flat of Compu (three sites)

	Flat	Source of variation	*F* ratio	*P* value	Results of ANOVA (Pelluco) and Tukey's HSD test (Compu)
No. of species	Pelluco	sites	103.22	<0.01	site 1 > site 2
		months	4.25	<0.01	
		interactions	7.10	<0.01	
	Compu	sites	32.89	<0.01	site 1 > site 2 > site3
		months	9.78	<0.01	
		interactions	5.25	<0.01	
Total abundance	Pelluco	sites	84.63	<0.01	site 1 > site 2
		months	3.52	<0.01	
		interactions	3.86	<0.01	
	Compu	sites	18.39	<0.01	site 1 = site 2 > site 3
		months	11.25	<0.01	
		interactions	2.05	<0.01	

Similar to that found in the cluster analyses, most of sampling dates of sites 1 and 2 grouped together and apart from that samples of site 3 (Fig. 6).

Discussion

This study shows that species richness and specimen abundances of the total macroinfauna were significantly higher at those sites with higher abundances of bivalves as compared with sites with lower abundance or absence of bivalves. At first glance, these results appear to be just another case of macroinfaunal promotion by sedimentary bivalves as has been discussed for other intertidal zones (cf. Commito, 1987; Dittmann, 1990; Commito & Dankers, 2001; Dame et al., 2001; Commito et al., 2005). However, the evidence published so far show that promotion affects primarily invertebrates with direct development such as tubificid oligochaetes and not necessarily other macroinfaunal groups. For example, Commito (1987) found that in a mussel bed located in an intertidal flat of New England (USA), the population abundances of oligochaetes inside the bed were nearly five times higher than outside the bank. Similarly, Dittmann (1990) found that in a tidal flat of the North Sea, only oligochaetes showed a significant increases in population abundances inside a mussel bank. In contrast to these previous studies, the results of this one show that most macroinfauna had significantly higher specimen abundances at sites with higher bivalve densities. For example, in Pelluco, the population abundances of *Poligordius* sp. (one of the most common species at both flats studied), were nearly six times higher in that sediments with high densities of *T. dombeii* as compared with sediments with lower abundances of bivalves. Similarly, the abundances of this worm at Compu, was nearly two times higher in those sediments dominated by *M. edulis* and *V. antiqua* as compared with sediments without bivalves. Also, the abundance of another common taxon in Compu, the snail *Caecum chilense*, was nearly 13 and seven times higher in the sediments dominated by *M. edulis* and *V. antiqua*, respectively, as compared with sediments without bivalves. It must be noted, however, that comparisons between the results of this study and those carried out in the Northern Hemisphere (mentioned above) cannot be directly comparable, since we studied bivalve assemblages living underneath the sediment surface, while the other studies refer to bivalves living above the sediment or at the sediment water interface. In other words, those studies refer to mussel beds with a three-dimensional architecture which may well not promote similar types of macroinfauna. The general structure of the mussel beds also acts as a sediment trap, which generally produces anoxic conditions in the sediments underneath the bivalves—which in turn may adversely affects the macroinfauna (cf. Ragnarsson & Raffaelli, 1999).

Our results are similar to those reported by Reise (1983), who found that experimental aggregates of the bivalve *Macoma balthic*a (L.)

Fig. 5 Dendrograms produced from the Bray–Curtis Similarity Index for sampling dates of Pelluco and Compu

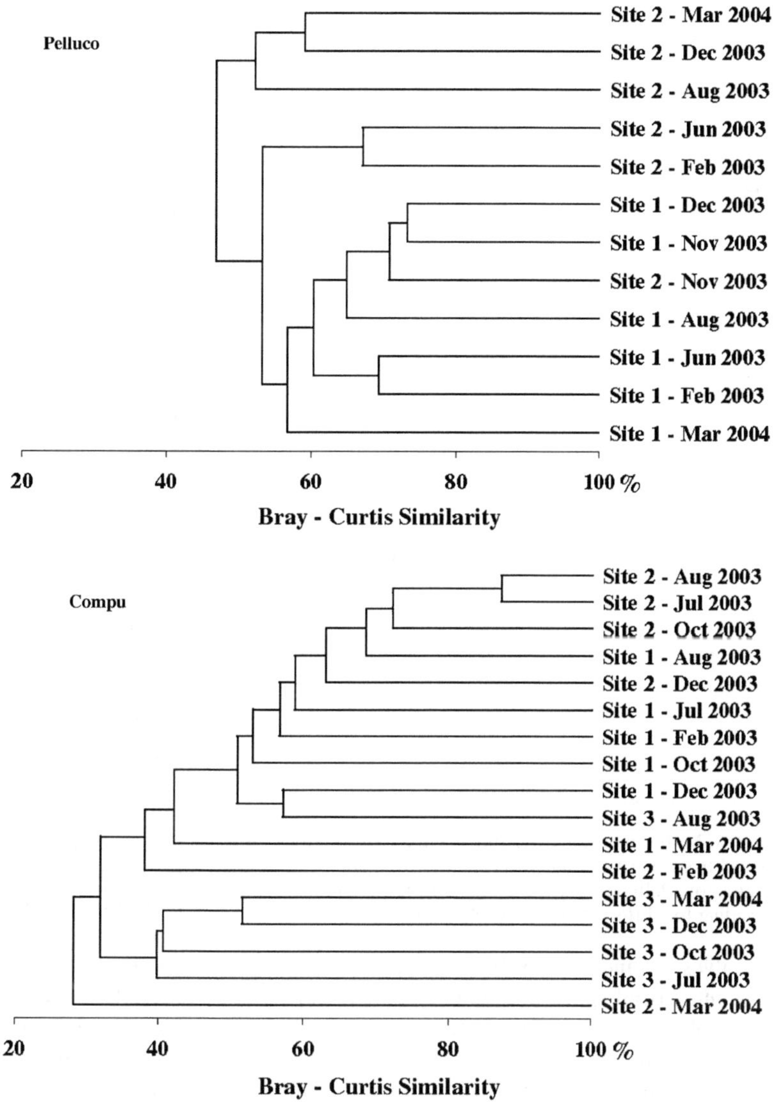

increased the abundance of meiofauna (Nematoda, Turbellaria and small Polychaeta) inhabiting a tidal flat of the North Sea. Based upon the feeding habits of some of that meiofauna, Reise (1983) concluded that sediments located around the aggregates of *Macoma* increased their amount of food in form of microorganisms, a situation that would explain the higher amount of meiofauna in sediments located around bivalves. We have no experimental data on the mechanisms by which bivalves promote species richness and specimen abundances of the macroinfauna on the flats studied on the Nord-Patagonic archipelagos. However, based on previous work elsewhere, it is possible to propose some alternative scenarios: (i) since bivalves are important recyclers of nitrogen in marine coastal habitats, the release of ammonium and dissolved organic nitrogen can be beneficial for microphytobenthos (Dame, 1996) and thus, actual or potential food resources for the macroinfauna would increase; (ii) biogenic reworking of sediments produced by

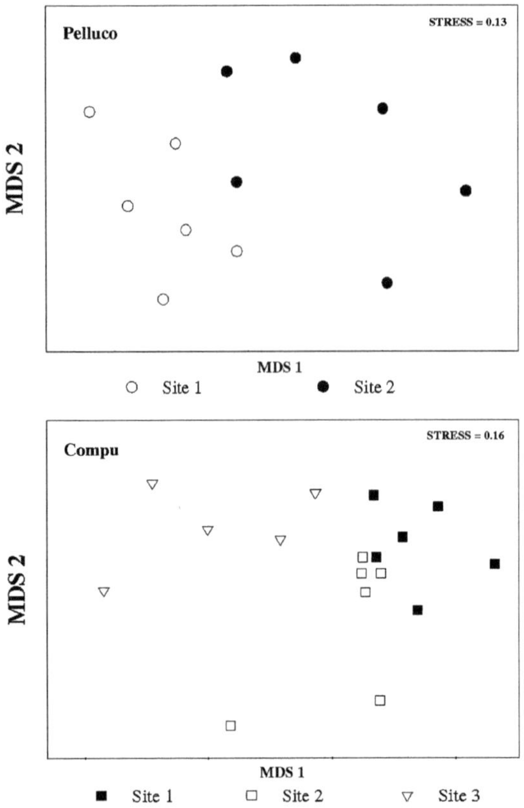

Fig. 6 Multidimensional scaling (MDS) ordination derived from macroinfaunal species data from the sampling dates of Pelluco and Compu. Different symbols and shading for different sites at each tidal flat

bivalves can increase flushing of water throughout the interstitial space and, consequently, penetration of oxygen into the sediments stimulating microbial metabolism (Dame, 1996), thus leading to higher macroinfaunal production; (iii) higher population abundances of bivalves result in higher production of faecal pellets and pseudofaeces, which can be used by microbial communities in the remineralization of the organic matter (cf. Black, 1980; Rhoads & Boyer, 1982).

The highest species richness and specimen densities of the intertidal macroinfauna occurred in those sediments that had the highest population abundances of the deep burrower *T.dombeii*. This suggests, that burrowing depth of these bivalves and, consequently, the depth of biogenic reworking may be an important factor affecting the abundance and distribution of the macroinfana. Since both studied tidal flats are included in the same zoogeographic zone of the Chilean coast (cf. Viviani, 1979), some simple comparisons can be done to illustrate this assertion. Overall, the mean species richness of the macroinfauna at the site of Pelluco with high abundances of *T. dombeii* (19 species) was 2.4–2.7 higher than that found at the sites of Compu dominated by the mid burrower *M. edulis* (8 species) and near-surface burrower *V. antiqua* (7 species). Similar comparisons for abundances of the total macroinfauna are as follows: the whole mean of macroinfaunal abundances at the site of Pelluco with high abundances of *T. dombeii* (210 individuals per 0.004 m^2), was nearly 4 times higher than that found at the site of Compu dominated by the mid burrower *M. edulis* (55 individuals per 0.004 m^2), and nearly 3 times higher than that calculated for the site of Compu with dominance of the near-surface burrower *V. antiqua* (71 individuals per 0.004 m^2). These simple analyses also suggest that macroinfaunal promotion by bivalves in the flats studied affect abundances more than species richness.

The above conclusions are in contrast with earlier conceptualizations for community organization of the soft bottom macroinfauna inhabiting intertidal flats and related to biological interactions occurring among different phyletic groups (i.e. Woodin, 1976). One of the predictions of that conceptual framework, was that suspension feeding bivalves (such as *T. dombeii* and *V. antiqua*, Lardies et al. 2001) will negatively affect the recruitment of species with planktonic larvae, by filtering them before they become established in the substrate. Having in mind that many of the macroinfaunal species studied here have indirect development, we should have expected a similar pattern to that predicted by Woodin (1976): i.e. lower population abundances of the macroinfauna at those sediments with higher densities of bivalves; however, we found just the opposite. One possible explanation for the higher population abundances of the macroinfauna in the sediments with higher densities of bivalves, is that the recorded densities were not high enough to produce any detrimental effect on the macroinfauna. Thus, we can hypothesize, that the pattern found in this study is consistent with the intermediate disturbance hypothesis; in other words, we found higher species richness and population abundances of the macroinfauna in sediments with the higher densities of bivalves, because at those densities beneficial effects of bivalves are still active, while at densities higher than those studied here, the effects might be overcome by detrimental ones such as that predicted by Woodin (1976). On the other hand, the following aspects which can account for the potential differences from other studies must be taken into account: (i) our results show important differences in sediment characteristics among sites; thus, it may well be possible that differences in macroinfaunal abundances are not only related to bivalve densities but also to other environmental differences; and (ii) due to the highly dynamic features of the tidal flats studied, many macroinfaunal organisms may settle on the studied sediments by bedload transport and not from the water column, thus avoiding consumption by suspension feeder bivalves. Clearly, experimental studies are needed for further comprehension of the interactions between bivalves and macroinfaunal organisms in tidal flats of southern Chile and elsewhere.

Acknowledgements To M. González for sorting faunistical samples and sedimentary analyses. To two anonymous reviewers who greatly improved an earlier manuscript. This research was supported by CONICYT, Chile (Fondecyt Project No. 1030335). Funds given to EJ to present this study during the 39th European Marine Biology Symposium (Genoa, Italia; July 2004) came from Dirección de Investigación y Desarrollo, Universidad Austral de Chile and Fondecyt Project No. 1030335.

References

Anderson, F., L. Black, L. Mayer & L. Watling, 1981. A temporal and spatial study of a mud flat texture. Northeastern Geology 3: 184–196.

Black, L. F., 1980. The biodeposition cycle of a surface deposit-feeding bivalve, *Macoma balthica* (l.). In Kennedy V. S. (ed.) Estuarine Perspectives. Academic Press, New York: 389–402.

Carr, M. R., 1997. PRIMER User Manual. Plymouth Marine Laboratory, Prospect Place, Plymouth PL1 3 DH, United Kingdom, 40 pp.

Clarke, K. R., 1993. Non-parametric multivariate analyses of changes in community structure. Australian Journal of Ecology 18: 117–143.

Commito, J. A., 1987. Adult–larval interactions: predictions, mussels and cocoons. Estuarine Coastal and Shelf Science 25: 599–606.

Commito, J. A. & N. M. J. A. Dankers, 2001. Dynamics of spatial and temporal complexity in European and North American soft-bottom mussel beds. In Reise, K. (ed.), Ecological Comparisons of Sedimentary Shores. Ecological Studies 151: 39–59.

Commito, J. A., E. A. Celano, H. J. Celico, S. Como & C. P. Johnson, 2005. Mussels matter: postlarval dispersal dynamics altered by a spatially complex ecosystem engineer. Journal of Experimental Marine Biology and Ecology 316: 133–147.

Dame, R. F., 1996. *Ecology of Marine Bivalves*: an ecosystem approach. CRC Press, New York.

Dame R. F., D. Bushek & T. C. Prins, 2001. Benthic suspension feeders as determinants of ecosystem structure and function in shallow coastal waters. In Reise, K. (ed.), Ecological Comparisons of Sedimentary Shores. Ecological Studies 151: 11–37.

Dittmann, S., 1990. Mussel beds – amensalism or amelioration for intertidal fauna ? Helgoländer Meeresuntersuchungen 44: 335–352.

Emery, K. O., 1938. A simple method of mechanical analysis of sands. Journal of Sedimentary Petrology 8: 105–111.

Gutiérrez, D., V. A. Gallardo, S. Mayor, C. Neira, C. Vásquez, J. Sellanes, M. Rivas, A. Soto, F. Carrasco & M. Baltazar, 2000. Effects of dissolved oxygen and fresh organic matter on the bioturbation potential of macrofauna in sublittoral sediments off Central Chile during the 1997/1998 El Niño. Marine Ecology Progress Series 202: 81–99.

Hall, S. J., 1994. Physical disturbance and marine benthic communities: life in unconsolidated sediments. Oceanography and Marine Biology: an annual review 32: 179–239.

Kanaya, G., E. Nobata, T. Toya & E. Kikuchi, 2005. Effects of different feeding habits of three bivalve species on sediment characteristics and benthic diatom abundance. Marine Ecology Progress Series 299: 67–78.

Lardies, M. A., E. Clasing, J. M. Navarro & R. A. Stead, 2001. Effects of environmental variables on burial depth of two infaunal bivalves inhabiting a tidal flat in southern Chile. Journal of the Marine Biological Association of the U.K. 81: 809–816.

Legendre, P., S. F. Thrush, V. J. Cummings, P. K. Dayton, J. Grant, J. E. Hewitt, A. H. Hines, B. H. McArdle, R. D. Pridmore, D. C. Schneider, S. J. Turner, R. B. Whitlatch & M. R. Wilkinson, 1997. Spatial structure of bivalves in a sandflat: scale and generating processes. Journal of Experimental Marine Biology and Ecology 216: 99–128.

Michaud, E., G. Desrosiers, F. Mermillod-Blondin, B. Sundby & G. Stora, 2005. The functional group approach to bioturbation: I. The effects of biodiffusers and gallery-diffusers of the *Macoma balthica* community on sediment oxygen uptake. Journal of Experimental Marine Biology and Ecology 326: 77–88.

Michaud, E., G. Desrosiers, F. Mermillod-Blondin, B. Sundby & G. Stora, 2006. The functional group approach to bioturbation: II. The effects of the *Macoma balthica* community on fluxes of nutrients and dissolved organic carbon across the sediment–water interface. Journal of Experimental Marine Biology and Ecology 337: 178–189.

Nowell, A. R. M., P. A. Jumars & J. E. Eckman, 1981. Effects of biological activity on the entraiment of marine sediments. Marine Geology 42: 133–153.

Peterson, C. H., 1977. Competitive organization of the soft-bottom macrobenthic communities of Southern California Lagoons. Marine Biology 43: 343–359.

Ragnarsson, S. A. & D. Raffaelli, 1999. Effects of the mussel *Mytilus edulis* L. on the invertebrate fauna of sediments. Journal of Experimental Marine Biology and Ecology 241: 31–43.

Reise, K., 1983. Biotic enrichment of intertidal sediments by experimental aggregates of the deposit-feeding bivalve *Macoma balthica*. Marine Ecology Progress Series 12: 229–236.

Rhoads, D. C. & D. K. Young, 1970. The influence of deposit-feeding organism on sediment stability and community trophic structure. Journal Marine Research 28: 150–178.

Rhoads, D. C. & L. F. Boyer, 1982. The effects of marine benthos on physical properties of sediments: a successional perspective. In McCall P. L. & M. J. S. Tevesz (eds) Animal-Sediment Relations – The Biogenic Alteration of Sediments. Plenum Press, New York: 3–52.

Seward-Thompson, B. & J. Hails, 1973. An appraisal on the computation of statistical parameters in grain size analysis. Sedimentology 11: 83–98.

Sokal, R. R. & F. J. Rohlf, 1995. Biometry: the principles and practice of statistics in biological research. Freeman, W.H., New York, 1–877.

Stead, R. A., E. Clasing, M.A. Lardies, L.P. Arratia, G. Urrutia & O. Garrido, 2002. The significance of contrasting feeding strategies on the reproductive cycle in two coexisting tellinacean bivalves. Journal of the Marine Biological Association of the U.K. 82: 443–453.

Vaughn, C. C. & C. C. Hakenkamp, 2001. The functional role of burrowing bivalves in freshwater ecosystems. Freshwater Biology 46: 1431–1446.

Vaughn, C. C. & D. E. Spooner, 2006. Unionid mussels influence macroinvertebrate assemblage structure in streams. Journal of the North American Benthological Society 25: 691–700.

Viviani, C. A., 1979. Ecogeografia del litoral chileno. Studies on Neotropical Fauna and Environment 14: 65–123.

Woodin, S. A., 1976. Adult–larval interactions in dense infaunal assemblages: Patterns of abundance. Journal of Marine Research 34: 25–41.

Hydrobiologia (2007) 580:97–108
DOI 10.1007/s10750-006-0462-z

BIODIVERSITY IN ENCLOSED SEAS

Response of zoobenthic communities to changing eutrophication in the northern Baltic Sea

Jonne Kotta · Velda Lauringson · Ilmar Kotta

Abstract The relationships between the concentration of water nutrients and the biomass of benthic invertebrate feeding guilds were examined at 46 sites in the northern Baltic Sea during 1993–2003. We analysed whether and how degree of exposure, presence of fronts, salinity, hypoxia, nutrient concentrations, depth, sediment type and structure of invertebrate communities contributed to these relationships. In general macrozoobenthos did not respond to the changing nutrient concentrations in the areas that were regularly impacted by fronts (river estuaries, bank slopes, straits). Macrobenthic species diversity, depth, 11-year average of nutrient concentration and sediment type explained best how strong the nutrient-invertebrate relationships were. The deposit feeders, that inhabited more diverse communities, were less sensitive to the increased concentration of nutrients than those in less diverse communities. On the other hand, the sensitivity of suspension feeders to rising nutrient load increased with benthic diversity. The response of macrozoobenthos to nitrogen level decreased with increasing depth. Our data did not support the hypothesis that there was a significant difference in the occurrence of nutrient-invertebrate relationships between hypoxic and normoxic conditions. The probability of finding negative nutrient-invertebrate relationships increased with depth. The results pointed to nitrogen limitation in the coarse and fine sediments and phosphorus limitation in the mixed sediments. Increased nitrogen values strengthened the response of suspension feeders to the concentration of phosphorus. Increasing phosphorus level dampened the relationships between benthic functions and concentration of phosphorus. This study confirmed that depth and sediment type were the best regularly monitored abiotic variables that could be used to determine the type areas within the northern Baltic Sea in sensu the European Community Water Framework Directive. As the nutrient-invertebrate relationships were significantly modified by macrobenthic diversity, the environmental classification should incorporate specific biological measures such as benthic diversity in order to better describe the quality status of the water body.

Keywords Baltic · Eutrophication · Front · Functional diversity · Hypoxia · Macrozoobenthos

Guest editors: G. Relini & J. Ryland
Biodiversity in Enclosed Seas and Artificial Marine Habitats

J. Kotta (✉) · V. Lauringson · I. Kotta
Estonian Marine Institute, University of Tartu, Mäealuse 10 a, 12618 Tallinn, Estonia
e-mail: jonne.kotta@sea.ee

V. Lauringson
Institute of Zoology and Hydrobiology, University of Tartu, Vanemuise 46, 51014 Tartu, Estonia

Introduction

Eutrophication is ranked among the most serious threats to the stability of marine ecosystems worldwide (Gray et al., 2002). The effect of eutrophication is more pronounced in the coastal areas which receive large amounts of organic and mineral nutrients from municipal wastes, agricultural and industrial effluents (Cloern, 2001; Gray et al., 2002). In order to protect coastal waters most European countries have to implement the European Water Framework Directive. According to the directive all waterbodies must obtain good water quality prior to 2015. To implement the directive, the reference (i.e. pristine) conditions and criteria for ecological status need to be defined (European Union, 2000).

Benthic invertebrate communities represent an intermediate trophic level and nutrient additions affect them in many ways. According to the Pearson-Rosenberg model (Pearson & Rosenberg, 1978) the biomass of benthic invertebrates increases gradually to a maximum as the load of organic matter increases. Then the biomass falls and often shows a secondary peak but lower than the first maximum. Increasing nutrient loads enhance the production of benthic and/or pelagic microalgae (Granéli & Sundbäck, 1985; Howarth, 1988) and, hence, increase available food for benthic grazers, suspension feeders or deposit feeders. As a consequence abundance and growth responses of invertebrates are observed from an initial increase of nutrients (Posey et al., 1999). Further eutrophication leads to hypoxia, appearance of ammonia and hydrogen sulphide, and consequently to the disappearance of benthic invertebrates. Thus, benthic communities are highly sensitive to eutrophication, which makes them a good indicator of water quality (Pearson & Rosenberg, 1978; Grall & Chauvaud, 2002; Gray et al., 2002).

The Baltic Sea is a semi-enclosed system with a small exchange of water. Due to the considerable river input of fresh water, the Baltic Sea may be classified as a large estuary with steep environmental gradients of salinity, temperature and oxygen (Segerstråle, 1957; Kullenberg, 1981). As compared to other waterbodies the effects of discharged nutrients are very acute here, especially at deeps where natural periodic low oxygen concentrations make the system highly vulnerable to the load of organic matter (Cloern, 2001; HELCOM, 2002; Karlson et al., 2002).

Eutrophication has been an increasing ecological threat in the Baltic Sea during the past 50 years. During this time the load of nutrients has grown fourfold for nitrogen and 8 times for phosphorus leading to an increased production at all trophic levels in the ecosystem (Elmgren, 2001). Increasing nutrient concentrations have also severely affected macrozoobenthos, being expressed as the loss of species and increase in abundance and biomass (Karlson et al., 2002). Due to low salinity and a short geological history, relatively few macrobenthic species have adapted to the Baltic conditions (Segerstråle, 1957; Hällfors et al., 1981). In the less saline northern Baltic Sea only a few feeding guilds can be found with one or two representatives within each group (Bonsdorff & Pearson, 1999). Thus, the loss of one species may lead to a whole function being lost.

Because of the strong seasonality, the concentration of nutrients in the water column during winter can be used as a proxy for eutrophication in the Baltic Sea area (HELCOM, 2002). Inorganic nutrients which have accumulated during winter are assimilated during the spring bloom. The new production is used directly by suspension feeders and benthic grazers and indirectly via sedimentation by deposit feeders (Rönnberg & Bonsdorff, 2004).

In this paper we study how different macrozoobenthos feeding guilds respond to changing nutrient concentrations and how environmental variables modify these relationships in the northern Baltic Sea. Our hypotheses are the following. (1) The relationship between water nutrients and the biomass of macrozoobenthos is strong in the shallow and less exposed areas due to high fluctuations in nutrient load and relatively weak water exchange (Josefson & Rasmussen, 2000). (2) Similarly, the benthic invertebrates respond to the increased nutrient levels in deep areas below the halocline owing to decreased oxygen concentrations following nutrient-mediated organic enrichment (Gray et al., 2002; Karlson et al., 2002). (3) The relationships between nutrients and macrozoobenthos biomass are positive in the shallow areas and negative in the deep areas described above. (4) Macrozoobenthic assemblages remain indifferent to changing nutrient concentrations in hydrodynamically active areas, e.g. river fronts, bank slopes, and straits through the wide range of nutrient concentrations (Cloern, 2001). (5) Different responses to the changing eutrophication level are expected for different feeding guilds.

Materials and methods

Within the framework of the Estonian National Coastal Water Monitoring Programme (ENCWMP) 28 areas were studied during summers 1993–2003 (Fig. 1). An average of 2 sites were sampled within an area (a total of 46 sites and 527 samples). The distance between sites within an area was on average 3 nautical miles. Generally one sample was taken per site and sampling occasion. Due to very homogenous environment and low number of species, one sample is representative of the macrofauna at a site. Replicate samples in the test areas (three sites in each water basins) showed less than 5% variability (this study and Kotta et al., 1999). Study areas differed considerably from one another in degree of exposure, water depth, sediment grain size and nutrient concentrations (Fig. 2). Based on literature (Kullas et al., 2000; Otsmann et al., 2001) the sampling sites were classified into frontal areas (including river fronts, straits and bank slopes) and hydrographically less active areas. Within hydrographically less active areas nutrient concentrations did not differ by more than two times whereas in frontal areas the concentrations varied by a factor of 4–5. Distance from shore was used as a proxy for exposure. Using the database of ENCWMP (available at the Estonian Marine Institute) the areas, wherein oxygen concentration at the nearbottom layer occasionally or regularly dropped below 2 ml l^{-1}, were defined

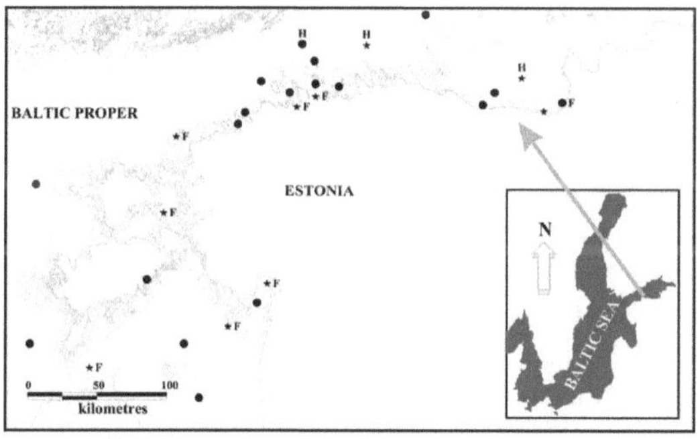

Fig. 1 Study area. Locations, where statistically significant nutrient-invertebrate relationships were found, are indicated by filled circles and those not impacted by stars, respectively. "F" refers to frontal areas and "H" shows hypoxic areas. Isobaths (2, 5, 10, 20 m) are indicated for the shallower study areas

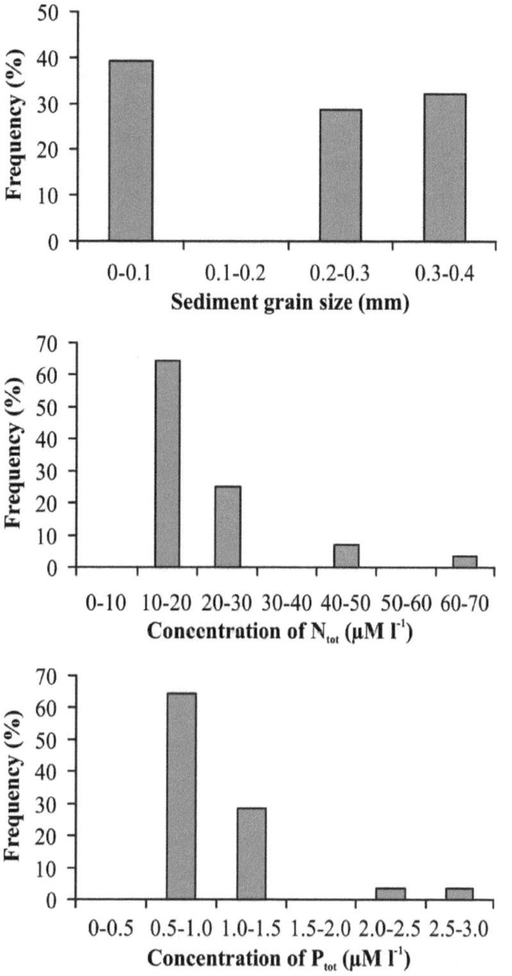

Fig. 2 Frequencies of mean sediment grain size and concentration of N_{tot} and P_{tot} in the water column in the study area

as hypoxic. Other areas were referred to as normoxic.

A van Veen bottom grab (0.1 m^{-2}) was used for bottom sampling. Sediment samples were sieved in the field on 0.25-mm mesh screens. The residuals were stored in a deep freezer and subsequent sorting, counting and determination of biomass of invertebrate species (dry weight g m^{-2}, 60°C at 48 h) were performed in the laboratory using a stereomicroscope. Bivalves were weighed with shells. Macrozoobenthos was classified into feeding guilds – deposit feeders, suspension feeders, herbivores and carnivores based on literature (Bonsdorff & Pearson, 1999) and field observations. Data on salinity and concentration of nutrients in the water column at the near-bottom layer were collected at each site during winter. Due to intensive mixing of seawater, however, there are no significant differences in nutrient concentrations between bottom and surface layers. The concentration of total phosphorus (P_{tot}) and nitrogen (N_{tot}), phosphate (PO_4), nitrate (NO_3) and ammonia (NH_4) were estimated using standard methods (Grasshoff, 1976; Solorzano & Sharp, 1980; Raimbault & Slawyk, 1991).

For univariate analysis the statistical programme "Statistica" was used (StatSoft Inc., 2004). Multivariate data analyses were performed by the statistical program "Primer" (Clarke & Warwick, 2001). Shannon-Wiener diversity index (H′) was calculated using the biomasses of invertebrate species (Pielou, 1975). We employed correlation and linear regression analyses to describe the relationships between abiotic and biotic environmental variables (Sokal & Rohlf, 1981). Our earlier investigations (Kotta & Kotta, 1995; Kotta, 2000; Kotta et al., 2000; Kotta et al., 2003) indicate that the macrozoobenthos is at the initial enrichment phase (Gray, 1992) in the Estonian coastal sea and, hence, the linear regression analysis is appropriate to describe the relationship between the concentration of nutrients in the water column and the biomass of benthic invertebrates.

Prior to regression analysis, autocorrelations were removed from the time series where appropriate (e.g., Chatfield, 1984; Viitasalo et al., 1995). Remaining residuals were considered to represent the numerical response of each series to the surrounding environment. Correlation and linear regression analyses were performed on data from sites to describe relationships between nutrient concentration in the water during non-productive winter months and the biomass of invertebrate feeding guilds during summer. All results were checked for their statistical significance (Sokal & Rohlf, 1981). Only the significant coefficients of determination were used in further uni and multivariate analyses. Zero values were used to indicate the absence of significant nutrient-macrofauna relationships.

The effect of fronts (presence, absence), sediment type (sand, sandy-silt, sandy-clay, silt, clayey-silt) and hypoxia (hypoxic, normoxic) on the proportion of significant nutrient-invertebrate relationships was estimated by analysis of variance at $P < 0.05$ (Sokal & Rohlf, 1981). As the design was incomplete the first order non-interactive effects of multiple categorical independent variables were analysed by "Main effects ANOVA" analysis.

Non-metric multidimensional scaling analysis (MDS) on fourth-root transformed data of macrobenthic biomasses was used to quantify the dissimilarities between study areas. The Bray–Curtis similarity measure on the biomasses of invertebrate species was used to construct the similarity matrices (Bray & Curtis, 1957). Macrozoobenthic communities were compared in areas with and without significant nutrient-invertebrate relationships (later referred to as impacted and nonimpacted areas) using the ANOSIM permutation test (Clarke, 1993). The contribution of different species and/or feeding guilds in the differences was calculated by SIMPER procedure (Clarke, 1993).

Another MDS analysis on fourth-root transformed data of the coefficients of determination was run to quantify the dissimilarities in nutrient-invertebrate relationships between study areas. A Spearman rank correlation (ρ) was computed between environmental data and the similarity matrices of the coefficients of determination (separate analyses for positive, negative and pooled values). The analysis shows which environmental variables predict best the functional response of invertebrate feeding guilds to the

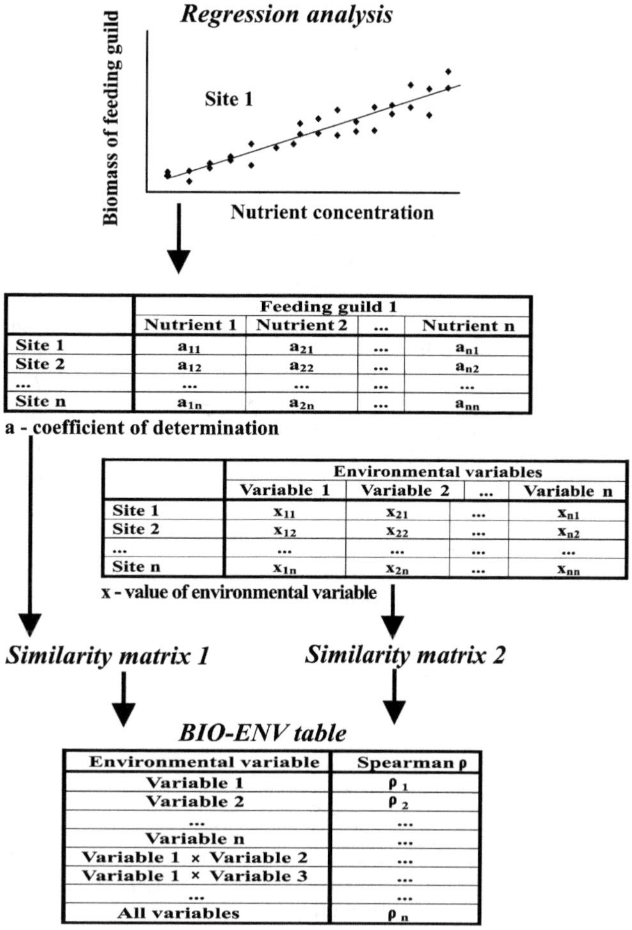

Fig. 3 Schematic procedure of statistical analyses

changing nutrient concentrations in water column (BIOENV procedure, Clarke & Ainsworth, 1993). The significance of the correlation was determined by the programme RELATE (Clarke & Warwick, 2001). The contribution of different nutrients and feeding guilds in the differences was calculated by SIMPER procedure. A permutation procedure is essential here because classical statistical approaches to significance testing are not valid for typical community matrices (Fig. 3).

Results

There were significant relationships between water depth, degree of exposure, salinity and macrozoobenthic diversity measured on the biomass of macrobenthic species (linear regression analyses, $P < 0.05$). There were significant differences in macrobenthic communities between most sediment types (ANOSIM, $P < 0.05$). Only sandy silt bottoms were not distinguished from sandy clay and silty bottoms. No significant relationship between depth and the grain size of sediment was found (correlation analysis, $P > 0.05$) (Figs. 4, 5).

Altogether 46 taxa were identified in the study area. Benthic invertebrate communities were dominated by bivalve molluscs. *Mytilus trossulus* Lamark, *Mya arenaria* L. and *Cerastoderma glaucum* Bruguière were observed in the shallower (<20 m) and more saline (>5 psu) parts of the study area. In the shallower and diluted environments either *Dreissena polymorpha* Pallas

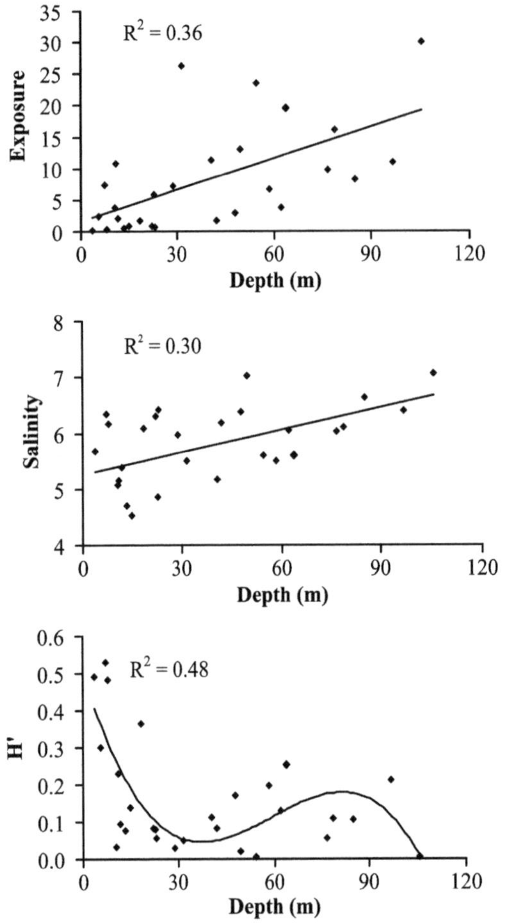

Fig. 4 Relationship between water depth, degree of exposure (distance from shore, nautical miles), salinity and diversity of macrozoobenthos in the study area. The coefficients of determination of linear and polynomial linear regression analyses are shown ($P < 0.05$)

and/or *Macoma balthica* L. were found. *M. balthica* was also the characteristic species of deep soft bottom communities. The five bivalve species comprised more than 90% of total dry weight of benthic invertebrates. Different feeding guilds dominated in different areas. Deposit feeders dominated at lower salinity but also in deeper areas. Higher species diversity was found in the shallower and more saline areas. Dominant carnivores were *Saduria entomon* (L.) and *Halicryptus spinulosus* von Siebold. Carnivores were mainly found in depths below 20 m. Dominant herbivores were *Hydrobia* spp. and *Theodoxus fluviatilis* (L.). Herbivores were confined to depths above 10 m. Although most sediment types had significantly different macrobenthos the most distinct communities were found on sandy and clayey-silty bottoms. Sandy bottoms were characterised by suspension feeding bivalves and clayey-silty bottoms by deposit feeding bivalves, respectively.

The relationships between the concentration of nutrients and macrozoobenthic biomass were non-significant in 10 (36%) and significant in 18 (64%) studied areas (linear regression analysis, $P < 0.05$) for at least one nutrients. Among nutrients, macrozoobenthos responded least to P_{tot}. The response to other nutrients was approximately twice as frequent as for P_{tot}. The relationships between nutrients and biomasses of faunal feeding guilds were strongest among deposit feeders followed by suspension feeders, carnivores and herbivores (Table 1).

Macrozoobenthos responded significantly less to the changing nutrient concentrations in the areas that were regularly impacted by fronts (river estuaries, bank slopes, straits) as compared to hydrodynamically less active areas (Table 2, Fig. 6). The nutrient-invertebrate relationships were somewhat weaker in areas impacted by hypoxia. However, the differences were not statistically significant. Thus, our data did not support the hypotheses that the response of invertebrates to changing nutrients was stronger in hypoxic than normoxic conditions. Degree of exposure, salinity, 11-year average and variability of nutrients, depth and sediment characteristics were not related to the presence or absence of the nutrient-invertebrate relationships (correlation analysis, ANOVA, $P > 0.05$) (Fig. 6, Table 2).

The biomass structure of macrozoobenthic communities was not statistically different between impacted and nonimpacted areas (ANOSIM, $P > 0.05$). However, the sites with significant nutrient-invertebrate relationships were characterised by lower biomasses of *M. trossulus*, *C. glaucum* and *M. arenaria* and higher biomass of *Saduria entomon* (L.) as compared to the sites where the relationship was non-significant (SIMPER).

BIOENV analysis suggested that macrobenthic diversity was the best predictor of the coefficients

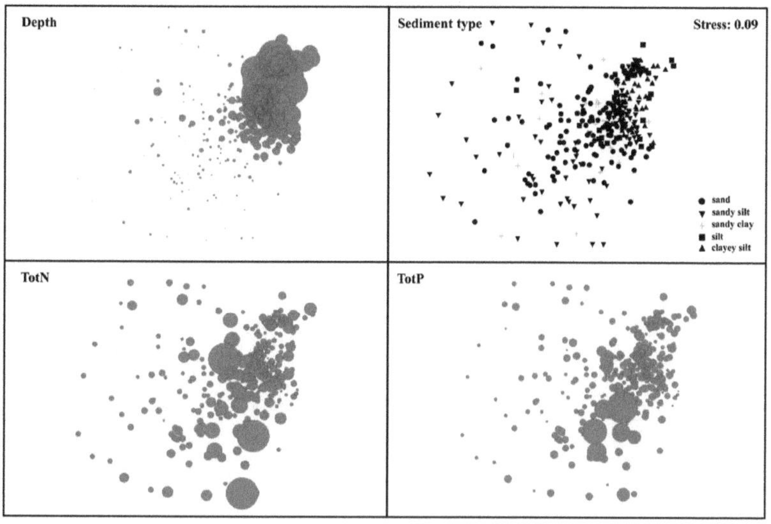

Fig. 5 MDS ordination of fourth-root macrozoobenthos species biomass data. The relative size of bubbles indicates depth and nutrient concentrations (total nitrogen and phosphorus). The code of sediment type is as follows: – sand, – sandy silt, – sandy clay, – silt, – clayey silt

Table 1 Percentage of study areas where the relationships between the concentration of nutrients and the biomass of invertebrate feeding types were significant (linear regression analysis, $P < 0.05$)

Function	NO_3	NH_4	N_{tot}	PO_4	P_{tot}	Total
Deposit feeder	11	14	11	18	7	39
Suspension feeders	0	7	4	4	4	11
Herbivores	0	0	4	4	0	4
Carnivores	14	4	7	7	4	18
Total	25	25	25	32	14	

of determination from nutrient-invertebrate relationships (BIOENV, $\rho = 0.215$). When all environmental variables were pooled together the relationships were best described by the combination of depth, 11-year-average of N_{tot} and P_{tot}, sediment type and macrobenthic diversity (BIOENV, $\rho = 0.430$) (Table 3).

The response of macrozoobenthos to the changing concentration of nutrients varied between different feeding guilds. The deposit feeders that inhabited more diverse communities were less sensitive to the increased concentration of N_{tot} and P_{tot} than those in less diverse communities. On the other hand, the sensitivity of suspension feeders to the rising concentration of nutrients increased with benthic diversity. As zero values were excluded from the analysis, the lack of suspension feeders in deeper areas had no impact on the results of analysis (correlation analysis, $P < 0.05$).

Table 2 Effects of fronts, hypoxia and sediment type on the occurrence of significant nutrient-invertebrate relationships

Source	df	MS	*F*	*P*
Front	1	30036	22.8	<0.001
Hypoxia	1	3219	1.371	0.252
Sediment	4	1146	0.441	0.777

Benthic invertebrates were more sensitive to the changing N_{tot} in the shallower areas than in deeper areas (correlation analysis, $P < 0.05$). Together with the increase in P_{tot} the relationships between the concentration of P_{tot} and the biomass of benthic functions became weaker. The concentration of N_{tot} modified the response of suspension feeders to the concentrations of PO_4 and P_{tot}. With increasing N_{tot} the relationships between PO_4, P_{tot} and the biomass of suspension feeders became stronger (correlation analysis, $P < 0.05$).

Sediment type significantly contributed to the nutrient-invertebrate relationships. Sandy-silty bottoms were different from sandy and silty-clay bottoms. In sandy-silty bottoms suspension feeders were related to the concentration of PO_4 and P_{tot} and not to NH_4, NO_3 and N_{tot} whereas deposit feeders inhabiting sandy and silty-clay bottoms were related to NH_4, NO_3 and N_{tot} and not to PO_4 and P_{tot} (ANOSIM, $P < 0.05$, SIM-

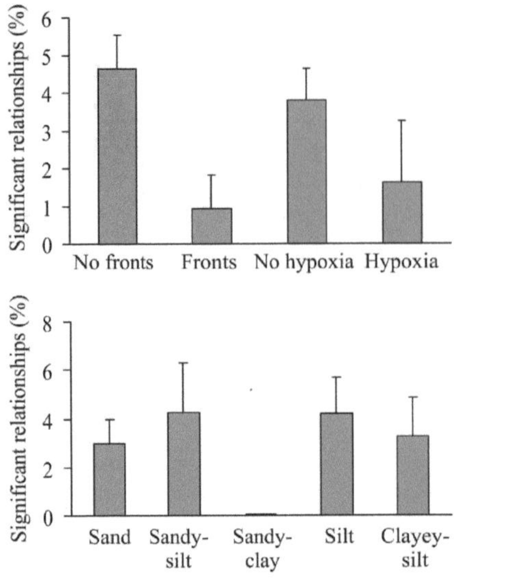

Fig. 6 Effect of fronts, hypoxia and sediment types on the occurrence of nutrient-invertebrate relationships. Average percentage and S.E. values are shown. For the details of ANOVA see also Table 2

PER). The differences in the other functional responses were not statistically significant (ANOSIM, $P > 0.05$).

In cases of significant relationship, carnivores and herbivores were always positively influenced by the addition of nutrients (positive values of coefficients of determination). However, the biomass of deposit and suspension feeders either increased or decreased with rising nutrient concentrations. The positive nutrient-invertebrate relationships were found in areas not affected by fronts (ANOVA, df = 1, $F = 16.01$, $P < 0.001$). The combination of depth, 11-year-average of N_{tot} and P_{tot} and macrobenthic diversity explained the best the strength of positive nutrient-invertebrate relationships (BIOENV, $\rho = 0.492$). The effects of environmental variables on the functional responses were similar as described above (linear regression analysis, $P < 0.05$).

There were significant differences in depth between the areas where the negative nutrient-invertebrate relationships were found and where the relationships were insignificant or positive. The negative relationships were more likely found in deeper areas (ANOVA, df = 1, $F = 6.48$, $P = 0.017$). However, when all other environmental variables were included, sediment type best explained the negative nutrient-invertebrate relationships (BIOENV, $\rho = 0.602$). The addition of nutrients (both phosphorus and nitrogen) reduced the biomass of deposit feeders on silty sediments and suspension feeders on sandy sediments (ANOSIM, $P < 0.05$, SIMPER).

Discussion

This study showed that the biomasses of invertebrate feeding guilds were related to the concentrations of nutrients in the water column and in most cases the relationships were positive. This indicates that benthic communities are food limited in our study area. Following an increased but moderate load of limiting nutrients in marine ecosystems the enhanced primary production and increased biomass of consumers have been often observed (Pearson & Rosenberg, 1978; Bonsdorff et al., 1991; Gray, 1992). However, our study demonstrated large differences among sites in the sensitivity of invertebrates to nutrient enrichment, reflecting system specific attributes and

Table 3 The combination of the best environmental variables that predicts the nutrient-invertebrate relationships. Only the significant coefficients of determination were used in the analyses

Similarity matrix	Significant environmental variables	Spearman ρ
Absolute values of coefficients of determination	Depth 11-year-average of N_{tot} and P_{tot} Sediment type Zoobenthic species diversity	0.430
Positive coefficients of determination	Depth 11-year-average of N_{tot} and P_{tot} Zoobenthic species diversity	0.492
Negative coefficients of determination	Sediment type	0.602

direct and indirect responses that act as a filter to modulate the responses to enrichment (Cloern, 2001). In hydrodynamically active areas there are no simple relationships between horizontal flows of nutrients and vertical flux of particulate organic matter that affect the benthic communities. This may explain why macrozoobenthos responded less to the changing nutrient concentrations in these areas.

Macrobenthic diversity best explained the nutrient-invertebrate relationships. However, the extent of the effect differed among locations due to the changing abiotic environment. Suspension feeders produce faeces and pseudofaeces and thus import large quantities of organic matter to sediment (Cloern, 1982; Kautsky, 1995; Dame, 1996; Peterson & Heck, 2001; Kotta & Møhlenberg, 2002; Kotta et al., 2005) and therefore support higher densities of deposit feeders (Tenore et al., 1985; Grant et al., 1995; Kautsky, 1995). This may explain why nutrients had a less acute effect on deposit feeders in more diverse communities with suspension feeding bivalves. The lack of deposit feeders, i.e., less intense bioturbation, leads faster to hypoxic/anoxic conditions in cases of high sedimentation of organic matter (Anderson & Meadows, 1978; Hansen & Kristensen, 1998). Thus, at high load of organic matter the oxygen level may control the population dynamics of suspension feeders in less diverse communities (without deposit feeders), whereas food may limit suspension feeders in more diverse communities (with deposit feeders). However, due to the relatively exposed environment, oxygen deficiency was rarely observed in the areas inhabited by suspension feeders. More likely the cohabitation of deposit and suspension feeders results in more intense food competition. The prevailing deposit feeder *Macoma balthica* may use different feeding modes including suspension feeding (Ólafsson, 1986). Owing to this variety of feeding modes the species is competitively superior to obligatory suspension feeders. Similarly Weigelt (1991) has shown that nonselective deposit feeders are favoured over suspension feeders with increasing eutrophication.

Alternatively, high macrobenthic diversity develops in the moderate parts of the prominent environmental gradient. It has been shown that diverse benthic assemblages respond more to the changes in nutrient load than functionally poor communities (Beukema & Cadée, 1997). In that case low diversity communities are under the stress of abiotic factors (e.g., low salinity, lack of suitable substrate, instability due to wave action) that is stronger than the food limitation (Herkül et al., 2006). Hence, the effect of eutrophication becomes negligible. However, as for deposit feeders, our data did not agree with this hypothesis as stronger effects of nutrients was observed in functionally poor communities.

Depth and salinity have an effect on macrobenthic diversity (Bonsdorff & Pearson, 1999). The proportion of suspension feeders progressively decreases with depth and increases with salinity while the opposite is true for deposit feeders. Thus, the changes in the relative dominance of feeding guilds along salinity and depth gradient also modify the response of benthic invertebrates to nutrient load.

The nutrient-invertebrate relationships suggested that nitrogen is the limiting factor for algal growth at least in shallower study areas. This is in agreement with earlier experimental observations (Granéli et al., 1990; Kivi et al., 1993). On the other hand phosphorus may limit algal growth at high N_{tot} values. Similarly, our data indicated the phosphorus (P_{tot}) saturation. Those observations support earlier results on phosphorus and/or nitrogen co-limitation of algal growth at low salinity areas (Granéli et al., 1990). The distribution of sediment types was related to the distribution pattern of nutrient limitation. Both coarser and finer sediments seemed to be limited by nitrogen whereas in mixed sediment types phosphorus was the limiting nutrient. The importance of sediment in the nutrient-invertebrate relationships suggests that our study area may be characterised as a high-productivity system contrasting to low-productivity systems where only food determines benthic biomass and diversity values (Grebmeier et al., 1989).

In the deeper sea below the halocline oxygen concentration is one of the most prominent structuring factors for benthic faunal communities (Laine et al., 1997; Karlson et al., 2002). Species and functions show a great variation in the sensitivity to low oxygen concentration.

Opportunists such as polychaetes are favoured in unstable oxygen conditions whereas the amphipods *Monoporeia affinis* Lindström and *Pontoporeia femorata* Kröyer reduce their density with moderate hypoxia and the bivalve *Macoma balthica* with severe hypoxia (Gray et al., 2002; Karlson et al., 2002). With increasing eutrophication organic matter is continuously accumulated in the bottom layer. Oxygen is consumed in its decomposition causing a decline in oxygen content in these deep waters (Grasshoff & Voipio, 1981). Although many studies have demonstrated the decrease of bottom fauna in these deep areas (Andersin et al., 1978; Cederwall & Elmgren, 1980; Laine et al., 1997), we did not observe any significant differences in the response of macrozoobenthos to changing eutrophication level between hypoxic and normoxic conditions. However, the occurrence of negative nutrient-invertebrate relationships was more frequent in the deeper areas than in shallower areas. As compared to the previous studies the Estonian coastal sea is very exposed and flushed by strong currents. Hence, hypoxic events occur at low frequency and persist in short periods. Moreover, the filtering capacity of such open coastal areas is small and the link from nutrients through primary producers to macrozoobenthos is weak.

Exponential growth of human populations in coastal areas in recent years leads to increasing concern about the marine ecosystems. This study showed that, in general, benthic invertebrates were a good indicator of nutrient enrichment except for frontal areas (river estuaries, bank slopes, straits). The response of macrozoobenthos to changing nutrients varied along the depth gradient and between different sediment types. The nutrient-invertebrate relationships were significantly modified by macrobenthic diversity. It is likely that interactions between and within the feeding guilds affect the flows and storage of matter in coastal ecosystems. Thus, besides total biomass values, the environmental assessments should incorporate biological measures such as benthic diversity in order to better describe the quality status of the water body.

Acknowledgements This study was financed by the Estonian Target Financing Programmes Nos 0182578s03 and 0182579s03, Estonian Science Foundation grant No 6015, 6016 and EU CHARM project.

References

Andersin, A. B., J. Lassig, L. Parkkonen & H. Sandler, 1978. The decline of macrofauna in the deeper parts of the Baltic Proper and the Gulf of Finland. Kieler Meeresforschungen. Sonderheft. 4: 23–52.

Anderson, J. G. & P. S. Meadows, 1978. Microenvironments in marine sediments. Proceedings of the Royal Society of Edinburgh 76B: 1–16.

Beukema, J. J. & G. C. Cadée, 1997. Local differences in macrozoobenthic response to enhanced food supply caused by mild eutrophication in a Wadden Sea area: food is only locally a limiting factor. Limnology and Oceanography 42: 1424–1435.

Bonsdorff, E., K. Aarnio & E. Sandberg, 1991. Temporal and spatial variability of zoobenthic communities in archipelago waters of the northern Baltic Sea – consequences of eutrophication. International Review of Hydrobiology 76: 433–450.

Bonsdorff, E. & T. H. Pearson, 1999. Variation in the sublittoral macrozoobenthos of the Baltic Sea along environmental gradients: a functional-group approach. Australian Journal of Ecology 24: 312–326.

Bray, J. B. & J. T. Curtis, 1957. An ordination of the upland forest communities of Southern Wisconsin. Ecological Monographs 27: 325–349.

Cederwall, H. & R. Elmgren, 1980. Biomass increase of benthic macrofauna demonstrates eutrophication of the Baltic Sea. Ophelia, supplement 1: 287–304.

Chatfield, C. 1984. The Analysis of Time Series. An Introduction. 3rd edn. Chapman and Hall, London, 286 pp.

Clarke, K. R., 1993. Non-parametric multivariate analysis of changes in community structure. Australian Journal of Ecology 18: 117–143.

Clarke, K. R. & M. Ainsworth, 1993. A method of linking multivariate community structure to environmental variables. Marine Ecology Progress Series 92: 205–219.

Clarke, K. R. & R. M. Warwick, 2001. Change in Marine Communities: An Approach to Statistical Analysis and Interpretation, 2nd edn. Plymouth Marine Laboratory, Plymouth

Cloern, J. E., 1982. Does the benthos control phytoplankton biomass in South San Francisco Bay? Marine Ecology Progress Series 9: 191–202.

Cloern, J. E., 2001. Our evolving conceptual model of the coastal eutrophication problem. Marine Ecology Progress Series 210: 223–253.

Dame, R. F., 1996. Ecology of Marine Bivalves: An Ecosystem Approach. CRC Press, Boca Raton, FL, 254 pp.

Elmgren, R., 2001. Understanding human impact on the Baltic ecosystem: changing views in recent decades. Ambio 30: 222–231.

European Union, 2000. Directive of the European Parliament and the Council Establishing a Framework for Community Action in the Field of Water Policy. Legislative acts and other instruments. ENV221 CODEC 513. European Union.

Grall, J. & L. Chauvaud, 2002. Marine eutrophication and benthos: the need for new approaches and concepts. Global Change Biology 8: 813–830.

Granéli, E. & K. Sundbäck, 1985. The response of planktonic and microbenthic algal assemblages to nutrient enrichment in shallow coastal waters, southwest Sweden. Journal of Experimental Marine Biology and Ecology 85: 253–268.

Granéli, E., K. Wallström, U. Larsson, W. Granéli & R. Elmgren, 1990. Nutrient limitation of primary production in the Baltic Sea area. Ambio 19: 142–151.

Grant, J., A. Hatcher, D. B. Scott, P. Pocklington, C. T. Schafer & G. Winter, 1995. A multidisciplinary approach to evaluating benthic impacts of shellfish aquaculture. Estuaries 18: 124–144.

Grasshoff K., 1976. Methods of Seawater Analysis. Chemie, New York, 317 pp.

Grasshoff, K. & A. Voipio, 1981. Chemical oceanography. In Voipio, A. (ed.), The Baltic Sea, Vol. 30. Elsevier Oceanography Series, Amsterdam, 183–218.

Gray, J. S. 1992. Eutrophication in the sea. In Colombo, G., I. Ferrari, V. U. Ceccherelli & R. Rossi (eds), Marine Eutrophication and Population Dynamics. Proceedings of the 25th European Marine Biology Symposium, Olsen & Olsen, Fredensborg, 3–15.

Gray, J. S., R. S. Wu & Y. Y. Or, 2002. Effects of hypoxia and organic enrichment on the coastal marine environment. Marine Ecology Progress Series 238: 249–279.

Grebmeier, J. M., H. M. Feder & C. P. McRoy, 1989. Pelagic-benthic coupling on the shelf of the northern Bering and Chukchi Seas 2. Benthic community structure. Marine Ecology Progress Series 51: 253–268.

Hällfors, G., Å. Niemi, H. Ackefors, J. Lassig & E. Leppäkoski, 1981. Biological oceanography. In Voipio, A. (ed.), The Baltic Sea, Vol. 30. Elsevier Oceanography Series, Amsterdam, 219–274.

Hansen, K. & E. Kristensen, 1998. The impact of the polychaete *Nereis diversicolor* and enrichment with macroalgal (*Chaetomorpha linum*) detritus on benthic metabolism and nutrient dynamics in organic-poor and organic-rich sediment. Journal of Experimental Marine Biology and Ecology 231: 201–223.

HELCOM, 2002. Environment of the Baltic Sea area 1994–1998. Baltic Sea Environmental Proceedings 82B: 1–215.

Herkül, K., J. Kotta, I. Kotta & H. Orav-Kotta, 2006. Effects of physical disturbance, isolation and key macrozoobenthic species on community development, recolonisation and sedimentation processes. Oceanologia 48: 1–16.

Howarth, R. W., 1988. Nutrient limitation of net primary production in marine ecosystems. Annual Review of Ecology and Systematics 19: 89–110.

Josefson, A. B. & B. Rasmussen, 2000. Nutrient retention by benthic macrofaunal biomass of Danish estuaries: importance of nutrient load and residence time. Estuarine Coastal and Shelf Science 50: 205–216.

Karlson, K., R. Rosenberg & E. Bonsdorff, 2002. Temporal and spatial large-scale effects of eutrophication and oxygen deficiency on benthic fauna in Scandinavian and Baltic waters – a review. Oceanography and Marine Biology. An Annual Review 40: 427–489.

Kautsky, U., 1995. Ecosystem Processes in Coastal Areas of the Baltic Sea. Ph.D. thesis. Stockholm University.

Kivi, K., S. Kaitala, H. Kuosa, J. Kuparinen, E. Leskinen, R. Lignell, B. Marcusse & T. Tamminen, 1993. Nutrient limitation and grazing control of the Baltic plankton community during annual succession. Limnology and Oceanography 38: 893–905.

Kotta, I., H. Orav-Kotta & J. Kotta, 2003. Macrozoobenthos assemblages in highly productive areas of the Estonian coastal sea. Proceedings of the Estonian Academy of Sciences. Biology. Ecology 52: 149–165.

Kotta, J., 2000. Impact of eutrophication and biological invasions on the structure and functions of benthic macrofauna. Dissertationes Biologicae Universitatis Tartuensis, 63, Tartu University Press: 1–160.

Kotta, J. & I. Kotta, 1995. The state of macrozoobenthos of Pärnu Bay in 1991 as compared to 1959–1960. Proceedings of the Estonian Academy of Sciences. Ecology 5: 26–37.

Kotta J., I. Kotta & J. Kask, 1999. Benthic animal communities of exposed bays in the western Gulf of Finland (Baltic Sea). Proceedings of the Estonian Academy of Sciences. Biology. Ecology 48: 107–116.

Kotta, J., I. Kotta & I. Viitasalo, 2000. Effect of diffuse and point source nutrient supply on the low diverse macrozoobenthic communities of the northern Baltic Sea. Boreal Environmental Research 5: 235–242.

Kotta, J. & F. Møhlenberg, 2002. Grazing impact of *Mytilus edulis* L. and *Dreissena polymorpha* (Pallas) in the Gulf of Riga, Baltic Sea estimated from biodeposition rates of algal pigments. Annales Zoologici Fennici 39: 151–160.

Kotta, J., H. Orav-Kotta & I. Vuorinen, 2005. Field measurements on the variability in biodeposition and grazing pressure of suspension feeding bivalves in the northern Baltic Sea. In Dame R. & S. Olenin (eds), The Comparative Roles of Suspension Feeders in Ecosystems. Springer, The Netherlands, Dordrecht: 11–29.

Kullas, T., M. Otsmann & Ü. Suursaar, 2000. Comparative calculation of flows in the straits of the Gulf of Riga and the Väinameri. Proceedings of the Estonian Academy of Sciences. Engineering 6: 284–294.

Kullenberg, G., 1981. Physical oceanography. In Voipio, A. (ed.), The Baltic Sea, Vol. 30. Elsevier Oceanography Series, Amsterdam, 135–181.

Laine, A. O., H. Sandler, A. B. Andersin & J. Stigzelius, 1997. Long-term changes of macrozoobenthos in the eastern Gotland basin and the Gulf of Finland (Baltic Sea) in relation to the hydrographical regime. Journal of Sea Research 38: 135–159.

Ólafsson, E. B., 1986. Density dependence in suspension-feeding and deposit-feeding populations of the bivalve *Macoma balthica*: a field experiment. Journal of Animal Ecology 55:517–526.

Otsmann, M., Ü. Suursaar & T. Kullas, 2001. The oscillatory nature of the flows in the system of straits and small semienclosed basins of the Baltic Sea. Continental Shelf Research 21: 1577–1603.

Pearson, T. H. & R. Rosenberg, 1978. Macrobenthic succession in relation to organic enrichment and pollution of the marine environment. Oceanography and Marine Biology. An Annual Review 16: 229–311.

Peterson, B. J. & K. L. Heck Jr, 2001. Positive interactions between suspension-feeding bivalves and seagrasses – a facultative mutualism. Marine Ecology Progress Series 213: 143–155.

Pielou, E. C., 1975. Ecological Diversity. John Wiley and Sons ed., NY, 165 pp.

Posey, M. H., T. D. Alphin, L. Cahoon, D. Lindquist & M. E. Becker, 1999. Interactive effects of nutrient additions and predation on infaunal communities. Estuaries 22: 785–792.

Raimbault, P. & G. Slawyk, 1991. A semiautomatic, wet-oxidation method for the determination of particulate organic nitrogen collected on filters. Limnology and Oceanography 36: 405–408.

Rönnberg, C. & E. Bonsdorff, 2004. Baltic Sea eutrophication: area-specific ecological consequences. Hydrobiologia 514: 227–241.

Segerstråle, S., 1957. Baltic Sea. Geological Society of America. Memoir 67: 757–800.

Sokal, R. R. & F. J. Rohlf, 1981. Biometry. The Principles and Practice of Statistics in Biological Research, 2nd edn. W. H. Freeman, San Francisco, California, USA, 859 pp.

Solorzano, L. & J. H. Sharp, 1980. Determination of total dissolved nitrogen in natural waters. Limnology and Oceanography 25: 751–754.

StatSoft, Inc., 2004. Electronic Statistics Textbook. Tulsa, OK: StatSoft. WEB: http://www.statsoft.com/textbook/stathome.html.

Tenore, K. R., J. Corral, N. Gonzalez & E. Lopez-Jamar, 1985. Effects of intense mussel culture on food chain patterns and production in coastal Galicia, NW Spain. Proceedings of the International Symposium on Utilization of Coastal Ecosystems: Planning, Pollution and Productivity. Rio Grande, Brazil 1: 321–328.

Viitasalo, M., I. Vuorinen & S. Saesmaa, 1995. Mesozooplankton dynamics in the northern Baltic Sea: implications of variations in hydrography and climate. Journal of Plankton Research 17: 1857–1878.

Weigelt, M., 1991. Short and long term changes in the benthic community of the deeper part of Kiel Bay (Western Baltic) due to oxygen depletion and eutrophication. Meeresforschung/Report marine Research 33: 197–244.

Hydrobiologia (2007) 580:109–115
DOI 10.1007/s10750-006-0461-0

Diversity of juvenile fish assemblages in the pelagic waters of Lebanon (eastern Mediterranean)

M. Bariche · R. Sadek · M. S. Al-Zein · M. El-Fadel

Abstract The opening of the Suez Canal resulted in the introduction of Indo-Pacific organisms (Lessepsian) to the eastern Mediterranean. Available information on the Levantine ichthyofauna concerns mainly necto-benthic species, while pelagic ones remain mostly uncharacterized. This paper presents a preliminary assessment of biodiversity and its temporal changes on the Lebanese coast (eastern Mediterranean) using species composition and abundance of pelagic communities as indicators. For this purpose, a total of 11,192 fishes, representing 32 species and 19 families were collected with purse seines. Lessepsian species represented 40.9% of the species number in the purse-seine catches but only 0.48% in abundance of individuals and 1.57% in biomass. The families most represented in terms of abundance were the Clupeidae (49.28%), the Engraulidae (41.69%) and the Scombridae (7.01%); in terms of biomass these families represented 56.76, 22.04 and 9.72%, respectively. Abundance and biomass exhibited clear temporal fluctuations with *Sardina pilchardus* and *Scomber japonicus* dominating the catches between May and June, and then replaced by *Sardinella aurita* in July and *Engraulis encrasicolus* in August. The highest values of species richness (12 species) and diversity indices ($H' = 1.37$; $D = 0.71$) were recorded in the last two weeks of June while the lowest values (5 species, $H' = 0.26$; $D = 0.11$) were recorded in early August. While Lessepsian fishes represented a minor part in terms of landings, they contributed considerably to the diversity of pelagic fish assemblages in Lebanese waters.

Keywords Lessepsian migration · Biodiversity · Pelagic fish · Eastern Mediterranean · Lebanon

Guest editors: G. Relini & J. Ryland
Biodiversity in Enclosed Seas and Artificial Marine Habitats

M. Bariche (✉) · R. Sadek · M. S. Al-Zein
Department of Biology, American University of Beirut, 11-0236 Beirut, Lebanon
e-mail: michel.bariche@aub.edu.lb

M. El-Fadel
Department of Civil and Environmental Engineering, American University of Beirut, 11-0236 Beirut, Lebanon

Introduction

The Mediterranean Sea is characterized by a relatively high biodiversity which is indicated by noting that it constitutes less than 1% of the world's ocean surface, while holding as much as 6% of all marine species (Quignard & Tomasini, 2000). This richness is probably due to numerous historical, ecological or paleogeographical, as well as other factors (Bianchi & Morri, 2000). The connection established by the Suez Canal in 1869 resulted in the introduction of Indo-Pacific marine organisms to the eastern Mediterranean Sea (Por, 1978, 1990). This phenomenon was termed "Lessepsian migration" and is considered among the most spectacular biological invasions witnessed in contemporary oceans (Boudouresque, 1999). Although 62 Lessepsian fish species have so far been reported from the eastern Mediterranean (Golani et al., 2002, 2004), very little is known about the effects of these alien species on the area.

The bulk of the available information regarding Lessepsian fishes concerns mainly reef-associated or demersal species (e.g. Golani & Ben-Tuvia, 1985; Golani & Diamant, 1991; Golani, 1993; Kaya et al., 1999; Bariche et al., 2003, 2004) while pelagic fish assemblages remain largely understudied (Ben-Tuvia, 1957, 1983). Assessing species composition in the pelagic environment is of great importance for understanding the contribution of Lessepsian species to the diversity of pelagic fish assemblages in the eastern Mediterranean. Schools of pelagic juvenile fishes, locally named *Bizree*, are present during the warm season close to the coast. They are attracted by floating lights and traditionally captured with purse seine nets over shallow waters (0–50 m). In this context, about two third of fish landings in Lebanon results from such an old destructive fishing method (Mouneimné, 1978).

Despite its long history, and contrary to the western Mediterranean basin, the entire area lacks long term fishery monitoring and similar studies in different localities. As such, this study assesses fish diversity in the pelagic environment of Lebanon in terms of (1) the specific composition, abundance and biomass; (2) the proportions of exotic versus native species and (3) the temporal change in these parameters. While the assessment is conducted along the coast of Lebanon, the results are of greater significance since they could be extended to the overall Levantine basin, which shares a common pelagic assemblage.

Materials and methods

Study area and fish collection

The study was conducted in the coastal waters of Lebanon, along the northern side of Beirut

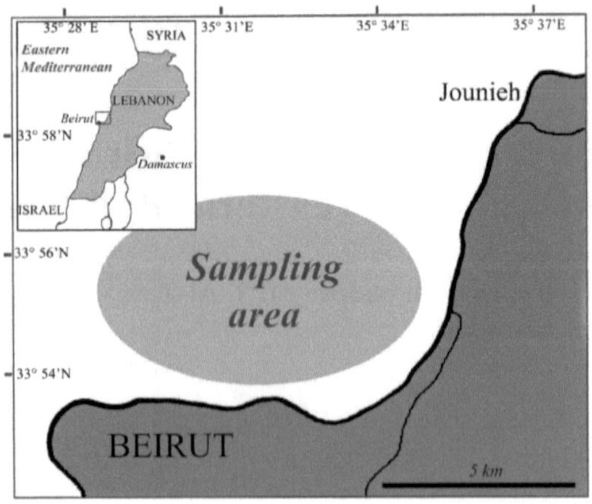

Fig. 1 Map of Lebanon showing the investigated area

(Fig. 1). It was initiated in May 2003, when the purse seine fishing season starts in Lebanon, and lasted for 4 months. Weekly samples were collected totaling 16 sampling cruises (5, 6, 14, 20, 22, 27 May; 4, 16, 29 June; 6, 17, 21, 31 July and 7, 20, 26 August 2003). A set of 4–5 floating lamps were lit at sunset. Fishing was carried out from midnight until the early hours of the morning whereby the seine was dropped around juvenile fish that regroup under each lamp. The purse seine (5 mm mesh, 170 m long and 40 m deep) was laid from the vessel in a circle around the light at about 20 m radius. The estimated volume covered by each seine was between 40,000 and 50,000 m^3. The seine was thereafter pulled manually by 6–9 fishermen. The entire process required approximately 1–1.5 h. A sample of fish was removed randomly from each seine haul by filling a three-liter bucket while the purse-bag was being emptied on board. For each cruise, 3–5 seines were hauled resulting in 54 samples collected between May and August. Once the fishing ended, samples were immediately transported on crushed ice to the laboratory for examination. After identification at the species level, the total length, L_T, to the nearest mm and total body mass, M_T, to the nearest 0.01 g were measured for all fish samples which were then preserved in 10% buffered formalin for further examination. Scarce species that were found in the catches but not in the random samples were

Table 1 Species list of fish found in the catches but not in the random samples

Family	Species	Environment	Status
Belonidae	*Belone belone*	Pelagic	Native
Exocoetidae	*Parexocoetus mento*	Pelagic	Exotic
Gobiidae	*Gobius niger*	Demersal	Native
	Gobius paganellus	Demersal	Native
	Oxyurichthys petersi	Demersal	Exotic
Gymnuridae	*Gymnura altavela*	Demersal	Native
Serranidae	*Serranus hepatus*	Demersal	Native
Siganidae	*Siganus luridus*	Reef-associated	Exotic
Sparidae	*Lithognathus mormyrus*	Demersal	Native
	Oblada melanura	Benthopelagic	Native

recorded (Table 1) but not included in any statistics. Invertebrates (e.g., squids, swimming crabs) were very rare in the landings and were disregarded in this study. Two trips (5 and 20 May, 2003) were excluded from the study because no fish were caught.

Data analysis

To test positive or negative associations between the most common fish species caught, we calculated the Pearson correlation coefficient between the abundance of each pair of fish in the original ungrouped (per sampling cruise) sample data. Considering problems regarding the assumptions needed for calculating Pearson's coefficient (r) in relation to the sampling procedure, the non-parametric Spearman's rho (ρ) was also calculated. Data were then grouped by 15-day intervals to provide a better representation of population dynamics. Species composition between each two consecutive sampling cruises was compared by χ^2 heterogeneity tests. Species richness (number of species collected), abundance (number of individuals per species) and biomass (wet weight per species) were determined. Two diversity indices were calculated (1) Simpson Diversity Index: $D = 1 - \sum_{i=1}^{S} (p_i)^2$; and (2) Shannon–Wiener Diversity Index: $H' = -\sum_{i=1}^{S} (p_i)(\ln p_i)$; where S is the number of species and p_i the proportion of total sample belonging to ith species (Magurran, 1998). All statistical analyses were performed using SPSS for windows (11.0.0) ©, SPSS Inc. or Microsoft® Excel 2002.

Results

Composition of the catches

Thirty-two fish species distributed among 29 genera and 19 families were identified (Tables 1 and 2). The number of species per fishing expedition ranged from three to ten, with an average of 6.5 ± 2.35 (SD). In all fish samples, the Carangidae and Clupeidae were the families most represented in terms of number of species. These were, *Trachurus trachurus*, *Trachurus mediterraneus*, *Alepes djedaba*, *Seriola dumerili*, *Sardina pilchardus*, *Sardinella aurita*, *Etrumeus teres* and *Herklotsichthys punctatus.* These families were followed in species number by Scombridae and Myctophidae (two species each). In contrast, the remaining families were all represented by one species each (Table 2). While some were typically pelagic (e.g. *Sardina pilchardus*, *Engraulis encrasicolus*) or mesopelagic species (e.g. *Hygophum benoiti*), others considered as reef-associated, benthopelagic or demersal were also captured (Table 2).

Nine Lessepsian species distributed among eight families were recorded, out of which two families (Siganidae, Monacanthidae) were not represented in the Mediterranean Sea prior to the opening of the Suez Canal (Table 2). Other families native to the Mediterranean, such as the Atherinidae, Callionymidae, Mullidae and Sphyraenidae, were represented in the samples only by Lessepsian species. In contrast, the Carangidae and Clupeidae comprised both native and exotic members (Table 2). Lessepsian species

Table 2 Species list of fish recorded between May and August 2003 from Lebanon showing their status in the Mediterranean, the overall number of species and biomass

Family, Species	Environment	Status	No. of specimens	%	Biomass	%
Atherinidae						
Atherinomorus lacunosus	Reef-associated	Exotic	2	0.02	23.18	0.060
Bothidae						
Arnoglossus laterna (juvenile)	Demersal	Native	3	0.03	1.58	0.004
Callionymidae						
Callionymus filamentosus	Demersal	Exotic	1	0.01	1.78	0.005
Carangidae						
Alepes djedaba	Reef-associated	Exotic	10	0.09	349.24	0.898
Trachurus trachurus	Pelagic	Native	20	0.18	198.62	0.511
Trachurus mediterraneus	Pelagic	Native	18	0.16	171.80	0.442
Seriola dumerili	Reef-associated	Native	1	0.01	36.40	0.094
Clupeidae						
Sardina pilchardus	Pelagic	Native	3,098	27.68	11,054	28.417
Etrumeus teres	Pelagic	Exotic	23	0.21	117.84	0.303
Sardinella aurita	Pelagic	Native	2,384	21.30	10,838	27.862
Herklotsichthys punctatus	Pelagic	Exotic	10	0.09	67.92	0.175
Engraulidae						
Engraulis encrasicolus	Pelagic	Native	4,666	41.69	8,575.34	22.045
Monacanthidae						
Stephanolepis diaspros	Demersal	Exotic	1	0.01	0.12	0.0003
Mullidae						
Upeneus moluccencis	Reef-associated	Exotic	1	0.01	18.16	0.047
Myctophidae						
Hygophum benoiti	Mesopelagic	Native	1	0.01	1.16	0.003
Myctophum punctatum	mesopelagic	Native	1	0.01	6.48	0.017
Scombridae						
Auxis rochei	Pelagic	Native	16	0.14	86.28	0.222
Scomber japonicus	Pelagic	Native	769	6.87	3,694	9.496
Siganidae						
Siganus rivulatus	Reef-associated	Exotic	1	0.01	0.32	0.001
Sparidae						
Boops boops	Reef-associated	Native	83	0.74	753.40	1.937
Sphyraenidae						
Sphyraena chrysotaenia	Pelagic	Exotic	3	0.03	33.56	0.086
Trichiuridae						
Trichiurus lepturus	Benthopelagic	Native	80	0.71	2,870.12	7.378
Total			11,192	100	38,899.30	100

The habitat of each species was formulated according to Froese & Pauly (2003)

represented 40.9% in species number of the purse-seine catches.

Variation in abundance and biomass

A total mass of 38.9 kg, representing 11,192 fish specimens, was collected (Table 2). The sample size per fishing cruise ranged from 153.5 to 270.2 individuals (214.9 ± 39.5), while the sample biomass ranged from 536.6 g to 1315.5 g (794.3 ± 249.8). In terms of abundance (A) and biomass (B), the Clupeidae (A: 49.3%; B: 56.8%) along with the Engraulidae (A: 41.7%; B: 22.0%) and the Scombridae (A: 7.01%; B: 9.72%) accounted for the greatest part of the landings. All remaining families accounted for only 1.99% (A) and 11.4% (B) (Table 2). Native species were the most represented (A: 99.5%; B: 98.4%), while Lessepsian species constituted only a small share of the landings.

Abundance showed clear temporal fluctuations, as *Sardina pilchardus* and *Scomber japonicus* were caught in relatively large proportions at the beginning of the season until June. They were gradually replaced by *Sardinella aurita* and *Engraulis encrasicolus*, which dominated the landings

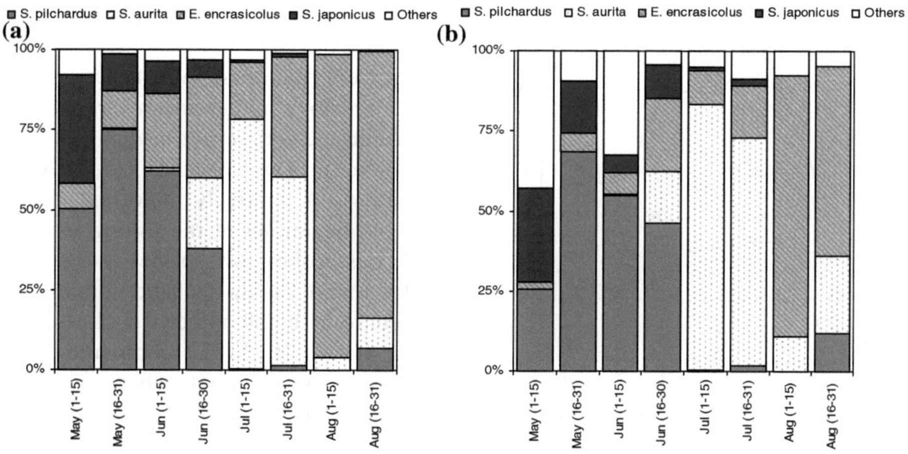

Fig. 2 Temporal variations in (**a**) fish abundance and (**b**) fish biomass grouped by most common species in the catches

in July and August, respectively (Fig. 2a). The χ^2 heterogeneity test showed significant differences ($P \ll 0.001$) in species composition between each consecutive pair of sampling cruises.

Except between *S. pilchardus* and *S. japonicus* ($r = 0.569$; $n = 14$; $P < 0.05$), no significant correlation existed between the abundance of various species using the Pearson's correlation coefficient. Spearman's non-parametric test showed a highly significant positive correlation between *S. pilchardus* and *S. japonicus* ($\rho = 0.817$; $n = 14$; $P < 0.05$), indicating that the two species coexisted in the same period (May–June) (Table 3). The presence of *S. japonicus* was, however, negatively correlated with *S. aurita* ($\rho = -0.697$; $n = 14$; $P < 0.05$) and *E. encrasicolus* ($\rho = -0.704$; $n = 14$; $P < 0.05$). Similarly, *S. pilchardus* and *S. aurita* were also found to be negatively correlated ($\rho = -0.638$; $n = 14$; $P < 0.05$), while no significant correlations existed between *E. encrasicolus* and either of *S. pilchardus* or *S. aurita* (Table 3). The biomass exhibited similar fluctuation patterns with differences mainly due to fish size for each species and to temporal increase in size (Fig. 2b). The remaining species (Others, Fig. 2b) constituted important proportions especially in the first 2 weeks of May and June.

Variation in species richness and diversity

A similar fluctuation pattern for indices of species richness and diversity was recorded throughout the study, except for early June (Fig. 3). The highest values calculated were in the last two weeks of June (16–30) while the lowest were in early August (1–15).

Discussion

Composition of the catches

Some reef-associated (Siganidae, Mullidae) or demersal species (e.g. Monacanthidae, Calli-

Table 3 Pearson (lower left) and non-parametric Spearman (upper right) correlations coefficients between the abundances of each pair of species

Species	*S. pilchardus*	*S. aurita*	*E. encrasicolus*	*S. japonicus*
S. pilchardus	1	–0.638 (0.014)	–0.377 (0.183)	0.817 (<0.001)
S. aurita	–0.415 (0.140)	1	0.384 (0.174)	–0.696 (0.005)
E. encrasicolus	–0.312 (0.278)	–0.061 (0.836)	1	–0.704 (0.004)
S. japonicus	0.569 (0.034)	–0.424 (0.131)	–0.354 (0.215)	1

Values of *P* are bracketed

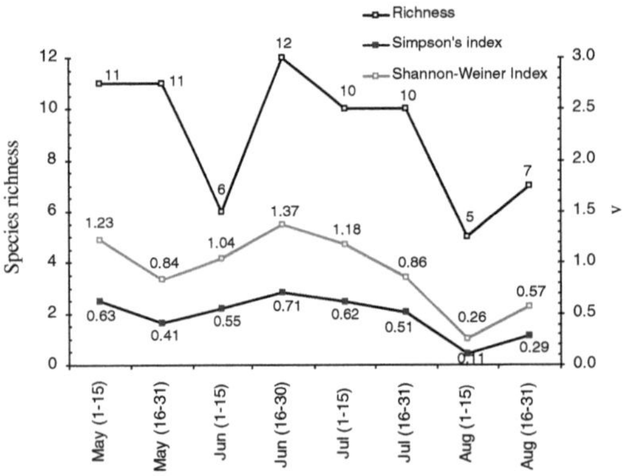

Fig. 3 Temporal variations in species richness and diversity indices during the sampling period

onymidae) were caught accidentally when the lower part of the seine touched the bottom. This is not the case for (1) *Seriola dumerili* (juveniles <15 cm TL), which seems to be attracted by the floating light, and (2) *Alepes djedaba* (juveniles <15 cm TL) or *Trichiurus lepturus* (>30 cm TL), which appear to be attracted to the prey available under the floating lights. Nonetheless, it remains difficult to assess whether it is light or prey that attracts other species (e.g. *Trachurus* spp., *Boops boops*) considering that numerous invertebrates, more particularly copepods and crustacean larvae, are also attracted to light.

As indicated earlier, Lessepsian species comprised 40.9% of the catches in terms of number of species. This proportion is certainly biased by rare Lessepsian species caught accidentally, since the three most common fishes in abundance (90.6%) and biomass (78.32%) were all native to the Mediterranean Sea (Table 2).

Two Clupeidae (*Dussumieria elopsoides*, *Spratelloides delicatulus*) and two Scombridae (*Rastrelliger kanagurta*, *Scomberomorus commerson*) are known to have been introduced to the eastern Mediterranean (Golani et al., 2002) but were not captured in the current study. This may presumably be due to their small populations size, low positive phototropism or possibly fast swimming capacity, the latter being observed in the native *Euthynnus alletteratus* (Scombridae) (MB, pers. obs.).

Variation in abundance and biomass

It is clear that purse seine fishery is dominated by the anchovy (*Engraulis encrasicolus*) and two sardines (*Sardina pilchardus*, *Sardinella aurita*) (Table 2). The mackerel (*Scomber japonicus*) compose a relatively important share especially at the beginning of the fishing season. Although 11 Lessepsian fish species were recorded (Tables 1 and 2), the pelagic community does not seem to be highly affected by Lessepsian invasion (A: 0.5%; B: 1.6%) compared to the necto-benthic environment where Lessepsian fishes represented larger proportions in commercial fisheries, i.e. 47% in weight of trawl catches and 25% of trammel nets catches of Israel (Ben-Tuvia & Golani, 1995). In their study on a beach seine fishery, George & Athanassiou (1967) reported 16 Lessepsian species, mostly necto-benthic of which *Upeneus moluccensis*, *Upeneus pori*, *Siganus rivulatus*, *Sphyraena chrysotaenia* and *Saurida undosquamis* contributed significantly to the fishery. Abundance and biomass of Lessepsian fish in the beach seine catch were A: 13.0% and B: 12.7%, respectively in Lebanon (George & Athanassiou, 1967) and are probably different at this time since *Scomberomorus commerson* and more recently *Fistularia commersonii* seem to contribute significantly to this fishery (MB, pers. obs.). Temporal fluctuations observed in abundance and biomass (Fig. 2), whereby *Sardina pilchardus* and *Scomber japonicus* dominated the beginning of the season and were then replaced by *Sardinella aurita* and later by *E. encrasicolus*, are possibly due to an adaptation to reduce competition for similar resources. The relatively large proportions of rare fishes (Others, Fig. 2b) in both early May (B: 43.0%) and early June (B: 32.6%) are due to schools of *Trichiurus lepturus* of large size (30.7–44.6 cm TL) captured on two different seines during the sampling period. *Boops boops* (5.4–14.3 cm TL) constituted the main share of the biomass during the rest of the sampling period (Fig. 2b).

Variation in species richness and diversity

Samples from the first part of June 2003 contained a large number of species that influenced diversity indices. Only one sampling trip (August 7th, 2003) was completed in early August due to technical difficulties with the fishing vessel. This relatively small sample (four seine hauls) contained a low number of species mainly *Engraulis encrasicolus*, which explains the low values of both richness and diversity indices observed for this date (Fig. 3).

Lessepsian migration affected eastern Mediterranean fish communities. However, while many necto-benthic and benthic Lessepsian fish are thriving, the inshore pelagic fish communities do not seem much affected by the dozen Lessepsian species inhabiting this area (Golani et al., 2002).

Acknowledgements We thank Joseph Chehwan for his unfaltering help in fish sampling, Houssein Darwish for his technical laboratory assistance and Charbel Bahout along with his crew for making us feel welcome on their boat during the long hours of fieldwork. Special thanks are extended to the United States Agency for International Development (USAID) for its continuous support to the Water Resources Center and the Interfaculty Graduate Environmental Sciences Program at the American University of Beirut.

References

Bariche, M., M. Harmelin-Vivien & J. -P. Quignard, 2003. Reproductive cycles and spawning periods of two Lessepsian siganid fishes on the Lebanese coast. Journal of Fish Biology 62: 129–142.

Bariche, M., Y. Letourneur & M. Harmelin-Vivien, 2004. Temporal fluctuations and settlement patterns of native and Lessepsian herbivorous fishes on the Lebanese coast (eastern Mediterranean). Environmental Biology of Fishes 70: 81–90.

Ben-Tuvia, A., 1957. Pelagic Fisheries in Israel, Les Pêches Pélagiques en Israël. General Fisheries Council for the Mediterranean. Proceedings and technical Papers, FAO, Rome 4: 383–391.

Ben-Tuvia, A., 1983. Tuna and Mackerel in Israel coastal waters. Israel Land and Nature 8: 152–157.

Ben-Tuvia, A. & D. Golani, 1995. Temperature as the Main Factor Influencing the Lessepsian Migration. La Méditerranée: Variabilités climatiques, environnement et biodiversité. Actes du Colloque Scientifique, Montpellier France, 159–161.

Bianchi, C. N. & C. Morri, 2000. Marine biodiversity of the Mediterranean Sea: situation, problems and prospects for future research. Marine Pollution Bulletin 40: 367–376.

Boudouresque, C. F., 1999. The Red Sea-Mediterranean link: unwanted effects of Canals. In Sandlund, O. T., P. J. Schei & Å. Viken (eds), Invasive Species and Biodiversity Management. Kluwer Academic Publishers, Dordrecht, 213–228.

Froese, R. & D. Pauly (eds), 2003. FishBase. World Wide Web electronic publication. www.fishbase.org, version 21 April 2004.

George, C. J. & V. Athanassiou, 1967. A two-year study of the fishes appearing in the seine fishery of St George Bay, Lebanon. Annali Del Museo Civico Di Storia Naturale de Genova 76: 237–294.

Golani, D., 1993. The biology of the Red Sea migrant, *Saurida undosquamis* in the Mediterranean and comparison with the indigenous confamilial *Synodus saurus* (Teleostei: Synodontidae). Hydrobiologia 271: 109–117.

Golani, D. & A. Ben-Tuvia, 1985. The biology of the Indo-Pacific squirrelfish, *Sargocentron rubrum* (Forsskål), a Suez Canal migrant to the Eastern Mediterranean. Journal of Fish Biology 27: 249–258.

Golani, D. & A. Diamant, 1991. Biology of the sweeper *Pempheris vanicolensis* (Cuvier & Valenciennes) a lessepsian migrant in the Mediterranean with comparison with the original Red Sea population. Journal of Fish Biology 38: 819–827.

Golani, D., L. Orsi-Relini, E. Massuti & J. -P. Quignard, 2002. CIESM Atlas of exotic species in the Mediterranean. In Briand, F. (ed.), Fishes. CIESM Publishers, Monaco 1: 254 pp.

Golani, D., L. Orsi-Relini, E. Massuti, J. -P. Quignard, 2004. Dynamics of fish invasions in the Mediterranean: update of the CIESM fish atlas. Rapports Commission internationale pour l'exploration scientifique de la Mer Méditerranée, Monaco 37: 367 pp.

Kaya, M., H. A. Benli, T. Katagan & O. Ozaydin, 1999. Age, growth, sex-ratio, spawning season and food of golden bend goatfish, *Upeneus moluccensis* Bleeker (1855) from the Mediterranean and south Aegean Sea cost of Turkey. Fisheries Research 41: 317–828.

Magurran, A. E., 1998. Ecological Diversity and its Measurement. Princeton University Press, Princeton, New Jersey.

Mouneimné, N., 1978. Poissons des côtes du Liban (Méditerranée Orientale), biologie et pêche. Thèse de Doctorat d'Etat ès-Sciences Naturelles. Université Pierre et Marie Curie, France.

Por, F. D., 1978. Lessepsian migration. The influx of Red Sea biota into the Mediterranean by way of the Suez Canal. Ecological Studies 23: 1–228.

Por, F. D., 1990. Lessepsian migration. An appraisal and new data. Bulletin de l'Institut Océanographique, Monaco. Special 7: 1–10.

Quignard, J. P. & J. A. Tomasini, 2000. Mediterranean fish biodiversity. Biologia Marina Mediterranea 7: 66 pp.

Hydrobiologia (2007) 580:117–124
DOI 10.1007/s10750-006-0460-1

BIODIVERSITY IN ENCLOSED SEAS

Stability of spatial pattern of fish species diversity in the Strait of Sicily (central Mediterranean)

G. Garofalo · F. Fiorentino · M. Gristina ·
S. Cusumano · G. Sinacori

Abstract In this paper, species diversity of demersal fish communities was analysed over an area covering about 45,000 km^2 of the Italian side of the Strait of Sicily (central Mediterranean). Fish abundance data come from a 10-year series (1994–2003) of experimental bottom trawl surveys carried out within the framework of the international program MEDITS. A simple GIS-based method was proposed to identify areas supporting high or low values of diversity and evaluate their temporal stability. A well-defined spatio-temporal pattern in diversity emerged from the analysis, with some areas of great ecological relevance being identified. Importantly, the greatest diversity within the fish communities was consistently seen at the offshore bank on the western part of the south Sicilian shelf (Adventure Bank). The site also supports high total biomass of demersal resources and shows the presence of species of great concern to fisheries. Results suggest that Adventure Bank represents a priority site for investigating the possibility of innovative management of marine ecosystems and demersal fisheries in offshore zones.

Guest editors: G. Relini & J. Ryland
Biodiversity in Enclosed Seas and Artificial Marine Habitats

G. Garofalo (✉) · F. Fiorentino · M. Gristina ·
S. Cusumano · G. Sinacori
IAMC – CNR, Section of Mazara del Vallo, Marine Living Resources Assessment (MaLiRA) Group, Via L. Vaccara, 61, 91026 Mazara del Vallo, Trapani, Italy
e-mail: germana.garofalo@irma.pa.cnr.it

Keywords Fish diversity · Mediterranean · Spatial pattern · Stability · Strait of Sicily

Introduction

The adverse effects of both human impacts and climate changes on marine ecosystems are increasing dramatically and threatening many habitats in the Mediterranean basin (Bianchi & Morri, 2000) as well as throughout the world. Marine biodiversity is being compromised and, because the process is largely irreversible, its conservation acquires an increasingly important concern in scientific and public discussions (Agardy, 2003). In particular, much debate has recently focused on the adverse effects of biodiversity loss on the stability and functioning of ecosystems (Chapin III et al., 2000; McCann, 2000; Tilman, 2000; Loreau et al., 2001; Duffy, 2003). Although the mechanisms underlying these processes are not yet well understood, it is by now recognized that preservation of biodiversity requires strategies alternative to those exclusively based on protection of diversity hotspots (species-rich areas under threat), which aim to maximize the number of protected species over small areas. New approaches are advocated based on the broader objective of maintaining the functioning of whole ecosystems. From this perspective, Kareiva & Marvier (2003) suggest that plans for biodiversity conservation may also include preservation of what they call biodiversity coldspots (species-poor areas under threat) if their degradation is supposed to affect ecosystem sustainability.

In marine ecosystems, fisheries activity plays a major role in biodiversity and productivity decline, especially when coupled to environmental degradation and climate changes (Agardy, 2000). In the last decade, plenty of scientific evidence has been provided about the adverse effects, both direct and indirect, of fisheries exploitation on function and structure of marine ecosystems (Jennings & Kaiser, 1998). As a result, a new approach to fishery management has been delineated that would balance objectives of fisheries sustainability, biodiversity conservation and overall health of ecosystems (Sinclair & Valdimarsson, 2003; Pikitch et al., 2004). Management options to reduce the threat to marine ecosystems include the institution of marine protected areas or no-take zones (Agardy, 2003; Kaiser et al., 2004). The localization of these areas is far from simple and requires an improving knowledge of the distribution of critical habitats, species richness, species of great concern, species at risk, or any other indicator of potential ecological value.

In this paper, we analyse across space and time the pattern of biodiversity at species level in the Strait of Sicily (central Mediterranean sea). In particular, we propose a simple GIS-based method to identify areas persistently supporting high or low values of fish diversity.

Materials and methods

Data source

The study area is the Italian side of the Strait of Sicily, where scientific bottom trawl surveys have been carried out annually in spring since 1994, in the framework of the international MEDITS program (Abelló et al., 2002), with the aim of monitoring demersal stocks. The area covers about 45,000 km^2 and a range of water depths from 10 m to 800 m. The program adopted a stratified random sampling design with allocation of hauls proportional to strata extension. The surveys were carried out using the same vessel, equipment and protocol throughout the entire period. A total of 56 hauls were made each year keeping fixed their location (Fig. 1). Although sampling effort has been more than doubled since 2002, only the fixed stations over the 10-year period from 1994 to 2003 were included in this study. Catches by haul were processed and sorted to species level. Abundances, standardized to 1 km^2 (N/km^2), were calculated for each species caught. Only demersal fish species (Osteichthyes and Chondrichthyes) were considered in the present study.

Data analysis

In order to choose the appropriate measure of diversity, some of the most common diversity indices were preliminarily evaluated. They were selected so as to capture the various aspects of biodiversity: species richness (number of species, *S*; Margalef's *D*), evenness (Pielou's *J′*) and diversity (Shannon–Wiener's *H′*; Simpson's $1 - \lambda$). The Spearman rank test was performed to analyse the correlations among the different indices (Table 1). The indices were significantly correlated with each other, with the exception of number of species versus Pielou index. Particularly, the Shannon diversity index showed the strongest positive correlations with the other indices. It accounts for both species richness and evenness and, although it suffers from some drawbacks, it is widely used to characterize species diversity in fishery studies. Accordingly, the Shannon diversity index was chosen and used in all successive analyses.

The resulting geo-referenced data set of each year was interpolated into a regular grid using inverse distance weighted interpolation (IDW) (Isaaks & Srivastava, 1989). Hence, in order to have an overall picture of the diversity distribution across time and space, the maps for the 10 years were combined to estimate the "average" geographical distribution of species diversity over the study area. In addition, the local coefficient of variation (CV) of H′ was calculated

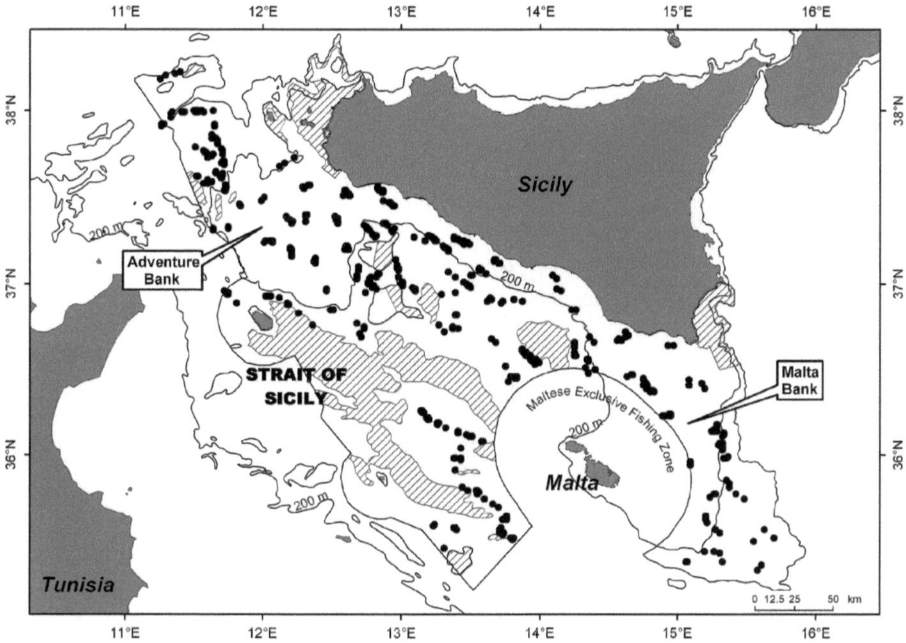

Fig. 1 Map of the study area showing the location of the sampling stations from 1994 to 2003. The dashed zones indicate untrawlable areas and water depths of more than 800 m

Table 1 Spearman rank correlations (r_s) among diversity indices, with significant ($P < 0.01$) coefficients in bold: number of species, *S*; Margalef's *D*; Pielou's *J'*; Shannon–Wiener's *H'*; Simpson's $1 - \lambda$

	S	*D*	*J'*	*H'*	$1 - \lambda$
S	1.00				
D	**0.75**	1.00			
J'	0.01	**0.17**	1.00		
H'	**0.52**	**0.53**	**0.80**	1.00	
1-λ	**0.37**	**0.41**	**0.87**	**0.97**	1.00

to assess the variability of the spatial pattern over the entire period.

To exactly delineate the areas of high diversity index and quantifying their time stability, each map was categorized into four classes corresponding to the quartiles of its value distribution. An index of persistence was defined to assess at which degree a given area was characterized by diversity values belonging to the fourth quartile for each year. The index is the ratio between the count of the years in which each grid point belonged to the fourth quartile and the total number of years. Hence, it ranged between 0 and 1, where 1 indicates persistent presence of the highest values throughout the years and 0 total absence of high values. The same procedure was followed to identify areas of consistently low diversity index, considering the first quartile. The resulting areas of both high and low diversity were showed in the same map.

Results

A total of 169 different species of fish (bony fish and elasmobranchs) was identified in the study area over the entire time series. On average 108 different species were found per survey. The number of species per sample was highly variable and ranged from 2 to 34 species. Only two species (*Merluccius merluccius* and *Phycis blennoides*) were widespread, being found at more than 50% of sampling stations. On average, 18% of species were rare, being represented by one or two individuals found in one or two sampling stations.

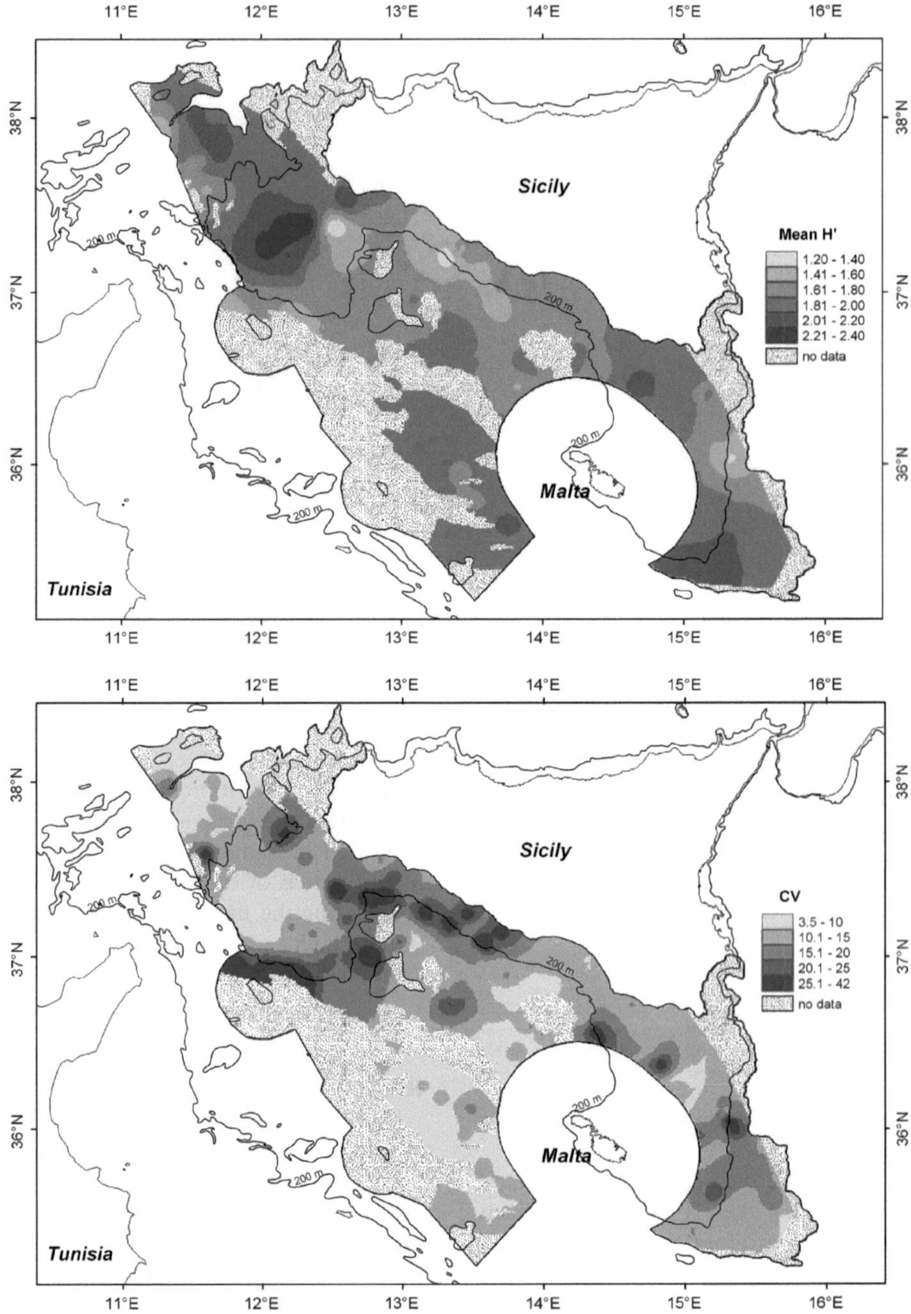

Fig. 2 Map of the (top) mean Shannon diversity index and (bottom) coefficient of variation over 10 years (1993–2004)

Figure 2(top) shows the average distribution pattern of local species diversity relative to the 10-year period, whereas Fig. 2(bottom) depicts the corresponding coefficient of variation. The region appears quite heterogeneous with respect to both the diversity value and its inter-annual variability, exhibiting areas of high diversity and low variability, areas of low diversity and high variability and areas of intermediate diversity values and low variability.

The highest diversity within the fish assemblages was seen at the offshore bank on the western part of the south Sicilian shelf (Adventure Bank). The site was also characterized by the lowest CV (<10%). Other sectors showing high species diversity included the area along the south-eastern boundary of the Maltese Exclusive Fishing Zone (MEFZ) and the north-western corner of the study area. On the other side, areas poor in diversity were found in the eastern sector of the Adventure Bank and along the central sector of the Strait of Sicily mainly in the outer shelf—upper slope. These latter areas were also characterized by the highest values of CV (>20%). The deep bottoms (more than 600 m) on the western edge of the MEFZ showed the peculiar characteristic of intermediate values of diversity and rather low variability.

Figure 3 shows the areas which were identified after analysing the persistence of the highest values (fourth quartile) and lowest values (first quartile) of the yearly diversity maps. The visualization of the persistence index was restricted to values greater than 0.5. Of utmost importance is the area located on the Adventure Bank, because of its extent, the high degree of persistence (0.9–1) of its core and its species composition. Indeed, the analysis of the catches inside this area shows that 58 different fish species occur in this area, which is about 34% of the total number of fish species collected over the entire study area. Nineteen species, including valuable commercial species, constitute 90% of the catch. The entire area delineated was inside the 100 m isobath. Other areas consistently supported high diversity indices even if they were slightly less stable than the Adventure Bank. Indeed the persistence index ranged between 0.6 and 0.8. These areas were located along the eastern edge of the MEFZ (between 200 m and 400 m depth) and in the north-western corner of the study area (between 400 m and 600 m depth).

Three large areas persistently characterized by low diversity indices were identified. The widest was located in the outer shelf—upper slope along

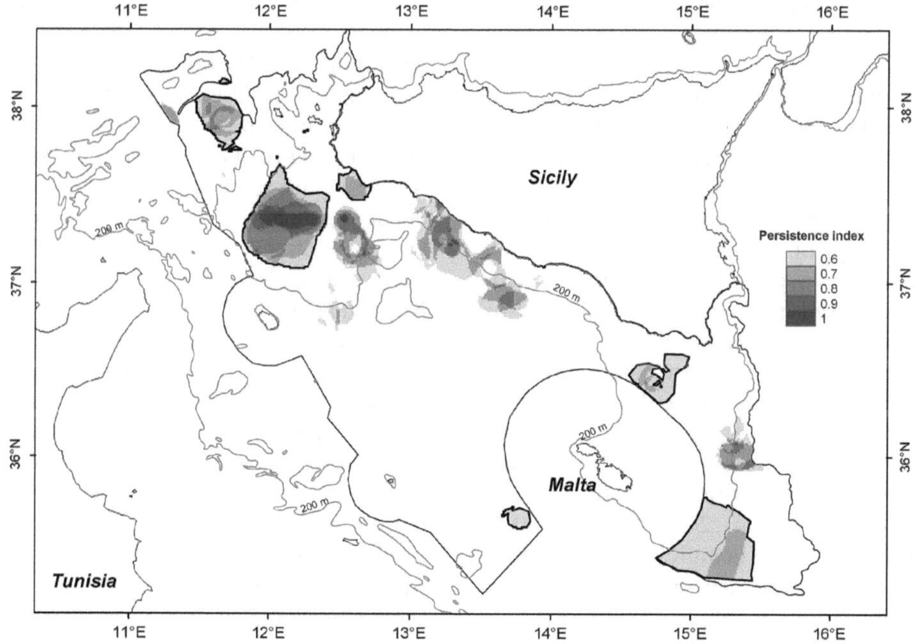

Fig. 3 Areas consistently supporting high (contoured areas) and low values of diversity. The degree of stability over the temporal series is measured by the persistence index

the central sector of the south Sicilian coast. The others were located on the eastern side of Adventure Bank and of Malta Bank, respectively.

Discussion

The many studies investigating the multivariate structure of species' assemblages in the Mediterranean (Ungaro et al., 1999; Biagi et al., 2002; Gaertner et al., 2002; Colloca et al., 2003; Labropoulou & Papaconstantinou, 2004) have all described patterns of demersal fish assemblages rather stable through time and strongly organized along depth gradients. Even though classical assemblage descriptions represent a useful basis of knowledge for the management of multispecies fisheries and by-catch problems, of greater importance is the capability to accurately describe spatial variability of exploited demersal resources and evaluate temporal changes which may be driven by anthropogenic or environmental factors. To address this issue, spatially explicit approaches of analysis need to be developed. Indeed, as advocated by Pauly et al. (2003), maps represent a major and intuitive tool to investigate the fisheries relevant ecological processes and to incorporate "ecosystem" considerations into fishery management.

In our study, the spatial organization of the multispecies demersal fishery in the Italian waters of the Strait of Sicily has been explored using a univariate feature of the species assemblages, the Shannon–Wiener diversity index. In addition, the temporal stability of the spatial pattern has been investigated over a period of 10 years. Specifically, the employed methodology has allowed us to: (i) identify areas supporting high or low values of fish diversity; (ii) evaluate their level of stability over time; (iii) delineate them at a useful management scale.

Diversity is an inherent aspect of community structure, related to ecosystem functioning and stability (Chapin III et al., 2000; McCann, 2000, Loreau et al., 2002), the comprehension of which represents a main challenge in modern fisheries management.

Our results indicate that, despite the diversity index not exhibiting any evident trend along depth gradients, a well-defined spatio-temporal pattern can be identified. Indeed, most of the study area is characterized by diversity values rather stable through time, the CV being below 15%. Interestingly, the areas showing the greatest inter-annual variability of diversity are located mainly along the shelf edge, where topographically induced upwelling may occur (Lermusiaux & Robinson, 2001), and particularly along the average trajectory of the Atlantic Ionian Stream (AIS) (Robinson et al., 1991). The AIS is the Modified Atlantic Water (MAW) current which enters the Strait of Sicily from its western boundary, flows along the Adventure Bank, gets close to the middle southern coast of Sicily and moves offshore again when it encounters the Malta Bank. It represents the principal hydrodynamic feature of the region, influencing, with its great inter-annual variability, the extension of the upwelling and the formation of fronts (Patti et al., 2004). The high variability of the diversity index along these areas could reflect a correlation between the annual upwelling index and the density of some species in the fish assemblage, similarly to what observed by Farina et al. (1997) in the Galician shelf (NW Spain).

The area where the AIS approaches the Sicilian coast is known to be a permanent upwelling area (Lermusiaux & Robinson, 2001) and it has been identified in this study as one of the three areas persistently characterized by low diversity values. The relationship between diversity and ecosystem productivity has been widely debated in the last decade and continues to be largely controversial (Lehman & Tilman, 2000; Loreau et al., 2002; Naeem et al., 2002; Groner & Novoplansky, 2003). Although a general positive relationship has been accepted, high temporal variability associated with high levels of primary production may not necessarily result in high diversity in marine systems, giving rise to a hump-shaped diversity-productivity relationship at local scale (Gaston, 2000; Loreau et al., 2001). Such a pattern has been observed in our study and reported as well by Farina et al. (1997) and Mueter and Norcross (2002). However, whereas they observed that higher primary productivity correlates with lower species diversity but also with greater abundance of demersal species, we found that the low diversity area supports low demersal fish biomass (Gristina et al., 2004). The presence of strong currents, which produce high hydrological instability, the limited extent of the shelf and the intensive exploitation by trawl fisheries may be explanatory factors. Conversely, over the less productive adjacent Adventure Bank, we located a large persistent area of high fish diversity, which supports the highest biomass indices of demersal resources of the study area (Gristina et al., 2004). It also includes spawning grounds of red mullet (*Mullus barbatus*) (Garofalo et al., 2004) and a relatively high abundance of rays, different from the remaining region (Garofalo et al., 2003). The evidence of such a diverse and healthy community is consistent with favourable environmental and hydrographic conditions. Indeed, the Adventure Bank is a wide shallow offshore bank encircled by a large cyclonic vortex (Adventure Bank Vortex), which is permanent even though greatly variable in shape and dimensions (Lermusiaux & Robinson, 2001). The heterogeneous nature and rough morphology of the sea bottom makes trawl fishing difficult and favours the persistence of a much greater fish species diversity and abundance than that found in the neighbouring shelf grounds.

Concerning the other delineated areas of stability, it is worthy to note that previous studies identified the two persistent areas of low-diversity, located on the eastern side of the Adventure Bank and Malta Bank, respectively, as being stable nurseries of hake (*Merluccius merluccius*) (Fiorentino et al., 2003). This finding confirms that prioritizing conservation areas on the basis of high species diversity may lead to other areas critical for fishery management being disregarded.

In conclusion, by also considering results from previous studies, we suggest that the areas identified with the described methodology have great significance from both ecological and fisheries perspectives, whereas their character of persistence has clear implications for site-specific management options. Although further studies are needed to elucidate the relationships between diversity and functioning of exploited communities in these areas, specifically investigating the relevant ecological mechanisms and the role of fisheries, our work moves toward incorporating "ecosystem" considerations into fishery management, a necessary measure to deal with the reduced productivity and biodiversity loss of most exploited marine systems.

References

Abelló, P., J. A. Bertrand, L. Gil de Sola, C. Papaconstantinou, G. Relini & A. Souplet, 2002. Mediterranean marine demersal resources: the MEDITS International trawl survey (1994–1999). Scientia Marina 66: 1–280.

Agardy, T., 2000. Effects of fisheries on marine ecosystems: a conservationist's perspective. ICES Journal of Marine Science 57: 761–765.

Agardy, T., 2003. An environmentalist's perspective on responsible fisheries: the need for holistic approaches. In Sinclair, M. & G. Valdimarsson (eds), Responsible Fisheries in the Marine Ecosystem. FAO, Rome: 65–85.

Bianchi, C. N. & C. Morri, 2000. Marine biodiversity of the Mediterranean sea: situation, problems and prospects for future research. Marine Pollution Bulletin 40: 367–376.

Biagi, F., P. Sartor, G. D. Ardizzone, P. Belcari, A. Belluscio & F. Serena, 2002. Analysis of demersal assemblages off the Tuscany and Latium coast (northwestern Mediterranean). Scientia Marina 66: 233–242.

Chapin, III, F. S., E. S. Zavaleta, V. T. Eviner, R. L. Naylor, P. M. Vitousek, H. L. Reynolds, D. U. Hooper, S. Lavorel, O. E. Sala, S. E. Hobbie, M. C. Mack & S. Diaz, 2000. Consequences of changing biodiversity. Nature 405: 234–242.

Colloca, F., M. Cardinale, A. Belluscio & G. Ardizzone, 2003. Pattern of distribution and diversity of demersal assemblages in the central Mediterranean sea. Estuarine, Coastal and Shelf Science 56: 469–480.

Duffy, J. E., 2003. Biodiversity loss, trophic skew and ecosystem functioning. Ecology Letters 6: 680–687.

Farina A. C., J. Freire & E. Gonzalez-Gurriaran, 1997. Demersal fish assemblages in the Galician continental shelf and upper slope (NW Spain): Spatial structure and long-term changes. Estuarine, Coastal and Shelf Science 44: 435–454.

Fiorentino, F., G. Garofalo, A. De Santi, G. Bono, G. B. Giusto & G. Norrito, 2003. Spatio-temporal distribution of recruits (0 group) of *Merluccius merluccius* and *Phycis blennoides* (Pisces; Gadiformes) in the Strait of Sicily (Central Mediterranean). Hydrobiologia 503: 223–236.

Gaertner, J. C., J. A. Bertrand & A. Souplet, 2002. STATIS-CoA: a methodological solution to assess the spatio-temporal organization of species assemblages. Application to the demersal assemblages of the French Mediterranean Sea. Scientia Marina 66: 221–232.

Garofalo, G., M. Gristina, F. Fiorentino, F. Cigala Fulgosi, G. Norrito & G. Sinacori, 2003. Distribution pattern of rays in the Strait of Sicily in relation to fishing pressure. Hydrobiologia 503: 245–250.

Garofalo, G., F. Fiorentino, G. Bono, S. Gancitano & G. Norrito, 2004. Identifying spawning and nursery areas of Red mullet (*Mullus barbatus*, L., 1758) in the Strait of Sicily. In Nishida, T., P. J. Kailola & C. E. Hollingworth (eds), GIS/Spatial Analyses in Fishery and Aquatic Sciences, Vol. 2. Fishery-aquatic GIS Research Group, Saitama, Japan: 101–109.

Gaston, K. J., 2000. Global patterns in biodiversity. Nature 405: 220–227.

Gristina, M., G. Garofalo, M. L. Bianchini, M. Camilleri & F. Fiorentino, 2004. Evaluating the performance of an index of trawling impact in the Strait of Sicily. Biologia Marina Mediterranea 11: 230–241.

Groner, E. & A. Novoplansky, 2003. Reonsidering diversity–productivity relationships: directness of productivity estimates matters. Ecology Letters 6: 695–699.

Isaaks, E. H. & R. M. Srivastava, 1989. An Introduction to Applied Geostatistics. Oxford University Press, New York: 561 pp.

Jennings, S. & M. J. Kaiser, 1998. The effects of fishing on marine ecosystems. Advances in Marine Biology 34: 201–352.

Kaiser, M. J., M. C. V. Austen & H. Ojaveer, 2004. European biodiversity action plan for fisheries: issues for non target species. Fisheries Research 69: 1–6.

Kareiva, P. & M. Marvier, 2003. Conserving biodiversity coldspots. American Scientist 91: 344–351.

Labropoulou, M. & C. Papaconstantinou, 2004. Community structure and diversity of demersal fish communities: the role of fishery. Scientia Marina 68 (Suppl. 1): 215–226.

Lehman, C. & D. Tilman, 2000. Biodiversity, stability and productivity in competitive communities. The American Naturalist 156: 534–552.

Lermusiaux, P. F. J. & A. R. Robinson, 2001. Features of dominant mesoscale variability, circulation patterns and dynamics in the Strait of Sicily. Deep-Sea Research I 48: 1953–1997.

Loreau, M., S. Naeem, P. Inchausti, J. Bengtsson, J. P. Grime, A. Hector, D. U. Hooper, M. A. Huston, D. Raffaelli, B. Schmid, D. Tilman & D. A. Wardle, 2001. Biodiversity and ecosystem functioning: current knowledge and future challenges. Science 294: 804–808.

Loreau, M., A. Downing, M. Emmerson, A. Gonzalez, J. Hughes, P. Inchausti, J. Joshi, J. Norberg & O. Sala, 2002. A new look at the relationship between diversity and stability. In Loreau, M., S. Naeem & P. Inchausti (eds), Biodiversity and Ecosystem Functioning: Synthesis and Perspectives. Oxford University Press, Oxford, 79–91.

McCann, K. S., 2000. The diversity-stability debate. Nature 405: 228–233.

Mueter, F. J. & B. L. Norcross, 2002. Spatial and temporal patterns in the demersal fish community on the shelf and upper slope regions of the Gulf of Alaska. Fishery Bulletin 100: 559–581.

Naeem, S., M. Loreau & P. Inchausti, 2002. Biodiversity and ecosystem functioning: the emergence of a synthetic ecological framework. In Loreau, M., S. Naeem & P. Inchausti (eds), Biodiversity and Ecosystem Functioning: Synthesis and Perspectives. Oxford University Press, Oxford, 3–11.

Patti, B., A. Cuttitta, A. Bonanno, G. Basitone, G. Buscaino, C. Patti, J. Garcia Lafuente, A. Garcia & S. Mazzola, 2004. Coupling between the hydrographic circulation in the Strait of Sicily and the reproductive strategy of the European anchovy *Engraulis encrasicolus*: effects on distribution of spawning grounds. In MedSudMed 2004. Report of the MedSudMed Expert Consultation on Small Pelagic Fishes: Stock Identification and Oceanographic Processes Influencing their Abundance and Distribution. GCP/RER/010/ITA/MSM-TD-05. MedSudMed Technical Documents 5: 132 pp.

Pauly, D., R. Watson & V. Christensen, 2003. Ecological geography as a Framework for a transition toward responsible fishing. In Sinclair, M. & G. Valdimarsson (eds), Responsible Fisheries in the Marine Ecosystem. FAO, Rome: 87–101.

Pikitch, E. K., C. Santora, E. A. Babcock, A. Bakun, R. Bonfil, D.O. Conover, P. Dayton, P. Doukakis, D. Fluharty, B. Heneman, E. D. Houde, J. Link, P. A. Livingston, M. Mangel, M. K. McAllister, J. Pope & K. J. Sainsbury, 2004. Ecosystem-based fishery management. Science 305: 346–347.

Robinson, A. R., M. Golnaraghi, W. G. Leslie, A. Artegiani, A. Hecht, E. Lazzoni, A. Michelato, E. Sansone, A. Theocharis & U. Unluata, 1991, The Eastern Mediterranean general circulation: features, structure and variability. Dynamics of Atmospheres and Oceans 15: 215–240.

Sinclair, M. & G. Valdimarsson, 2003. Responsible Fisheries in the Marine Ecosystem. FAO, Rome: 426 pp.

Tilman, D., 2000. Causes, consequences and ethics of biodiversity. Nature 405: 208–211.

Ungaro, N., G. Marano, R. Marsan, M. Martino, M. C. Marzano, G. Strippoli & V. Volra, 1999. Analysis of demersal assemblages from trawl surveys in the South Adriatic Sea. Aquatic Living Resources 12: 177–185.

Hydrobiologia (2007) 580:125–133
DOI 10.1007/s10750-006-0459-7

BIODIVERSITY IN ENCLOSED SEAS

Recurrent high-biomass blooms of *Alexandrium taylorii* (Dinophyceae), a HAB species expanding in the Mediterranean

M. G. Giacobbe · A. Penna · E. Gangemi · M. Masò · E. Garcés · S. Fraga · I. Bravo · F. Azzaro · N. Penna

Abstract Summer outbreaks of the dinoflagellate *Alexandrium taylorii* Balech are recurrent events in nearshore waters of Sicily (Italy)—a central region in the Mediterranean Sea—producing dense yellowish–green patches. Beyond the local phenomenon, the problem covers a broader geographic scale, involving also other European localities, mostly in Spain.

Biological, environmental, and molecular data are reported here from a semi-closed bay of Sicily (Vulcano Island, Tyrrhenian Sea, 2000–2003), showing in summer the recurrence of high-biomass blooms and events of water discolouration. Without underestimating the setbacks to the tourism industry, the ecological impact of *A. taylorii* blooms may be important considering the high levels of biomass produced (West Bay, Vulcano: up to a magnitude order of 10^7 cells l^{-1}, 50–180 μg-Chl*a* l^{-1}, June 2002 and 2003) and coincident conditions of oxygen supersaturation of the waters (130–170%). Trophic trends in the Tyrrhenian site indicate high amounts of nutrients linked to the increased anthropogenic activity in summer, although recently there has been an apparent shift of the marked eutrophic conditions towards a slighter eutrophy.

Genetic data on isolates of *A. taylorii* from the Mediterranean Sea are also discussed. Molecular analyses implied the sequencing of target rDNA regions (5.8S rDNA and ITS regions) of several isolates from different Mediterranean localities, as well as the application of species-specific PCR assays for rapid species identification in preserved field samples. The confirmation of the specific identity provided new insights into the biogeography of this species and further evidence of the occurrence of *A. taylorii* in a number of Mediterranean localities, both in the western side (the Catalan coast of Spain) and the eastern area (Greece). Analyses of the molecular diversity of geographically distinct isolates of *A. taylorii* from Italy, Spain, and Greece based on the 5.8S rDNA-ITS region sequences showed a high level of similarity, indicating the existence of an unique Mediterranean population.

Guest editors: G. Relini & J. Ryland
Biodiversity in Enclosed Seas and Artificial Marine Habitats

M. G. Giacobbe (✉) · E. Gangemi · F. Azzaro
IAMC-CNR Istituto per l'Ambiente Marino Costiero, Sezione di Messina, Italy
e-mail: mariagrazia.giacobbe@iamc.cnr.it

A. Penna · N. Penna
CBA Centro di Biologia Ambientale, University of Urbino, Urbino, Italy

M. Masò · E. Garcés
CSIC-ICM Institut Ciències del Mar, Barcelona, Spain

S. Fraga · I. Bravo
IEO-COV Instituto Español de Oceanografia, Vigo, Spain

Keywords *Alexandrium taylorii* · Dinoflagellate blooms · Mediterranean · PCR · Phylogeny

Introduction

Blooms of phytoplankton organisms are recurrent events in European coastal waters and worldwide (Anderson, 1989), being potentially harmful to human health, marine ecosystems and resources such as tourism, fisheries and aquaculture. The socio-economic impact of HABs (Harmful Algal Blooms) highlights the need for ecological studies addressed to the understanding of HAB dynamics and processes of bloom formation and dispersal. Summer outbreaks of the dinoflagellate *Alexandrium taylorii* Balech produce dense yellowish–green patches of limited extension in coastal waters of Sicily, Italy (Giacobbe & Yang, 1999; Penna et al., 2002), a central island in the Mediterranean Sea. Beyond the local phenomenon, similar events are also observed in beach areas along the Costa Brava and Balearic Islands (Spain), causing in some localities water discolouration lasting two months (Garcés et al., 1999). Dense aggregates of cells are also found near the sediment, this representing a possible way of avoiding advection from the area and prevent population losses. As a whole, the observations reported in literature suggest that the magnitude of the event is increasing in some places and that this species is in a phase of expansion, probably due to ballast water release and other causes.

Some authors have shown the morphological variability of *A. taylori*, the high production of temporary cysts having a significant role in the bloom dynamics (Garcés et al., 2002), and the existence of a sexual resting stage as a part of its life cycle (Delgado et al., 1997, Giacobbe & Yang, 1999). Thus, the specific identity may be easily masked when dealing with encysted stages—vegetative or sexual resting cysts.

In this study, we examined a set of biological and environmental data obtained in the last four years (2000–2003, EU Project Strategy, EVK3-CT-2001–00046) from a Tyrrhenian locality of Sicily subject to bloom events. Sequence analyses of different strains of *A. taylorii* isolated from various Mediterranean areas, as well as results of the application of molecular probes (by PCR assays) to preserved field samples, are also provided.

Study area

The West Bay of Vulcano (Sicily), a tourist locality of the Aeolian Islands (Tyrrhenian Sea), is a shallow, semicircular area extending approximately 300 × 100 m. The site is located at 38°25′04″–10″ Lat N and 14°57′14″–23″ Long E and covers as a whole about 7 ha. Recurrent summer blooms of *A. taylorii* take place in some nearshore points of the bay with evident water discolouration due to the high-biomass produced by this dinoflagellate (Penna et al., 2002). Before 2000, there had been no studies on HABs in this Tyrrhenian locality, although popular references on events of red tide date back to 1995.

Materials and methods

Sample collection and analyses

Three nearshore stations (1 m depth) located in the West Bay of Vulcano (Stn 1: 38°25′4″ N, 14°57′14″ E; Stn 2: 38°25′6″ N, 14°57′19″ E; Stn 3: 38°25′10″ N, 14°57′23″ E) were sampled in spring-summer 2000–2003, on 7–19 occasions each year. 100–250 ml of surface seawater samples were collected and fixed with Lugol's solution (0.4% final concentration). Samples were taken at the same time of day (between 12 and 15 h). The bloom density was determined by cell counts of target species and other phytoplankton groups. The procedure for identification and quantification of phytoplantkon cells involved: sedimentation of a subsample in 10 or 50 ml settling chambers and counts of cells in an appropriate area (Throndsen, 1995) using an inverted Zeiss Axiovert 200 microscope equipped with epifluorescence. The identity of *A. taylorii* was confirmed by staining the cells with Calcofluor White M2R (Fritz & Triemer, 1985) and using a Zeiss UV filter set Fs 01.

Measures of environmental parameters such as water temperature, salinity and oxygen were taken by a mobile Multiprobe (YSI 556 MPS). Nutrient samples were filtered and frozen immediately after water collection. Concentrations of nitrate, nitrite and phosphate were determined following the Strickland and Parsons (1972) method, and ammonia according to Aminot & Chaussepied (1983).

Clonal cultures of *Alexandrium taylorii* were established at the Consiglio Nazionale delle Ricerche (CNR, Messina), Institut de Ciències del Mar (CSIC, Barcelona), and Instituto Español de Oceanografia (IEO, Vigo) from unfixed seawater samples taken at various sites of the Mediterranean. All marine cultures were maintained in F/2 or F/20 medium (see http://ccmp.bigelow.org/), at 18 ± 1°C and 12:12 h (light:dark) photoperiod. Illumination was provided by a photon irradiance of 100 μmol m^{-2} s^{-1}.

Molecular assays

Total DNA extractions of different *A. taylorii* isolates and mixed phytoplankton populations from seawater samples collected in different localities of the Mediterranean basin—Vulcano (Italy), St. Carles, Catalan Sea (Spain), and Saronikos Gulf (Greece)—were performed by using DNeasy Plant Kit (Qiagen, Valencia, CA, USA), according to the manufacturer's instructions. PCR amplification conditions of the 5.8S gene and ITS regions were performed as described in Penna & Magnani (1999). PCR amplified products of all *A. taylorii* isolates used in this study were directly sequenced. The sequences were deposited at EMBL Bank and given in Table 1.

Sequence alignment was done with Clustal X software and subsequently improved by eye. The rDNA-ITS regions were aligned with other ITS sequences of *Alexandrium* species, listed in GenBank. Sequence alignment analyses were also carried out by including other sequence data of *Alexandrium* species from GenBank (Table 2). Phylogenetic relationships, based on the 5.8S rDNA-ITS regions, were inferred using the Neighbor-Joining (NJ) (Saitou & Nei, 1987). The sequence of *Gymnodinium sanguineum* Hirasaka [=*Akashiwo sanguinea* (Hirasaka) Hansen & Moestrup] (GenBank AF131075) was used as outgroup. Robustness of the phylogenetic tree generated by NJ was tested by using the non-parametric bootstrap (Felsenstein, 1985) with 10,000 pseudoreplicates. The above analysis was performed with the software package MEGA ver. 3.0.

A. taylorii specific PCR assays were carried out on the genomic DNA of mixed phytoplankton population samples by using two species-specific designed primers (Penna et al., 2004). The PCR products were visualized on 2.0% agarose gel.

Results

Spring-summer *A. taylorii* densities in the last four years (2000–2003) showed exceptionally high values (Fig. 1), compared with the concentrations of phytoplankton usually found in Sicilian coastal waters. On various occasions (2001–2003), the bloom density reached a 10^7 order of magnitude, with maxima of 2–3×10^7 cells l^{-1} and amounts of Chl*a* as high as 50–180 μg l^{-1} in June 2002 and 2003. During these blooms, the water temperature was 27.5–32°C, salinity 37.1–37.2, dissolved inorganic nitrogen 2.42–2.88 μM DIN, phosphates 2.32–5.48 μM, with a corresponding condition of oxygen supersaturation of the waters (up to 170%). The highest concentrations of DIN were due to nitrates, or alternatively to ammonia, and phosphates in some cases reached values of ≈10 μM (2001, Fig. 1) with a slight decreasing trend through the years.

The water discolouration lasted 1–2 months, as confirmed again in 2004 and 2005 (data not shown), gradually disappearing in July or August–September (2001). The bloom persistence and discolouration was apparently favoured by optimal, summer weather conditions. The horizontal distribution of *A. taylorii* cells within the West Bay (Vulcano) was characterized in all cases by the assembling of cells nearshore, in the most sheltered part of the bay, as shown in Fig. 2, although in some cases a wind-driven advection of cells to NE of the bay was observed.

The percentage of temporary cysts versus motile cells of *A. taylorii* (Fig. 3), as determined

Table 1 List of *A. taylorii* strains, sample locations, origin, source and EMBL accession numbers from this study

Species	Strain	Sampling location and year	Origin cellular phase	Source	EMBL accession no.
A. taylorii	[a]CNR-AT4	Mediterranean, Ionian Sea, Siracusa, Italy, 1997	vegetative	M.G. Giacobbe	AJ251653
A. taylorii	[a]CNR-ATAYB2	Mediterranean, Tyrrhenian Sea, Vulcano, Italy, 2000	vegetative	M.G. Giacobbe	AJ300451
A. taylorii	Field sample	Mediterranean, Tyrrhenian Sea, Vulcano, Italy, 2000	temporary cysts	M.G. Giacobbe/A. Penna	AJ291785
A. taylorii	[b]CBA-1	Mediterranean, Aegean Sea, Kavala, Italy, 2001	vegetative	K. Koukavas	AJ416856
A. taylorii	[c]CSIC-AV8	Mediterranean, Catalan Sea, La Fosca, Spain, 1995	vegetative	E. Garcés/S. Fraga	AJ251654

[a] CNR = Consiglio Nazionale delle Ricerche, Messina, Italy

[b] CBA = Centro di Biologia Ambientale, Urbino, Italy

[c] CSIC-ICM = Institut de Cienciès del Mar, Barcelona, Spain

Table 2 List of *Alexandrium* species 5.8S rDNA and ITS region sequences used in this study

Species	Strain	Location	Source[a]	Accession No.	Toxicity
A. affine	H1	Harima Nada, Japan	Adachi et al. 1996	AB006995	No-PSP
A. catenella	M17	Harima Nada, Japan	Adachi et al. 1996	AB006990	Yes-PSP
A. catenella	ACC01	Chile	Marin et al. 2001	AJ272120	Yes-PSP
A. fraterculus	AF-1	Pusan, Korea	Cho 1999	AF208242	No-PSP
A. insuetum	S1	Shoudoshima, Japan	Adachi et al. 1996	AB006996	No-PSP
Alexandrium sp.	AS-1	Pusan, Korea	Cho 1999	AF208243	No-PSP
A. margalefii	[b]CNR-AM1	Oliveri, Italy	Penna et al. 1999	AJ251208	No-PSP
A. minutum	LAC 27	Trieste, Italy	Penna et al. 2000	AJ005050	Yes-PSP
A. minutum	[b]CNR-AMIA1	Siracusa, Italy	Penna et al. 2003	Vila et al. 2005	Yes-PSP
A. minutum	[b]CNR-AMIA4	Siracusa, Italy	Penna et al. 2003	Vila et al. 2005	Yes-PSP
A. minutum	[b]CNR-AMIA5	Siracusa, Italy	Penna et al. 2003	Vila et al. 2005	Yes-PSP
A. minutum	[c]IEO-AL8C	Arenys, Spain	Penna et al. 2003	Vila et al. 2005	Yes-PSP
A. minutum	[c]IEO-AL9C	Arenys, Spain	Penna et al. 2003	Vila et al. 2005	Yes-PSP
A. minutum	[d]CSIC-D1	Arenys, Spain	Penna et al. 2003	Vila et al. 2005	Yes-PSP
A. pseudogonyaulax	H1	Harima Nada, Japan	Adachi et al. 1996	AB006997	No-PSP
A. tamarense	CU-15	Gulf of Thailand, Thailand	Adachi et al. 1996	AB006992	Yes-PSP
A. tamarense	FK-788	Funka Bay, Japan	Adachi et al. 1996	AB006993	Yes-PSP
A. tamarense	AT-A	Chinhae Bay, Korea	Cho 2001	AF374224	PSP?
A. tamarense	AT-B	Chinhae Bay, Korea	Cho 2001	AF374225	PSP?
A. tamarense	AT-2	Chinhae Bay, Korea	Cho 2001	AF374227	PSP?
A. tamarense	AT-6	Chinhae Bay, Korea	Cho 2001	AF374228	PSP?
A. tamarense	AT-10	Chinhae Bay, Korea	Cho 2001	AF374226	PSP?
A. tamarense	[e]CCMP116	Vigo, Spain	Penna et al. 1999	AJ005047	Yes-PSP
A. tamarense	WKS-1	Kushimoto, Japan	Adachi et al. 1996	AB006991	Yes-PSP
A. tamiyavanichii	MS-01	Sebatu, Malaysia	Usup et al. 2002	AF145224	No-PSP

[a] who sequenced the strain

[b] CNR = Consiglio Nazionale delle Ricerche, Messina, Italy

[c] IEO = Instituto Español de Oceanografia, Vigo, Spain

[d] CSIC-ICM = Institut de Cienciès del Mar, Barcelona, Spain

[e] CCMP = Provasoli-Guillard National Centre for Culture of Marine Phytoplankton, Maine, US

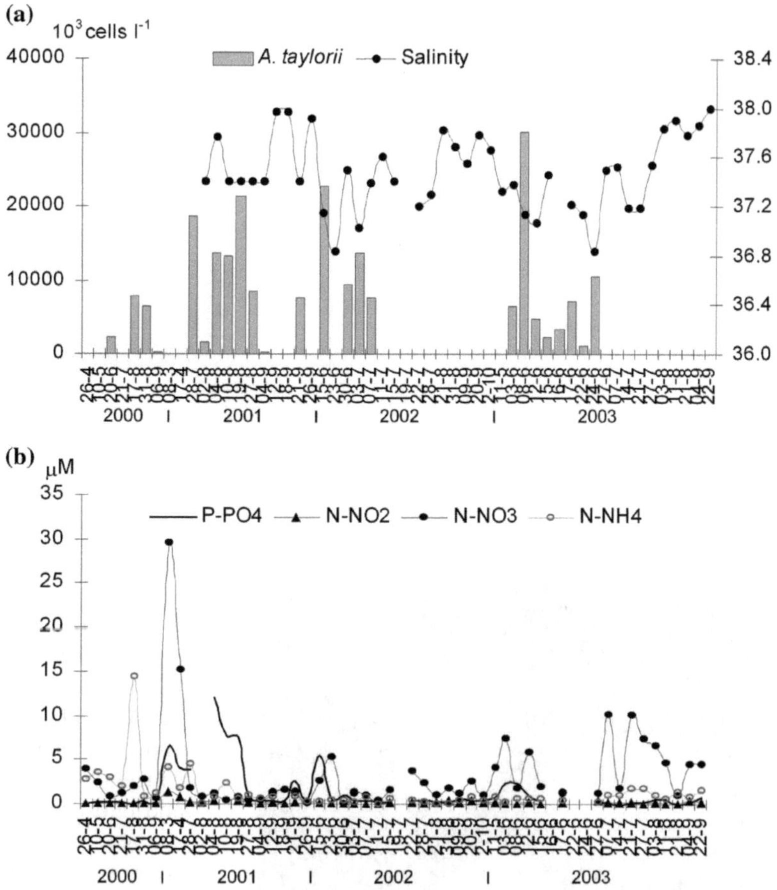

Fig. 1 Spring-summer values of (a) *Alexandrium taylorii* densities, salinities, and (b) nutrients in Vulcano-St.1, Sicily. Years 2000–2003

in 2001 in water samples, showed an incidence of resting vegetative stages of 7–11% over the total *A. taylorii* population, with the highest percentages near the sediment. These values increased to 22% in late summer when the bloom was declining.

The NJ phylogenetic tree of *A. taylorii* and other *Alexandrium* species based on the 5.8S rDNA-ITS regions is shown in Fig. 4. In respect to the other species of *Alexandrium* examined to date, *A. taylorii* appeared to diverge. Within the genus *Alexandrium*, the Mediterranean *A. taylorii* isolates formed a homogeneous clade, distinct from the others. These isolates shared identical nucleotide sequences.

Species-specific PCR assays of fixed field samples are shown in Fig. 5. By using species-specific primers designed for *A. taylorii*, PCR amplifications of the 5.8S rDNA-ITS regions yielded products of 297 base pair size for the field samples containing *A. taylorii* cells. Species-specific PCR amplifications for *A. taylorii* revealed positive results for the presence of this species in the Vulcano phytoplankton population samples in agreement with the microscopic examination.

Discussion

Without understimating the setbacks to the tourism industry, the ecological impact of *A. taylorii* blooms may be important considering the high levels of biomass produced, as in the case of the West Bay, Vulcano (up to 3×10^7 cells l^{-1}) and coincident conditions of oxygen supersaturation of the waters, that may threaten marine life. Trophic trends in the Tyrrhenian target site

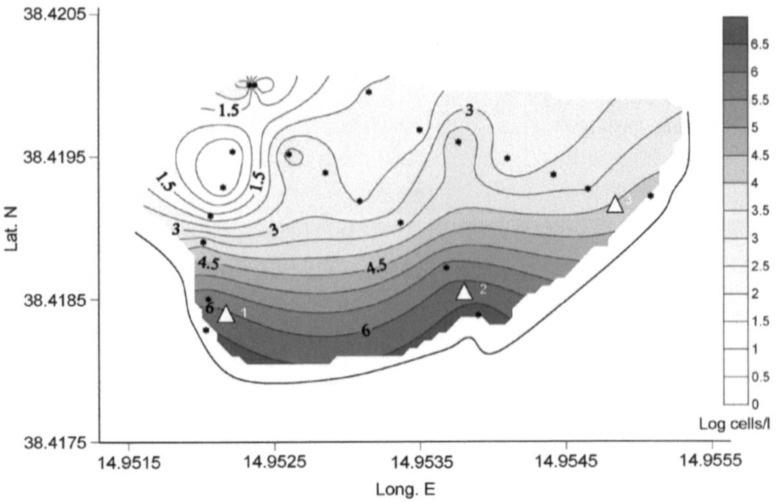

Fig. 2 Horizontal distribution of *A. taylorii* cells in the West Bay, Vulcano. Surfer plot of 22 July 2001. Triangles (1–3) indicate routine sampling stations. Asterisks indicate additional cruise stations

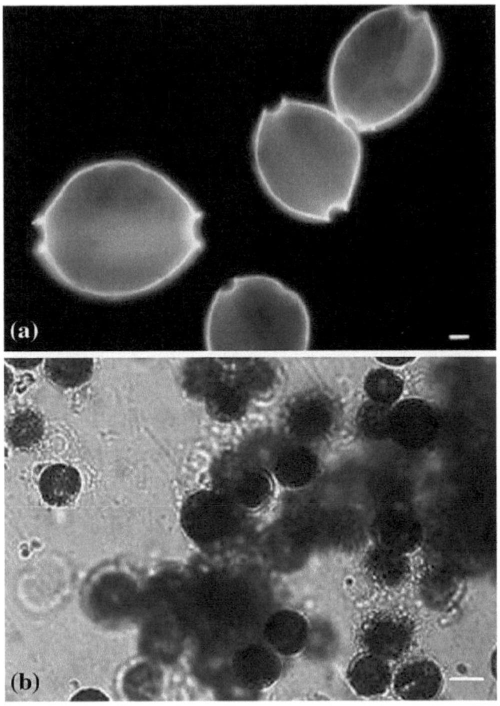

Fig. 3 General morphology of motile cells (fluorescence) and temporary cysts (bright field). Scale bars = 4 μm (**a**) and 25 μm (**b**)

indicate high amounts of nutrients linked to the increased anthropogenic activity in summer (see also Penna et al., 2001), although we observed an apparent shift of the marked eutrophic conditions towards a slighter eutrophy. This could be due to a recent improvement of the sewage treatment. Previous studies in this area suggested that nutrients play a relevant role in the development and maintenance of *A. taylorii* blooms, the high concentrations in summer of inorganic nitrogen and phosphorus being probably due to the hotel sewage, when the anthropogenic pressure increases in relation to the tourism (Penna et al., 2002). Then, climatic conditions such as high water temperature, calm sea and weak tide, seem to favour the bloom development. Garcés et al. (2002) suggested that under short-term environmental perturbations, for instance storms or swells, *A. taylorii* encysts to withstand the unfavourable conditions and restores the population after a few days of calm weather, thus functioning as a population stock. In addition, encystment would also reduce the population losses from the area, also through cell aggregation in clusters of cysts. This, together with the physical characteristics (relatively low water renewal) and favourable wind conditions, may favour population growth and persistence (Garcés et al., 1999, Basterretxea et al., 2005).

Many factors, natural or anthropogenic, are known to contribute to the spreading worldwide of HAB species. Such phenomena are amplified by human activities, such as the wrong

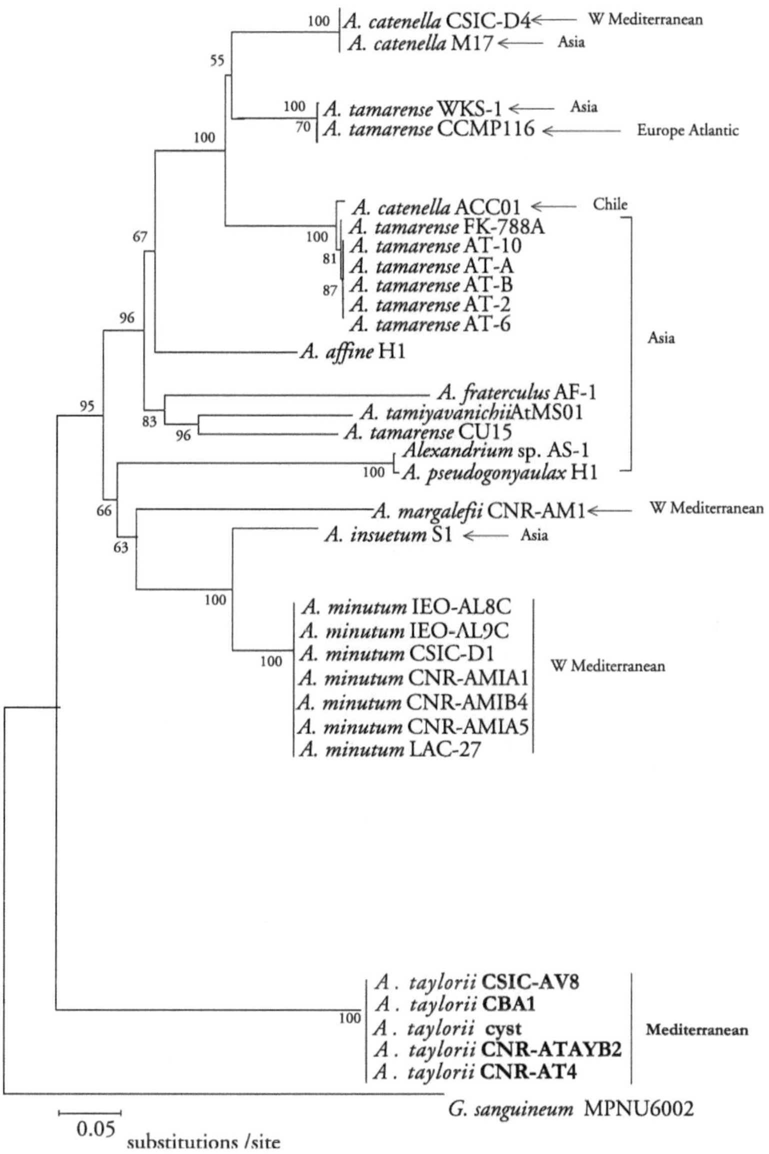

Fig. 4 Neighbor-Joining (NJ) tree inferred from the 5.8S rDNA and ITS sequences of *Alexandrium* species. Sequences of *A. taylorii* from this study are in bold. *G. sanguineum* was used as an outgroup. Numbers on the major nodes represent NJ (10000 pseudo-replicates) bootstrap values. Only bootstrap values ≥ than 50% are shown

management of the coastal area leading to increased amounts of nutrients in the waters and eutrophy, lack of seawage treatments, use of fertilizers in agriculture, the sea discharge of ship's ballast water favouring the introduction of non-indigenous organisms, the so-called "alien species". Cell dispersion can also be favoured by anthropogenic means, as plastics (Masò et al., 2003) or movement of recreational boats in summer. Therefore, areas previously free by harmful species can later become subject to HABs, if such species find in the new environment conditions favouring cell growth.

In the specific case of *A. taylorii*, we obtained further evidence for its spreading in the Mediterranean region, from the western side (the Catalan coast of Spain) to the eastern area (Greece).

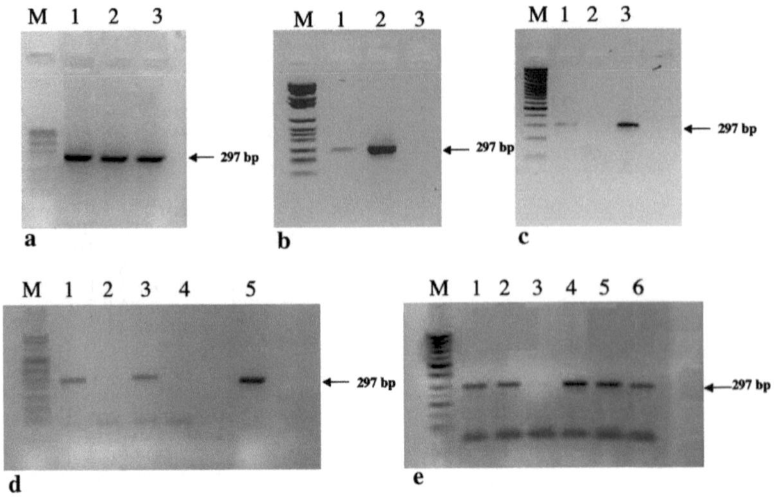

Fig. 5 *A. taylorii* specific PCR assays of mixed phytoplankton population samples from different Mediterranean localities. (**a**) template plasmid containing *A. taylorii* 5.8S rDNA-ITS regions at different concentrations: lane 1, 10 ng; lane 2, 1 ng; lane 3, 0.1 ng. (**b**) Saronikos Gulf (Greece) field sample: lane 1, phytoplankton template DNA (1 ng); lane 2, template plasmid containing *A. taylorii* 5.8S rDNA-ITS regions (1 ng); lane 3, negative control (no template DNA). (**c**) St. Carles field sample, Catalan Sea (Spain): lane 1, phytoplankton template DNA (1 ng); lane 2, negative control (no template DNA); lane 3, template plasmid containing *A. taylorii* 5.8S rDNA-ITS regions (1 ng). (**d**) Vulcano West Bay field samples, Tyrrhenian Sea (Italy), 15th June 2003: lane 1, Station1 phytoplankton template DNA (1 ng); lane 2, Station 3 phytoplankton template DNA (1 ng); lane 3, Station2 phytoplankton template DNA (1 ng); lane 4, Station Capo Testa Grossa phytoplankton template DNA (1 ng); lane 5, template plasmid containing *A. taylorii* 5.8S rDNA-ITS regions (1 ng). (**e**) Vulcano field sample, West Bay Station1: lane 1, phytoplankton template DNA (10 ng), 16th June 2003; lane 2, phytoplankton template DNA (0.1 ng), 16th June 2003; lane 3, phytoplankton template DNA (0.01 ng), 16th June 2003; lane 4, phytoplankton template DNA (10 ng), 17th June 2003; lane 5, phytoplankton template DNA (0.1 ng), 17th June 2003; lane 6, phytoplankton template DNA (0.01 ng), 17th June 2003. M, size standards; arrow indicates reactions that produced a single species-specific PCR product

Aiming at the improvement of the HAB detection tools, species-specific primers for *A. taylorii* were applied in this study in parallel to conventional techniques. This allowed the detection of the target organism also when it occurred as resting stage and the exact species identity was more difficult to be determined. The application of the PCR analysis to field seawater samples, evidenced the presence of *A. taylorii* in all mixed phytoplankton population samples collected in different sampling points in the Vulcano West Bay, with exception of the Station 3, where the cell abundance was often extremely low. Further, the species-specific primer application to the PCR analyses resulted also suitable for the Greek and Spanish field samples containing the target species. These results on the species-specific detection of *A. taylorii* by PCR assay were always in agreement with the microscopic examination.

The phylogenetic analyses of geographically distinct isolates of *A. taylorii* (Italy, Spain, Greece) indicated the existence of a unique, homogeneous Mediterranean population, constituted by isolates with identical ITS-5.8S rDNA sequences.

In this study, the confirmation of the specific identity through the coupling of microscopic and molecular analyses provided new insights into the biogeography of *A. taylorii*, showing the extension of its geographic range from the western Mediterranean to the eastern area. Future researches will give priority to the advancement in both control and prevention strategies as safeguard of the coastal marine environment and public health, and development of advanced technologies for the rapid detection and mitigation of HAB events through laboratory experiments and in situ application.

Acknowledgements This study was supported by the EU project STRATEGY (EVK3-2001-00046). Thanks to Mariagrazia Galletta, Antonella Marini, Franco Decembrini, Francesco Raffa for analyses of nutrients, biomass and technical cooperation, Dr. Francesca Andreoni and Dr. Elena Bertozzini for molecular analyses.

References

Anderson, D. M., 1989. Toxic algal blooms and red tides: a global perspective. In Okaichi T., D. M. Anderson & T. Nemoto (eds), Red Tides: Biology, Environmental Science and Toxicology. Elsevier, New York, 11–16.

Aminot, A. & M. Chaussepied 1983. Manuel des analyses chimiques en milieu marin. Centre National pour l'Exploration des Oceans, 1–1395.

Basterretxea, G., E. Garcés, A. Jordi, M. Masò & J. Tintoré, 2005. Breeze conditions as a favoring mechanism of *Alexandrium taylori* blooms at a Mediterranean beach. Estuarine, Coastal & Shelf Science 62: 1–12.

Delgado, M., E. Garcés, M. Vila & J. Camp, 1997. Morphological variability in three populations of the dinoflagellate *Alexandrium taylori*. Journal of Plankton Research 19: 749–757.

Felsenstein, J., 1985. Confidence limits on phylogenies: an approach using the bootstrap. Evolution 39: 783–791.

Fritz, L. & R. E. Triemer, 1985. A rapid simple technique utilizing calcofluor white M2R for the visualization of dinoflagellate thecal plates. Journal of Phycology 21: 662–664.

Garcés, E., M. Masò & J. Camp, 1999. A recurrent and localized dinoflagellate bloom in a Mediterranean beach. Journal of Plankton Research 21: 2373–2391.

Garcés, E., M. Masò & J. Camp, 2002. Role of temporary cysts in the population dynamics of *Alexandrium taylori* (Dinophyceae). Journal of Plankton Research 24: 681–686.

Giacobbe, M. G. & X. Yang, 1999. The life history of *Alexandrium taylori* (Dinophyceae). Journal of Phycology 35: 331–338.

Masó, M., E. Garcés, F. Pagès & J. Camp, 2003. Drifting plastic debris as a potential vector for dispersing Harmful Algal Blooms (HAB) species. Scientia Marina 67: 107–111.

Penna A., G. Fusco, E. Bertozzini, M. Vila, E. Garcés, M.G. Giacobbe, A. Luglié, M. Masò, L. Galluzzi & M.Magnani, 2004. Monitoring of HAB species in the Mediterranean Sea through a filter system PCR assay detection method. 11th International Conference on Harmful Algae. Cape Town, 15–19 November 2004. Abstract.

Penna, A., M. G. Giacobbe, F. Andreoni, E. Garcés, S. Berluti, R. Cantarini, N. Penna & M. Magnani, 2001. Blooms of *Alexandrium taylori* (Dinophyceae) in the Mediterranean: a preliminary molecular analysis of different isolates. In Hallegraeff G. M., S. I. Blackburn, C. J. Bolch & R. J. Lewis (eds), Harmful Algal Blooms 2000. UNESCO, Paris, 218–221.

Penna, A., M. G. Giacobbe, N. Penna, F. Andreoni & M. Magnani, 2002. Seasonal blooms of the HAB dinoflagellate *Alexandrium taylori* Balech in a new Mediterranean area (Vulcano, Aeolian Islands). Marine Ecology 23: 320–328.

Penna, A. & M. Magnani, 1999. Identification of *Alexandrium* (Dinophyceae) species using PCR and rDNA-targeted probes. Journal of Phycology 35: 615–621.

Saitou, N. & M. Nei, 1987. The Neighbour-Joining method: a new method for reconstructing phylogenetic trees. Molecular Biology and Evolution 4: 406–25.

Strickland, J. D. H. & T. R. Parsons, 1972. A Practical Handbook of Seawater Analysis. Fisheries Research Board of Canada, Ottawa.

Throndsen, J., 1995. Estimating Cell Numbers. In Hallegraeff G. M., D. M. Anderson & A. D. Cembella (eds), Manual on Harmful Marine Microalgae Intergovernmental Oceanographic Commission Manuals and Guides no 33. UNESCO, Paris, 63–80.

Vila, M., M. G. Giacobbe, M. Maso', E. Gangemi, A. Penna, N. Sampedro, F. Azzaro, J. Camp & L. Galluzzi, 2005. A comparative study on recurrent blooms of *Alexandrium minutum* in two Mediterranean coastal areas. Harmful Algae 4: 673–695.

Hydrobiologia (2007) 580:135–142
DOI 10.1007/s10750-006-0458-8

Lack of epifaunal response to the application of salt for managing the noxious green alga *Caulerpa taxifolia* in a coastal lake

K. M. O'Neill · M. J. Schreider · T. M. Glasby · A. R. Redden

Abstract Since the discovery of the green alga *Caulerpa taxifolia* in Lake Macquarie (New South Wales, Australia) in 2001, the New South Wales Department of Primary Industries (Fisheries) has attempted various control methods, including covering the alga with granulated sea salt to induce osmotic shock and cell lysis. In Lake Macquarie, *C. taxifolia* often occurs in patches within beds of the native seagrass *Zostera capricorni.* Although the effects of the salt treatment on blades of *Z. capricorni* and infauna have been shown to be minimal, there have been no tests of any effects on other native biota, including seagrass epifauna. In this study, we tested the general hypothesis that the abundance and diversity of epifauna would be reduced by salting. We used a 'Beyond BACI' experimental design whereby epifaunal invertebrates were sampled 3 months, 6 weeks and 6 days before and then again after salting. Epifaunal abundances at the putatively impacted (salted) location were compared to those at 4 control locations (where no salt was applied). Abundances of most organisms varied significantly among times and locations with no evidence of the consistent effect of salting on diversity or abundance of epifauna. The study represents an example of the use of large-scale managerial action as a scientific experiment.

Keywords Epifauna · *Caulerpa taxifolia* · Environmental impact · Salting · *Zostera capricorni*

Guest editors: G. Relini & J. Ryland
Biodiversity in Enclosed Seas and Artificial Marine Habitats

Electronic supplementary material Supplementary material is available for this article at <http://dx.doi.org/10.1007/s10750-006-0458-8> and accessible for authorised users

K. M. O'Neill (✉) · M. J. Schreider ·
A. R. Redden
School of Applied Sciences, Ourimbah Campus, University of Newcastle, PO Box 127, Ourimbah, NSW 2258, Australia
e-mail: karenon@optusnet.com.au

T. M. Glasby
NSW Department of Primary Industries, Port Stephens Fisheries Centre, Private Bag 1, Nelson Bay, NSW 2315, Australia

Introduction

Caulerpa taxifolia (Vahl) C. Agardh is a marine green alga endemic to sub-tropical and tropical regions of the world. *C. taxifolia* is capable of growing extremely quickly and vegetative growth seems to be the primary mode by which the alga has invaded large areas of seafloor in NSW and in other countries (Meinesz et al., 1993; Smith & Walters, 1999). Experiments in the Mediterranean have shown that small fragments of *C. taxifolia* can establish on the edges of beds of seagrass during the warmer months (Ceccherelli & Cinelli, 1999). These authors suggested that fragments may accumulate in beds of seagrass due to patterns of water flow and physical entrapment within the seagrass beds. To date, there is little published empirical evidence that *C. taxifolia* damages seagrass or is capable of reducing the size or density of beds of seagrass (de Villèle & Verlaque, 1995; Jaubert et al., 1999, 2003). However, given that *C. taxifolia* in the Mediterranean and in NSW grows amongst seagrass, there is this potential.

Native populations of *C. taxifolia* can be found in a number of northern subtropical states in Australia (Phillips & Price, 2002). *C. taxifolia* has, however, recently been detected in temperate Australian estuaries and coastal lagoons. *Caulerpa taxifolia* was first discovered in temperate New South Wales waters in 2000 and has since been found in ten other estuaries or coastal lakes of NSW (Glasby et al., 2005). The northernmost of these invasions is Lake Macquarie, where the alga was found in 2001, and grows in patches amongst beds of *Zostera capricorni* (Ascherson). Concerns regarding the potential loss of seagrass and potential reductions in biodiversity of fish and invertebrates led to management action by the New South Wales Government (see Creese et al., 2004).

Research into methods for controlling *Caulerpa taxifolia* has indicated that treatment by granulated salt (at a concentration of 50 kg/m^2) is effective in removing *C. taxifolia*, with limited short-term effects on *Z. capricorni* and benthic infauna (Glasby et al., 2005). The salting method was applied to *C. taxifolia* in Lake Macquarie in an attempt to control its growth and spread and is estimated to have lead to the removal of almost 5,200 m^2 of *C. taxifolia*. There is the potential for salting to affect seagrass epifauna and their food source (largely epiphytes) by extreme short-term changes in salinity and turbidity (coincident with dumping granulated salt onto soft sediments), so this study was designed to investigate possible short- (days) and medium-term (months) effects of salting on epifauna associated with seagrasses. It was predicted that the application of salt would lead to:

(i) a significant short- and medium-term change in the composition of epifaunal assemblages associated with adjacent beds of eelgrass *Zostera capricorni*.

(ii) a significant short and medium-term change in the abundances of some organisms (especially less mobile ones, such as gastropod molluscs).

Another (ongoing) component of the project examined infauna, the results of which will be presented elsewhere.

Materials and methods

Study area

Lake Macquarie is a relatively large and shallow, saline coastal lake situated in the Lower Hunter region about 150 km north of Sydney, Australia (Fig. 1). The study was undertaken at Mannering Park, which is located at the south-western end of Lake Macquarie. *C. taxifolia* was found in patches within beds of *Z. capricorni* and associated macroalgal species, close to shore, to a depth of approximately 1 m. The distribution and abundance of *C. taxifolia* at Mannering Park varies seasonally—fronds were generally not visible during late winter and early spring. New growth commenced in spring and patches increased in size rapidly throughout summer reaching maximum patch size (about 25 m^2) and density at this site in early winter.

Sampling and experimental design

The study used a Beyond BACI (Before-After-Control-Impact) asymmetrical design (Underwood, 1991, 1992) that comprised one putatively impacted location (where *Caulerpa* was found and salt was applied) and four control locations approximately 100 m apart (i.e. controls ranged from 100 m to about 200 m from the salted plots). Control locations did not contain *C. taxifolia*, but otherwise were the same as the salted location. There were two randomly selected sites (15 m apart) nested within each location and five replicate samples were taken at each site. Salt application to *C. taxifolia* patches was conducted by New South Wales Fisheries on October 16 2003. Epifaunal samples were taken from all sites at 3 months, 6 weeks and 6 days before salting (June, August and September 2003) and 6 days,

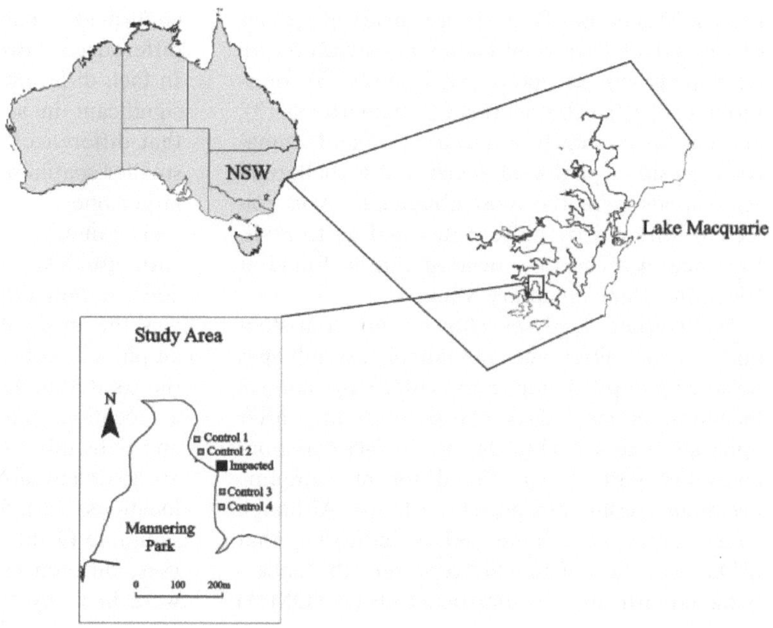

Fig. 1 Map of Lake Macquarie with sampling locations

6 weeks and 3 months after salting (October, November 2003 and February 2004).

A total of 300 epifaunal samples was collected using a specially designed sampler (BROMAR), which comprised a metal cylindrical tube (20 cm in diameter), with a cutting blade underneath and a mesh screen (0.5 mm) on top. The blade was used to cut seagrass at the base and ensured that only vegetation and associated epifauna were collected on the sieve. Samples were bagged and returned to the laboratory where the contents were rinsed in fresh water over 1 mm and 500 μm mesh sieves; both fractions were retained in 70% ethanol. Vegetation was dried at 60°C for 24 h and dry weight measured. Fauna was sorted, enumerated and identified to the lowest taxonomic level. Abundances were expressed as the number of organisms per g dry weight of seagrass.

Statistical Analyses

Analyses of patterns of difference in assemblages were performed using PRIMER v5 (Clarke & Warwick, 2001). As abundances of taxa within some samples varied from one to over 100 individuals, raw data were fourth-root transformed prior to analyses. Transformation of this kind increases the importance of rare taxa and decreases the importance of dominant or abundant taxa (Clarke & Green, 1988). Non-parametric multi-dimensional scaling (nMDS) using Bray–Curtis similarity matrices produced two-dimensional ordination plots of assemblages in each location for each time. To test for differences between assemblages in control and salted locations we used nested analyses of similarities (ANOSIM) with a two-factor design of sites nested within locations for each time.

Asymmetrical analyses of variance (Underwood, 1994; Glasby, 1997) were used to test for differences in total abundance of abundant taxonomic groups. Before vs After salting and Impact vs Control were treated as fixed orthogonal factors with Locations and Sites as random, nested factors. Cochran's test was used to test for homoscedasticity.

Results

A total of 59 epifaunal taxa were represented in the 300 epifaunal samples collected from *Zostera* beds at Mannering Park. These included species of gastropods (16), polychaetes (7), bivalves (6), amphipods (5), decapods (8), isopods (3), opisthobranchs (2), echinoderms (2), nemerteans (3), and one species each of anthuran, tanaid, cumacean, mysid, cephalopod, leech and turbellarian. Amphipods were the most abundant taxon collected, followed by polychaetes and gastropods. Presence of species by location can be found at Electronic Supplementary Material.

Multivariate analyses showed no consistent patterns of difference in faunal assemblages between putatively impacted (salted) and control locations, either 6 days before or 6 days after application of salt (Fig. 2a, b). In fact this non-consistent pattern was found for all sampling occasions (results not presented here). Although stress values were large (>0.2), indicating that MDS plots did not reliably represent the underlying similarity matrix, statistical tests (ANOSIM) confirmed that there were no consistent differences between salted and unsalted locations. In fact, differences between individual sites were significant in several cases (Table 1) indicating that differences among epifaunal assemblages at smaller spatial scales are often greater than at the larger ones.

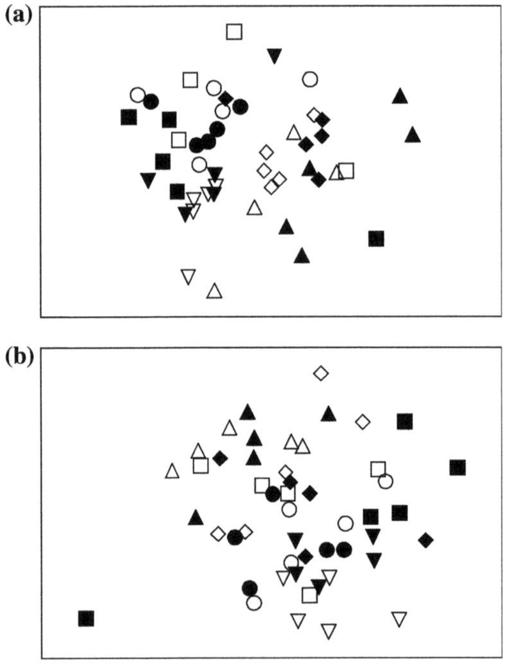

Fig. 2 nMDS plots of the assemblages associated with *Zostera capricorni* for Salted location (■), Control 1 (▲), Control 2 (▼), Control 3 (•) and Control 4 (♦) for a) 6 days before October 2003 (stress = 0.22) and (b) 6 days after October 2003 (stress = 0.21). Closed symbols represent Site 1 and Open Symbols Site 2

The number of amphipods varied through time with greatest abundances detected in October 2003, 6 days after salting (Fig. 3a). *Caprella* sp. was the most abundant of the eight species of amphipods collected. On each sampling occasion, the total abundance of amphipods was similar at all locations indicating that salting did not have any detectable impact on the abundance of this group of animals. Patterns among the four control locations varied significantly among times of sampling (Table 1). As for the multivariate analyses, differences between sites within locations were in many cases greater than the differences among the locations (Table 2).

Although the abundances of polychaetes varied among the control locations, there was no detectable pattern of difference between impacted and control locations at any time of sampling (Fig. 3b). Similarly to amphipods, abundances of polychaetes were more variable between sites within locations than among locations (Table 1).

Gastropods were the most diverse group of epifauna, with four species representing 75% of gastropod abundance. *Batillaria australis* (Quoy & Gaimard) and *Calthaliota* sp. were the dominant molluscs. Abundances of gastropods were lower in all locations for the three times after salting compared to the times before salting (Fig. 3c), indicating wide-spread temporal variability rather than any impact of salting. The patterns of difference among sites at control locations only varied through time (Table 1).

Discussion

Overall, there was no long-term effect of salting on abundance and composition of epifauna associated with *Z. capricorni*. Abundances were highly variable with differences at small spatial and temporal scales often more pronounced than

Table 1 Summary of one-way ANOSIM tests comparing assemblages at each site of the salted location (I) with each site in each of the four control locations (C1, C2, C3 and C4)

	C1-1	C1-2	C2-1	C2-2	C3-1	C3-2	C4-1	C4-2	I1
Before—(6 days before salting) September 2003 Global *R* = 0.392 **									
C1-2	0.0								
C2-1	0.50**	022*							
C2-2	0.69**	0.33**	0.02						
C3-1	0.66**	0.28**	0.38*	0.72**					
C3-2	0.62*	0.22*	0.51*	0.65**	0.16				
C4-1	0.03	0.04	0.40*	0.65**	0.47*	0.50*			
C4-2	0.14	0.05	0.13	0.49**	0.71**	0.60**	0.27		
I1	0.23	–0.13	0.25**	0.35**	0.18*	0.14	0.31*	0.34*	
I2	0.2	–0.11	0.17*	0.30**	0.18*	0.12	0.10	0.20	–0.12
After—(6 days after salting) October 2003 Global *R* = 0.318 **									
C1-2	–0.032								
C2-1	0.82*	0.90**							
C2-2	0.73**	0.94**	0.64**						
C3-1	0.59**	0.69**	0.25*	0.24					
C3-2	0.48*	0.62*	0.25*	0.24	–0.16				
C4-1	0.48**	0.55*	0.28*	0.65**	–0.07	–0.02			
C4-2	0.09	0.40*	0.23*	0.36**	–0.05	0.04	–0.04		
I1	0.50**	0.42*	0.38**	0.39*	0.31*	0.14	0.18	0.21	
I2	0.22	0.32*	0.54*	0.58**	0.21	0.14	–0.01	0.03	0.23*

Values shown are *R*-values for pair-wise comparisons between sites with significance level (*$P < 0.05$, **$P < 0.01$, Blank $P > 0.05$). C1-1, control location 1, site 1 etc

Table 2 Summary of asymmetrical analyses of variance comparing total abundances of amphipods, polychaetes and gastropods between salted and control locations

				Amphipods			Polychaetes			Gastropods		
Source of variation	df		*F*-ratio vs	MS	*F*	*P*	MS	*F*	*P*	MS	*F*	*P*
B/A	1		Irrelevant									
Time(B/A)	4		Irrelevant									
Location	4		Irrelevant									
Salt vs Controls		1	Irrelevant									
Among Controls		3	Irrelevant									
Site(Loc)	5		Irrelevant									
Site(Salt)		1	Irrelevant									
Site(Controls)		4	Irrelevant									
B/A × Loc	4		Irrelevant									
B/A × S vs C		1	B/A × Among Cs	4.26	0.33	0.61	8.24	0.34	0.60	0.02	0.40	0.57
B/A × Among Cs		3	Time(B/A) × Loc[a]	13.05	0.13	0.94	24.04	0.56	0.65	0.04	0.31	0.82
Time(B/A) × Loc	16		Irrelevant	98.20			42.79			0.13		
Time(B/A) × S vs C		4	Time(B/A) × Among Cs	11.81	0.14	0.97	0.52	0.01	1.00	0.01	0.04	1.00
Time(B/A) × Among Cs		12	Time(B/A) × Site(L)	86.39	1.09	0.42	42.27	5.45	**0.00**	0.13	1.32	0.28
B/A × Site(Loc)	5		Time(B/A) × Site(Loc)	19.09	0.24	0.94	2.22	0.29	0.91	0.11	1.00	0.44
B/A × Site(Salt)		1	Time(B/A) × Site(L)	3.29	0.04	0.84	0.17	0.02	0.88	0.01	0.06	0.81
B/A × Site(Controls)		4	Time(B/A) × Site(L)	22.38	0.28	0.89	2.39	0.31	0.09	0.11	1.19	0.35
Time(B/A) × Site(Loc)	20		Irrelevant	79.53			7.76			0.11		
Time(B/A) × Site(Salt)		4	Residual	9.88	0.13	0.97	0.01	0.00	1.00	0.01	0.28	0.89
Time(B/A) × Site(Controls)		16	Residual	69.65	0.94	0.52	7.75	0.61	0.87	0.10	2.26	**0.00**
Residual	240			73.96			12.72			0.05		

[a] This test created only when B/A × site(Loc) could be eliminated ($P > 0.25$)

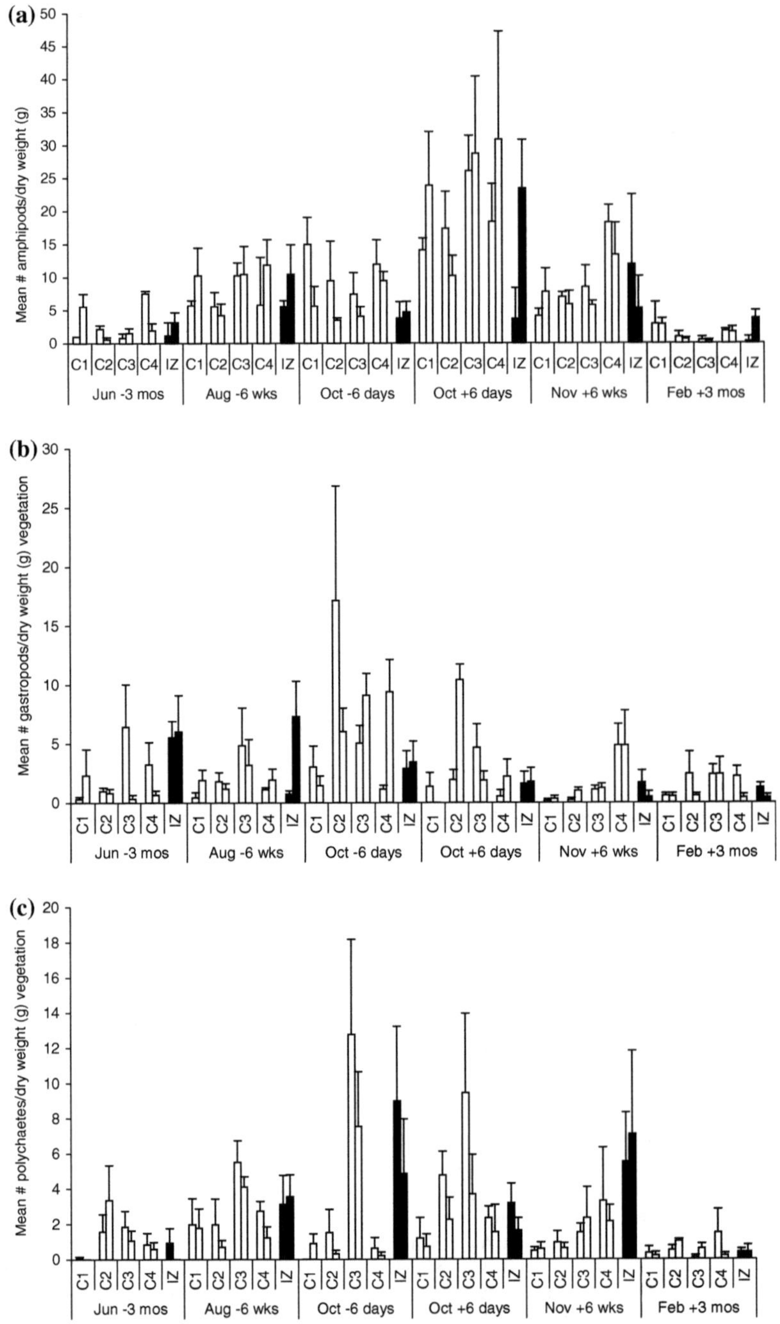

Fig 3. Mean (+SE, n = 5) number of (**a**) amphipods (**b**) polychaetes and (**c**) gastropods per dry weight (g) of vegetation per site per location for the six sampling occasions (Before salting—June 2003, August 2003, October 2003 and after salting—October 2003, November 2003 and January 2004) at Mannering Park, Lake Macquarie. (Each column represents an average number per site; sites are plotted along side each other, Site 1 followed by Site 2. The sites at Impacted locations are highlighted as solid black bars)

differences at larger scales. While amphipods did display some larger scale variability in numbers, this was not related to salting. Their numbers were highest in October and lowest in November and February and there was no consistent pattern of difference between controls and the salted location. This result is similar to that observed for benthic infauna (Glasby et al., 2005), which showed only a short-term decrease in abundance at salted locations.

The possibility that control locations were situated too close to the salted area, and so were also impacted, is highly unlikely, especially considering that the farthest control locations were at least 200 m away from the salted location (Fig. 1). Furthermore, Glasby et al. (2005) identified only a very localized effect of salt on benthic infauna in their study which used 'close' (2–5 m from the salting) and 'distant' (50 m from the salting) control locations. In the present study, the scale of the salting effect was not explicitly investigated, but measurements of salinity (at 25 cm intervals starting from just above the surface of the sediment at all locations) taken within 2 days of salting showed a negligible increase in salinity (i.e. <2ppt) and no change at the control locations.

Epifaunal abundances exhibit high spatial and temporal variability due to a number of factors, including differences in behaviour and rates of growth coupled with differences in population dynamics (Hanski, 1991). As well as spatial variability, changes in temperature, water currents, turbidity, salinity and other physico-chemical factors can lead to changes in patterns of recruitment and differences in local abundances of invertebrates (Connell & Keough, 1985). It is not surprising then that variability at small scales (between sites or between days) is often greater than at larger scales. The colonisation of available substrata by epifauna is also usually very rapid (e.g. Martin-Smith, 1994). Thus, even if there were short-term decreases in abundance immediately after salting, recolonisation would have occurred very quickly. Importantly, organisms inhabiting estuaries and semi-enclosed waterbodies already experience fluctuations in salinity and turbidity because of the dynamic nature of the environment in which they live; therefore, increases in turbidity and salinity associated with applications of large doses of salt may not represent a significant change for these animals.

This study has shown that using salt for the management of *Caulerpa taxifolia* does not have any significant effects on the abundance and diversity of mobile epifauna. Even if there were any immediate changes in epifaunal assemblages, these were not detectable after 6 days and, thus, are not relevant from a conservation point of view. Nevertheless, salting for management of *C. taxifolia* presented a unique opportunity to conduct a scientific 'experiment' at a relatively large scale that otherwise would not have been possible. The research outcomes of this study, in addition to those of Glasby et al. (2005), support the use of salting to control *C. taxifolia.*

Acknowledgements We would like to thank Vanessa Hannan for her valuable assistance in the field and NSW Department of Primary Industries (formerly NSW Fisheries) for assistance with co-ordination of salting and sampling. Bob Creese provided valuable comments on the manuscript.

References

Ceccherelli, G. & F. Cinelli, 1999. Effects of *Posidonia oceanica* canopy on *Caulerpa taxifolia* in a north-western Mediterranean bay. Journal of Experimental Marine Biology and Ecology 240: 19–36.

Clarke, K. R. & R. M. Warwick, 2001. Change in marine communities: an approach to statistical analysis and interpretation, 2nd edn. PRIMER-E, Plymouth.

Clarke, K. R. & R. H. Green, 1988. Statistical design and analysis for a 'biological effects' study. Marine Ecology Progress Series 46: 213–226.

Connell, J. H. & M. J. Keough, 1985. Disturbance and patch dynamics of subtidal marine animals on hard substrata. In Pickett S. T. A. & P. S. White (eds), The Ecology of Natural Disturbance and Patch Dynamics. Academic Press, Inc., New York, 125–151.

Creese, R. G., A. R. Davis & T.M. Glasby, 2004. Eradicating and Preventing the Spread of *Caulerpa taxifolia* in NSW. NSW Fisheries Final Report Series, No. 64, 110 pp.

de Villèle, X. & M. Verlaque, 1995. Changes and degradation in a *Posidonia oceanica* bed invaded by the introduced tropical alga *Caulerpa taxifolia* in the north western Mediterranean. Botanica Marina 38: 79–87.

Glasby, T. M., 1997. Analysing data from post-impact studies using asymmetrical analysis of variance: a case study of epibiota on marinas. Australian Journal of Ecology 22: 448–459.

Glasby, T. M., R. G. Creese & P. T. Gibson, 2005. Experimental use of salt to control the invasive marine alga *Caulerpa taxifolia* in New South Wales, Australia. Biological Conservation 122: 573–580.

Hanski, I., 1991. Single-species meta population dynamics: concepts, models and observations. Biological Journal of Linnaean Society 42: 17–38.

Jaubert, J. M., J. R. M. Chisholm, D. Ducrot, H. T. Ripley, L. Roy & G. Passeron-Seitre, 1999. No deleterious alterations in *Posidonia* beds in the Bay of Menton (France) eight years after *Caulerpa taxifolia* colonization. Journal of Phycology 35: 1113–1119.

Jaubert, J. M., J. R. M. Chisholm, A. Minghelli-Roman, M. Marchioretti, J. H. Morrow & H. T. Ripley, 2003. Re-evaluation of the extent of *Caulerpa taxifolia* development in the northern Mediterranean using airborne spectrographic sensing. Marine Ecology Progress Series 263: 75–82.

Martin-Smith, K. M., 1994. Short-term dynamics of tropical macroalgal epifauna: patterns and processes in recolonisation of *Sargassum fissifolium*. Marine Ecology Progress Series 110: 177–185.

Meinesz, A., J. de Vaugelas, B. Hesse & X. Nmarin, 1993. Spread of the introduced tropical green alga *Caulerpa taxifolia* in the northern Mediterranean water. Journal of Applied Phycology 5: 141–147.

Phillips, J. A. & I. R. Price, 2002. How different is Mediterranean *Caulerpa taxifolia* (Caulerpales: Chlorophyta) to other populations of the species? Marine Ecology Progress Series 238: 61–71.

Smith, C. M. & L. J. Walters, 1999. Fragmentation as a strategy for *Caulerpa* species: fates of fragments and implications for management of an invasive weed. Marine Ecology 20: 307–319.

Underwood, A. J., 1991. Beyond BACI: experimental designs for detecting human environmental impacts on temporal variations in natural populations. Australian Journal of Marine and Freshwater Research 42: 569–587.

Underwood, A. J., 1992 Beyond BACI: the detection of environmental impacts on populations in the real, but variable, world. Journal of Experimental Marine Biology and Ecology 161: 145–178.

Underwood, A. J., 1994. On Beyond BACI: sampling designs that might reliably detect environmental disturbance. Ecological Applications 412: 3–15.

Hydrobiologia (2007) 580:143–155
DOI 10.1007/s10750-006-0457-9

BIODIVERSITY IN ENCLOSED SEAS

Artificial habitats and the restoration of degraded marine ecosystems and fisheries

William Seaman

Abstract Artificial habitats in marine ecosystems are employed on a limited basis to restore degraded natural habitats and fisheries, and more extensively for a broader variety of purposes including biological conservation and enhancement as well as social and economic development. Included in the aims of human-made habitats classified as artificial reefs are: Aquaculture/marine ranching; promotion of biodiversity; mitigation of environmental damage; enhancement of recreational scuba diving; eco-tourism development; expansion of recreational fishing; artisanal and commercial fisheries production; protection of benthic habitats against illegal trawling; and research. Structures often are fabricated according to anticipated physical influences or life history requirements of individual species. For example, many of the world's largest reefs have been deployed as part of a national fisheries program in Japan, where large steel and concrete frameworks have been carefully designed to withstand strong ocean currents. In addition, the differing ecological needs of porgy and sea bass for shelter guided the design of the Box Reef in Korea as a device to enhance productivity of marine ranching. The effect of these and other structures on fisheries catch is positive. But caution must be exercised to avoid using reefs simply as fishing devices to heavily exploit species attracted to them. No worldwide database for artificial habitats exists. The challenge to any ecological restoration effort is to define the condition or possibly even the historic baseline to which the system will be restored; in other words, to answer the question: "Restoration to what?" Examples of aquatic ecosystem restoration from Hong Kong (fisheries), the Pacific Ocean (kelp beds), Chesapeake Bay (oysters) and the Atlantic Ocean (coral reefs) are discussed. The degree to which these four situations consider or can approach a baseline is indicated and compared (e.g., four plants per 100 m^2 are proposed in one project). Measurement of performance is a key factor in restoration planning. These situations also are considered for the ecosystem and fishery contexts in which they are conducted. All use ecological data as a basis for physical design of restoration structures. The use of experimental, pilot and modeling practices is indicated. A context for the young field of marine restoration is provided by reviewing major factors in ecosystem degradation, such as high stress on 70% of commercially valuable fishes worldwide. Examples of habitat disruption include an

Guest editors: G. Relini & J. Ryland
Biodiversity in Enclosed Seas and Artificial Marine Habitats

W. Seaman (✉)
Department of Fisheries and Aquatic Sciences, and Florida Sea Grant College Program, University of Florida, Gainesville, FL 32611-0400, USA
e-mail: seaman@ifas.ufl.edu

extensive hypoxic/anoxic zone in the Gulf of Mexico and nutrient and contaminant burdens in the North Sea. Principles of ecological restoration are summarized, from planning through to evaluation. Alternate approaches to facilitate ecological recovery include land-use and ecosystem management and determining levels of human population, consumption and pollution.

Keywords Artificial habitats · Reefs · Estuaries · Ocean · Restoration

Introduction

Degradation of coastal and ocean habitats, ecosystems and fisheries is a global concern. It motivated the content of the 2004 World Fisheries Congress, for example, where the issues of serial depletion of fisheries by size, area and trophic level, and impairment and destruction of ecological system structure and function were quantified. Examples of overharvest come from all seas, such as the collection of fewer and smaller sea horses and damage to their coral reef habitat in the Indo-Pacific to supply the world aquarium trade and certain medicinal markets. According to the Food and Agriculture Organization of the United Nations (FAO), 18% of major marine fish stocks or species groups are reported as overexploited (FAO, 2002). One of the voices calling attention to this condition is Daniel Pauly, keynote speaker at the World Fisheries Congress, whose "Sea Around Us" project quantifies the consequence of "fishing down the food chain" as top carnivores are decimated by fishing and landings shift to emphasize lower trophic levels (Pauly et al., 1998).

Aquatic habitats are characterized by impairments such as dredging, draining and damming of riverine floodplains and destruction of coastal wetlands. Worldwide, one-third of the world's coasts are at "high potential risk of degradation," according to the United Nations. Along the seacoasts of Europe, degradation may take the form of seagrass bed destruction, eutrophication or fishery overharvest. In the North Sea, for example, the impacts of fisheries activities, trace organic contaminants and nutrients are classified as "First Priority" by the Convention for the Protection of the Marine Environment of the North-East Atlantic (OSPAR Commission, 2000) of the Oslo and Paris Commissions. Despite improvements for certain pollutants/disturbances, the OSPAR Quality Status Report raises concerns for these three stressors and future loss or disturbance of "many sensitive habitats" in coastal areas.

As one type of response, when appropriate, ecological restoration aims to return a system to some level of pre-degraded state. One intent is "to establish a functional ecosystem of a designated type that contains sufficient biodiversity to continue its maturation by natural processes and to evolve over longer time spans in response to changing environmental conditions" (Clewell et al., 2000).

This paper addresses the role of artificial reef habitats in restoration of degraded marine systems. It first examines the overall context for ecological restoration in both terrestrial and aquatic environments, provides definitions for various objectives and practices, and directs the reader to relevant information resources. Some trends and guiding principles (e.g., establishment of measurable objectives) relevant generally to restoration and specifically to artificial marine habitat technology are indicated. This information is presented to assist the multiple disciplines and interests concerned with the use of artificial habitats to better assess their relevance and role in ecosystem and fishery science and management, and in return aid the practitioners. The proper role of artificial habitats in aquatic systems continues as an item of debate in scientific circles. Evidence for their role is presented in a brief analysis of four situations concerning restoration of kelp, coral reefs, oysters and fish populations.

The information presented complements the second theme of the 39th European Marine Biology Symposium, on "Artificial Habitats and Restoration of Degraded Systems," which contained 17 oral and 17 poster contributions. The approach to preparation of this paper was to review a predetermined number of organizations (10), journals (5), articles (ca. 20) and websites (15) representative of effort in this field in Asia, Africa, Australia and the Americas, and to a lesser degree in Europe.

Context for marine restoration

A brief indication of the nature of ecosystem degradation (including causes and effects) and restoration (practices and results) is given here. Some overall trends, definitions and efforts are noted. Coastal restoration is a growing endeavor, albeit limited. One example is the initiation of over 650 "community-based restoration projects" sites in the United States since 1996 under the auspices of the NOAA Restoration Center, established in the U.S. National Marine Fisheries Service, Office of Habitat Conservation (NOAA Restoration Center, 2004). This program relies heavily on local community participation, including volunteer efforts by citizens. Yet, as noted by Beck et al. (2003, p. 10), "our ability to restore ecosystems such as marshes and seagrass meadows is quite limited."

Definitions

Ecological Restoration is the "process of assisting the recovery of an ecosystem that has been degraded, damaged, or destroyed," according to the Society for Ecological Restoration (SER, 2004, p. 2). This source further identifies "deviations from the normal or desired state of an intact ecosystem" to include degradation, damage, destruction and transformation, recognized as overlapping and sometimes unclear terms that describe the degree of alteration. Degradation is defined as pertaining to "subtle or gradual changes that reduce ecological integrity and health." Recovery or restoration of an ecosystem is achieved when "it contains sufficient biotic and abiotic resources to continue its development without further assistance or subsidy" (SER, 2004, p. 3). Moreover, it is intentional, and "initiates or accelerates an ecological pathway through time towards a reference ecosystem or a target ecosystem condition" (CEM [Commission on Ecosystem Management, IUCN/The World Conservation Union, Gland, Switzerland], unpublished draft, Ecological Restoration).

According to the NOAA Restoration Center, restoration is "the process of reestablishing a self-sustaining habitat that closely resembles a natural condition in terms of structure and function...These habitats support fish and wildlife, and human uses such as swimming, diving, boating, and recreational and commercial fishing. Restoration usually does not focus on a single species but strives to replicate the original natural system to support numerous species. The goal is to expedite natural processes in rebuilding a healthy, functioning natural ecosystem that works like it did before it was polluted or destroyed."

Measures of restoration or recovery will determine, for example, progress in "returning a polluted or degraded environment as closely as possible to a successful, self-sustaining ecosystem with both clean water and healthy habitats" (NOAA Restoration Center). Thus, "An ecosystem is considered to be fully restored when it contains sufficient biotic and abiotic resources to sustain its structure, ecological processes and functions with minimal assistance or subsidy. It will demonstrate resilience to normal ranges of environmental stress and disturbance. It will interact with contiguous ecosystems in terms of biotic and abiotic flows and social and economic interactions. It will support, as appropriate, local social and economic activities. Such a state, however, is rarely achieved, even in the long-run. Nevertheless, significant environmental and social benefits can be realized even in the earliest stages of restoration" (CEM).

In the preceding definitions, both the terms ecosystem and habitat are used. "Habitat refers to the dwelling place of an organism or community that provides the requisite conditions for its life processes" (SER, 2004, p. 4). In turn, habitat is part of the ecosystem, defined by the Ecological Society of America (Beck et al., 2003, p. 3) as "characteristic assemblages of plants and animals and the physical environment they inhabit (e.g., marshes or oyster reefs). The term habitat refers to the area used by a species." This definition extends to include modifiers that identify "particular habitats used by an animal. For example, the blue crab...has a seagrass habitat and a marsh habitat...portions of seagrass and marsh ecosystems, respectively."

Extent

Habitat and ecosystem degradation is documented from the local to the global level. Among terrestrial organisms butterfly and songbird life cycles in North America are threatened by destruction of critical habitat along migration routes (Nature Conservancy; website: http://nature.org). Two hundred years ago the "Corps of Discovery" led by the so-called Lewis and Clark Expedition of 1804–1805 encountered herds of bison (buffalo) that were part of a species of 70,000,000 individuals. By 1881 the number of bison was reduced to 350, and in modern times it has been restored to 325,000, a small fraction of the original abundance.

"The state of the world's fisheries is poor, and continues to degenerate. 70% of commercially valuable fisheries have collapsed or are over-fished and en route to collapse. The biggest threats to fishery health worldwide include: Pollution from land-based sources; habitat alteration and destruction; non-sustainable and destructive fishing techniques; global climate change. The deteriorating state of the world's fisheries has social, economic and ecological implications: commercial and artisanal fishing is a source of income and a way of life for coastal populations, seafood is an important source of food and protein for the global population and demand for it is rising, and the depletion in stocks of commercially targeted fish, as well as the depletion of marine species that are incidentally caught (by-catch) with targeted species, has altered and unbalanced the food web of the world's oceans. The consequences of this destabilization are ecologically complex and only beginning to be understood." (International Oceanographic Commission; website: http://ioc.unesco.org/iocweb/ecosystems.php).

In marine systems, all major environments have been affected. Causes of degradation include land-based wastes including nutrients and toxic chemicals. Impacts include "dead zones" such as an area of as large as 20,000 square kilometers (Turner et al., 2004) in the Gulf of Mexico (an area about half the size of Switzerland). The United Nations Atlas of Fisheries states that "one of the greatest long-term threats to the viability of commercial and recreational fisheries is the continuing loss of marine, estuarine and other aquatic habitats." Further, "pathogens, toxic waste and toxins from Harmful Aquatic Blooms (HABs) have a major impact on fisheries, not only from the impact on human health, but from the closure of fisheries in contaminated areas. The losses in areas which are permanently polluted is hard to measure, however during periods of HABs, the economic losses have been calculated for a number of locations. A red tide in Hong Kong in 1998 caused losses of US$32 million from the closure of fish farms, whilst an algal bloom in Korea in 1991–1992 was estimated to cost US$133 million. Solid waste also has an impact on fisheries. The constant trawling and dredging operations have significant impact on the sea floor. It is similar to farming in that areas are cleared of rocks and obstacles, the terrain is leveled, and each succeeding year gear passes easier over the bottom. At the same time though, just as fields of wheat replace forests, trawlable bottom replaces coral heads and rock piles. The ecology of plants and animals is greatly changed." (United Nations Atlas of the Oceans; website: http://www.oceansatlas.org.)

Responses

The rationales for restoration include maintaining food supply, maintaining biodiversity, protecting nature, protecting human health, creating jobs and preserving ways of life (NOAA Restoration Center). Indigenous peoples and industrialized nations alike have responded to the growing loss of marine habitat function and structure. As reported by A. Vincent at the 2004 World Fisheries Congress, residents of artisanal fishing communities in the Philippines are keenly aware of options for fisheries restoration and adopting sustainable conservation practices concerning seahorses (personal communication; website: http://www.projectseahorse.org/). Similarly, in the United States local initiative has been responsible for coastal restoration nationwide, in part facilitated by programs such as the NOAA Restoration Center which seeks to study ecosystem structure, function and recovery, and develop restoration methods, success criteria and monitoring practices.

Efforts to develop and exchange information on a global basis include the development of restoration workshops and guidelines by the Commission on Ecosystem Management (CEM) of IUCN/The World Conservation Union. The mission of the Society for Ecological Restoration (SER) is "to promote ecological restoration as a means of sustaining the diversity of life on Earth and reestablishing an ecologically healthy relationship between nature and culture." It has published a primer on the subject, produces two journals and conducts an annual conference, for which it creates a lasting record on its website. Its membership of 2,300 is engaged in committees and working groups.

Principles and performance

Ecosystem restoration is young. Science-based policies and guidelines for effective restoration practices, including planning and evaluation, have been formulated in recent years. Among the international and national policies and laws concerning ecosystem restoration, the European Environment Agency (EEA) Strategy for 2004–2008 offers one approach to establishing a foundation for restoration and recovery of ecosystems. As part of its sixth environment action programme, the EEA lists as priorities halting biodiversity loss, assessment of marine ecosystem health and support for implementation of the EU marine strategy (EEA, 2003).

In North America, a partnership between the Estuarine Research Federation and Restore America's Estuaries produced a document entitled "Principles of Estuarine Habitat Restoration" (Waters, 1999) organized into four categories: (1) Context (four principles—preservation, stewardship, increasing scale, public participation); (2) Planning (two principles—ecosystem perspective, stakeholders/science); (3) Design (four principles—goals long-term and measurable, success criteria linked to reference habitats, impacts, monitoring); (4) Implementation (four principles—ecological engineering, adaptive management, protection, public access).

Among the science-based practices promulgated by organizations mentioned in this paper, and others, Clewell et al. (2000) identify and briefly describe 51 guidelines for ecological restoration, which address planning, organization, implementation and evaluation phases (Table 1). The extensive list of tasks would be especially useful to organizations just beginning to conceptualize restoration efforts in marine ecosystems.

The absolute need for establishment of measurable objectives in ecosystem restoration is emphasized by numerous authors and organizations. For example, two questions posed by Proffitt (2004) are: "What are appropriate time frames and measures for evaluating success? What do we establish as target restoration conditions?" Beck et al. (2003) note a consistent lack of effort to monitor restoration in nearshore ecosystems, thereby compromising efforts to gauge success or failure. Moreover, these authors encourage comprehensive evaluation to document returns of species, communities, and ecological functions.

Finally, a context for habitat restoration may be framed by asking: "Restoration to what?" It is essential that a "baseline" condition be defined by ecosystem, fishery and habitat scientists and managers, and other informed stakeholders, as a guide for design of an environmental restoration project and for its evaluation. This definition should be established prior to implementation. "Natural" conditions presumed for a given coastal system likely represent a "shifted" baseline given the ubiquity of historical overexploitation (Jackson, 2001). An example derives from what is possibly the largest restoration ever attempted in the world, the Greater Everglades Ecosystem Restoration, with costs estimated at U.S. $8 billion (USACOE, 1999) over 20 years. This program aims to return estuaries in South Florida, USA to earlier conditions by restoring natural quantities, qualities, timing, and distribution of freshwater inputs altered by drainage and flood control systems implemented over the past 50 years. Efforts include filling straightened channels of a river and restoring its original meandering course. However, elevation changes over the landscape due to soil oxidation and subsidence in drained wetlands make the historical condition of freshwater flow impossible to restore. Also, a substantial portion of the system is now in urban development.

Table 1 Excerpts[a]: Society for Ecological Restoration International "Guidelines for Ecological Restoration." (Note: Most annotations for the 51 steps listed are not reproduced here.)

The mission of every ecological restoration project is to reestablish a functional ecosystem of a designated type that contains sufficient biodiversity to continue its maturation by natural processes and to evolve over longer time spans in response to changing environmental conditions.

(A). CONCEPTUAL PLANNING (Reasons why restoration is needed, general strategy for conducting it.)

1. Identify the project site location and its boundaries.
2. Identify ownership.
3. Identify the need for restoration. (Tell what happened at the site that warrants restoration. State the intended benefits of restoration.)
4. Identify the kind of ecosystem to be restored and the type of restoration project.
5. Identify restoration goals, if any, that pertain to social and cultural values.
6. Identify physical site conditions in need of repair.
7. Identify stressors in need of regulation or re-initiation.
8. Identify biotic interventions that are needed.
9. Identify landscape restrictions, present and future.
10. Identify project-funding sources.
11. Identify labor sources and equipment needs.
12. Identify biotic resource needs.
13. Identify the need for securing permits required by government agencies.
14. Identify permit specifications, deed restrictions, and other legal constraints.
15. Identify project duration.
16. Identify strategies for long-term protection and management.

(B). PRELIMINARY TASKS (These tasks form the foundation for well-conceived restoration designs and programs.)

17. Appoint a restoration ecologist.
18. Appoint the restoration team.
19. Prepare a budget to accommodate the completion of preliminary tasks.
20. Document existing project site conditions and describe the biota. (Project evaluation depends in part upon being able to contrast the project site before and after restoration.)
21. Document the project site history that led to the need for restoration.
22. Conduct pre-project monitoring as needed. (Obtain baseline measurements.)
23. Gather baseline ecological information and conceptualize a reference ecosystem from it upon which the restoration will be modeled and evaluated.
24. Gather pertinent autecological information for key species.
25. Conduct investigations as needed to assess the effectiveness of restoration methods.
26. Decide if ecosystem goals are realistic or if they need modification.
27. Prepare a list of objectives designed to achieve restoration goals. (Objectives are the specific activities to be undertaken for the satisfaction of proper goals. Objectives are explicit, measurable, and have a designated time element.)
28. Secure permits required by regulatory and zoning authorities.
29. Establish liaison with other interested governmental agencies.
30. Establish liaison with the public and publicize the project.
31. Arrange for public participation in project planning and implementation.
32. Install roads and other infrastructure needed to facilitate project implementation.
33. Engage and train personnel who will supervise and conduct project installation tasks.

(C). INSTALLATION PLANNING (The care and thoroughness with which installation planning is conducted will be reflected by how aptly project objectives are realized.)

34. Describe the interventions that will be implemented to attain each objective.
35. State how much of the restoration can be accomplished passively.
36. Prepare performance standards and monitoring protocols to measure the attainment of each objective. (A performance standard [also called a design criterion] provides evidence on whether or not an objective has been attained. This evidence is gathered by monitoring. It is essential that performance standards and monitoring protocols be selected prior to any project installation activity.)
37. Schedule the tasks needed to fulfill each objective.
38. Procure equipment, supplies, and biotic resources.
39. Prepare a budget for installation tasks, maintenance events, and contingencies.

(D). INSTALLATION TASKS

40 Mark boundaries and secure the project area.
41. Install monitoring features.
42. Implement restoration of objectives. (Restoration tasks identified in Guideline #34.)

Table 1 continued

(E). POST-INSTALLATION TASKS
43. Protect the project site against vandals and herbivory.
44. Perform post-implementation maintenance.
45. Reconnoiter the project site regularly to identify needs for mid-course corrections.
46. Perform monitoring as required to document the attainment of performance standards.
47. Implement adaptive management procedures as needed.
(F). EVALUATION
48. Assess monitoring data to determine if performance standards are being met.
49. Describe aspects of the restored ecosystem that are not covered by monitoring data.
50. Determine if project goals were met, including those for social and cultural values. (Based on monitoring data and other documentation [Guidelines #46, #49], evaluate the restoration with respect to its project goals. These will include the primary goal to restore a functional ecosystem that emulates the reference ecosystem at a comparable ecological age [Guideline #4]).
51. Publish an account of the restoration project and otherwise publicize it. Publicity and documentation should be incorporated into every restoration project for the following reasons: Published accountings are fundamental for instituting the long-term protection and stewardship of a completed project site. Policy makers and the public need to be apprised of the fiscal and resource costs, so that future restoration projects can be planned and budgeted appropriately. Restoration ecologists improve their craft by becoming familiar with how restoration objectives were accomplished.

[a] Source: Clewell et al. (2000)

Artificial habitats in marine restoration

Over centuries the role of artificial habitats in aquatic environments has expanded from a relatively simple set of procedures applied at a small-scale and using natural materials designed to enhance success of local fishing harvest, to a more involved technology used more broadly in environmental management. From documented origins in Japan, these practices are employed in scores of nations in temperate and tropical areas. The historical goal of increased food production continues in both artisanal fisheries (e.g., India) and commercial fisheries (e.g., Taiwan) settings. In a more controlled situation, artificial reefs are used as a physical basis for aquaculture (e.g., Italy), which in its more complicated aspects is known as marine ranching due to the use of complementary manipulations such as introduction of hatchery-reared fingerlings to augment recruitment (e.g., Korea). In the last 10 years, another historical goal—enhancement of recreational fishing (e.g., Australia)—has been augmented by use of artificial habitats to promote recreational diving (e.g., Canada) and eco-tourism (e.g., Bahamas) and conservation of biodiversity (e.g., Monaco). It is likely that ecosystem restoration is the newest and least widespread application for artificial reefs.

The definition of an artificial reef as "a submerged structure placed on the seafloor deliberately, to mimic some characteristics of a natural reef" appears in the OSPAR Guidelines on Artificial Reefs in Relation to Living Marine Resources (OSPAR, 1999), having been adopted from the definition of the European Artificial Reef Research Network. This material derives from the Convention for the Protection of the Marine Environment of the North-East Atlantic, presented at the Ministerial Meeting of the Oslo and Paris Commissions in 1992. Among a diversity of applied and conservation purposes for artificial reefs, restoration of marine areas (including "regeneration of marine habitats") is identified. Considerations in the OSPAR Guidelines include: definition and purpose for artificial reefs; justification and impacts; materials, design and placement; monitoring to verify fulfillment of objectives and degree of benefit; and the role of pilot studies and experiments.

Principally due to a series of international and more recent regional and national scientific conferences on artificial habitats, a body of technical literature has developed for this field in the last 20 years. This material includes description of the uses of reefs noted above. The larger international meetings attracted as many as 350 individuals; programs in 1983, 1987, 1989, 1995 and 1999 resulted in five published volumes. Regional and national meetings, meanwhile, have occurred in Europe (Jensen et al., 2000), Korea, Canada, Brazil and elsewhere. Of 56 papers in the *ICES Journal of Marine Science* issue devoted to a proceedings of the Seventh International Conference on Artificial Reefs and Related Aquatic Habitats, 29 were produced by European authors (ICES, 2002). Over the approximate 20-year history of the development of this research literature, an initial body of descriptive work has been augmented by results from experimental and hypothesis-driven research designed for understanding and predicting ecological behavior of reefs.

The following four case studies are presented as a guide to current and emerging considerations for habitat restoration. The practices addressed in this paper involve purposeful placement of either human-made or natural materials in a benthic marine ecosystem, generally on the coastal shelf or in an estuary, with a goal of modifying ecological structure and function. These examples from Atlantic, Indian and Pacific Ocean biogeographic regions address restoration of plant habitats, coral reefs, bivalve mollusk systems and fisheries stocks. Restoration is just one tool available for the response to system degradation. Clearly, reduction of pollution, limits to fishing, regulation of coastal development, and dealing with both human population growth and consumption of natural resources all must be considered for application to aspects of the situations considered in this paper.

Kelp bed mitigation and restoration

A large kelp (*Macrocystis pyrifera* Linnaeus) bed is being created as mitigation for habitats destroyed in the coastal Pacific Ocean by the operations of the San Onofre Nuclear Generating Station in southern California, USA (Reed et al., 2002). The owners of the electrical power plant were mandated to create 61 ha (150 acres) of new kelp bed habitat. Placement of artificial reefs is part of this project (Fig. 1). Work began with a moderate sized 8.9 ha (22 acres) pilot project costing U.S. $4 to 6 million, to gain assurance that an appropriate design for the full-scale reef would yield the habitat characteristics and functions legally required by the mitigation permit.

Fig. 1 Deployment of material for kelp mitigation reefs is typical of worldwide practices that use barges for transportation and staging. (Photograph from Southern California Edison.)

Experimental reefs with different substrate characteristics (quarry rock vs. concrete and different coverage of hard substrate vs. sand) are undergoing extensive evaluation to determine the degree of habitat improvement for fishes and benthic communities provided. Recruitment and growth of giant kelp, or survival and growth of transplants, onto the artificial reef structure is a principal concern. Certain performance standards are in terms of an absolute historic baseline: the total amount of kelp that was lost, ultimately 61 ha, at a density of 4 adult plants/100 m^2 (Reed, 2002). Others are stated relative to current status of other similar habitats in the area. For example, fish assemblage, recruitment, and production should be "similar to natural reefs in the region" (Reed et al., 2002). Initial monitoring indicates that both kelp density, and fish recruitment as measured by young-of-year juvenile fish density, compare favorably with natural reference reefs. Deysher et al. (2002) conclude that a low relief structure with moderate sand cover between reefs has the best chance for success.

Coral reefs in site-specific situations

It is estimated that shallow coral reefs worldwide occupy some 284,300 square kilometers, less than 1.2% of the world's continental shelf area (Spalding et al., 2001). Indonesia possesses the largest amount of coral reef, followed by Australia and the Philippines. Reefs worldwide are degraded by over-fishing, coastal development, the introduction of sewage, fertilizer and sediment, and more recently by tsunami, with an estimated 27% lost.

The 1999 International Conference on Scientific Aspects of Coral Reef Assessment, Monitoring, and Restoration convened scientists and stakeholders from all oceans with 13 of 51 published papers addressing restoration, a field in its infancy but rooted in growing scientific understanding of coral reproduction, recruitment and physiology (National Coral Reef Institute, 2001). According to the United States Coral Reef Task Force (2000) the majority of experience for restoration is based on repair of vessel grounding sites, through creation of habitat and transplantation.

Coral reef damage results from ship groundings and fishing gear damage. In the Experimental Oculina Research Reserve in the Atlantic off Florida, USA the ivory tree coral, *Oculina*, was degraded by commercial and recreational fishing. Extensive areas were reduced to rubble by trawling or dredging. Reef fish populations were low. Restoration was attempted with concrete modules, cement blocks (Fig. 2) and PVC piping on first an experimental basis (1996–1999) and then more extensively (2000–2001) (Koenig, 2001). Live *Oculina varicosa* (Lesueur) colonies, approximately 15 cm (6") in diameter, and small *Oculina* fragments were attached to each reef ball with concrete and cable ties. Also, 450 patio stones with an *Oculina* fragment attached to the top of a 30 cm (1") PVC pipe were deployed. Koenig (2001) reported that high rates of coral transplant survival in pilot studies led to a larger restoration effort by the U.S. National Marine Fisheries Service, which led to an increase in numbers of individuals of groupers (Serranidae) and possible spawning and nursery functions for this habitat.

In 1993 a United States submarine ran aground on a coral reef off southeast Florida (Jaap, 2000). Physical damage to the reef substrate covered 2,310 m^2, with 1,205 m^2 totally destroyed. In 1997, the State of Florida was awarded a settlement of U.S. $750,000 by the Federal government for environmental damages caused by the submarine grounding. In experiments using artificial reefs, scientists are examining three restoration strategies: (1) enhancing coral recruitment through the use of coral larval attractants, (2) the effect of reef structure on fish assemblages, and (3) the interaction between fish assemblages and coral recruitment and survival (R. Dodge, Nova Southeastern University, personal communication). This is a good example of the advantages offered by artificial reefs for manipulation of ocean habitats for experimentation.

Oysters in ecosystem context

The Chesapeake Bay on the Atlantic coast of the United States represents a system with a drastically shifted baseline. Declines in oyster abundance and habitat and loss of seagrasses have

Fig. 2 Concrete building blocks are common in experimental manipulations of marine benthic habitat, such as in the study of restoring *Oculina* reefs. (Photograph from NOAA Restoration Center Image Catalog, image r0022703, U.S. National Oceanic and Atmospheric Administration [NOAA].)

been co-incident with increased turbidity, eutrophication and anoxia. As early as 1881, oyster (*Crassostrea virginica* Gmelin) beds had reduced structure, increased amounts of sand and mud, and were composed of 97% broken shell and debris as compared with 30% for unfished beds (Wilson, 1881, cited in Kennedy & Breisch, 1981). Oyster harvest in Chesapeake Bay has declined from its peak in 1874 (14 million bushels, about 50 million ton) to less than half a million bushels (1.8 million ton) in recent years. The current stock is approximately 2% of the historic baseline.

Because oysters are filter feeders, there has come to be wider acceptance that their loss may have been a factor in the decline of water quality (Coen & Luckenbach, 2000), and that restoration of habitat may depend on maintaining a certain biomass of filter feeders in the system. Traditional oyster habitat restoration approaches focused narrowly on providing low artificial shell reefs to attain fisheries goals, i.e., increasing harvestable oysters, but did not meet with success (Lenihan, 1999; Coen & Luckenbach, 2000).

A more recent scientific consensus (e.g., Coen & Luckenbach, 2000) is emerging that, for artificial oyster reefs to constitute effective habitat improvement, they need to be (1) taller in order to provide more structurally complex habitat and a potential refuge from bottom anoxia events and (2) protected from harvesting in order to provide for persistence of the reef structure and the maintenance of sufficient oyster biomass both for filtering and for reproductive capacity (Coen & Luckenbach, 2000). Models have predicted that maintaining an average oyster biomass of 25 g/m^2 would reduce turbidity by an order of magnitude. This would greatly increase the amount of light reaching the bottom and thereby expand the suitable area for seagrasses habitat (Newell et al., 2003). A small number of limited pilot restoration efforts are beginning.

Fisheries populations

The fourth case study of structural/physical responses to habitat and fishery degradation includes the most emphasis on simulation modeling of ecosystems, fishing and policy, and is the newest. In Hong Kong, China, high trawling effort during the last quarter of the 20th century produced declining catch, high fishing mortality, greater relative capture of low-value short-lived species, and virtual elimination of longer lived demersal species of higher value (Pitcher et al., 2002). After a peak fishery harvest of over 240,000 tons in 1989, catch in 1998 was under 145,000 tons (Wilson et al., 2002). In response a multi-faceted approach including fishing licenses, protected areas, and restoration and enhancement of habitats was proposed; a five-year Artificial Reef Programme started in 1996, funded at U.S. $13,000,000 (Wilson et al., 2002). These latter authors described preliminary results including juvenile fish recruitment for species of Sparidae and Lutjanidae, residence of adult Serranidae, and increased catch of small-scale fisheries for bream (Sparidae).

Habitat restoration structures included deployed vessels (including along park boundaries to prevent trawling), rock, tire units and concrete units (28,000 m^3 total) (Fig. 3) in two marine parks. This was according to a voluntary no-fishing arrangement made possible by placement of additional artificial reefs for fishing in open mud areas. An area of 10% of Hong Kong waters has been set aside as a "Fisheries Protection Area." According to predictions by Pitcher et al. (2001) the value of the fishery would increase by over 50% if 10–20% of waters were managed on a no-take basis.

Fig. 3 The trend in construction of artificial reefs is for increasing use of designed modules, such as this concrete structure deployed in Hong Kong. Dimensions are 4.0 × 4.4 × 1.6 m^3. (Photograph from Agriculture, Fisheries and Conservation Department of the Government of Hong Kong Special Administrative Region.)

The responses to this artificial reef-based fishery restoration project were forecast by Pitcher et al. (2002), using three ecosystem and resource models. Information from a variety of local databases and consultations allowed these authors to incorporate (1) diet, growth and mortality data for 255 reef-associated and non-reef fish species, sorted by size, and collected into 27 functional groups, and (2) descriptions of seven sectors of the Hong Kong fishery into "Ecosim" and "Ecopath" models. These provided the basis for dynamic "Ecospace" simulations to predict fishery performance according to fishery sector and habitat. An actual increase of harvest of large reef fish is forecast when artificial reefs are deployed, in contrast to a non-reefs scenario that depicted continuing depletion of the fishery and increase of lower-trophic level organisms. In one situation, the authors forecast a total catch of reef fish of 100 tons per year, including 60 tons of large demersal reef fish.

This early application of ecosystem simulation to artificial reef performance in a coastal fishery/habitat restoration situation has advantages including the capabilities for analysis of trade-offs among marine protected area and reef deployment design practices and for comparison of policy options (Pitcher et al., 2002). Potential concerns include levels of confidence and uncertainty.

Trends in success of habitat restoration

The preceding four case studies were selected because in aggregate they possess attributes useful to planning other marine ecosystem restoration efforts. In contrast with many typical artificial reef deployments that have relatively small areal "footprints," such as individual ships or "patch reefs" of concrete modules, three of the studies are being implemented on a relatively larger scale, from a 61-ha site in California to regional marine parks in Hong Kong to virtually the entire Chesapeake Bay. Comparison of the preceding project summaries with the 51 steps given in Table 1 indicates the thoroughness of planning and execution of the projects, such as in identifying need for restoration, gathering ecological information, monitoring, etc.

Selected attributes suggested as desirable for marine ecosystem restoration are summarized in Table 2. Each situation includes the measurable objectives necessary to successful implementation of aquatic ecosystem restoration. Both the Chesapeake Bay and San Onofre efforts specify units of oyster biomass (25 g/m^2) and plant density (4/

Table 2 Components for marine ecosystem restoration, as addressed in four situations using artificial reefs

Component of restoration	System			
	Kelp forests	Coral reefs	Oyster reefs	Reef fisheries
Goal/ performance measure	4 plants/100 m^2; monitoring in progress	Increased coral biomass/ structure; monitoring	Oyster biomass = 25 g/m^2; monitoring in progress	Increased fishery yield; monitoring in progress
Ecosystem context	Adjacent natural reefs as reference target and source of recruits	Site-specific	Oysters as critical component of ecosystem to enhance water quality; opportunity for recovery of other habitats (e.g., seagrass)	Considers adjacent natural reefs and open mud and sand
Ecological basis for design	Height, spacing of reefs; predators	Species suited to sites	Physical structure; anoxia events	Species diet, growth
One tool of many used	Kelp transplantation being evaluated	Not considered	Coupled to watershed management	Coupled to management of fishing effort
Advanced techniques	Experimental pilot study to ensure design most likely to attain targets	Compatible substrates for transplants; test hypotheses	Modeling to predict ecosystem benefits; water quality—seagrass linkages	Modeling forecasts of fishery response

100 m^2), respectively, which in fact are derived from historical baseline values. The Hong Kong program is more general in seeking increased fishery catch, and the coral reef situations were smaller but focused in increasing structure. In all cases monitoring to acquire data for measurement of performance is in place. Further, the ecology of organisms has been used to direct design of reef structures, such as in defining height of kelp reefs to minimize both scour by currents and grazing by herbivores.

In three cases, reefs are being used in two broader contexts. As a fishery management tool, for example, they are coupled with new fishing license measures in Hong Kong. In a broader ecosystem context, management of nutrients from the Chesapeake Bay watershed along with oyster reef restoration and protection to enhance filtering are expected to improve water quality and increase opportunity for seagrass bed recovery. The southern California kelp bed project is explicitly quantifying recruitment of kelp, other benthic species, and fishes in a spatially explicit way, cognizant of the importance of the mosaic of surrounding habitats for reference and as a source of recruits. Finally, the use of pilot studies to test reef designs (kelp), ecological modeling to predict reef function (oysters, Hong Kong), and testing of hypotheses (coral) represent effective steps in maximizing success of the projects through rigorous scientific study design.

In conclusion, the technology for marine ecosystem restoration and application of artificial reefs to it are young. As indicated by the case studies above, there is a valid role for artificial reefs in marine ecosystem restorations. Even before reefs can be used in a restoration setting, though, the nature and extent of degradation must first be established, particularly in terms of characterizing and quantifying the pre-existing (baseline) condition. Other solutions that may be more appropriate must be evaluated, ranging from control of pollution or land-use practices to management of the ecosystem. In defining the utility of artificial habitats in restoration, the question of habitat-limitation in the ecosystem must be considered. As discussed by Frid and Clark (1999), scientific knowledge of ecosystems must be integrated with economic and social forces impacting the environment.

Acknowledgements Information about habitat restoration for oysters and kelp was developed for the 2004 World Fisheries Congress in cooperation with M. Miller of U.S. NOAA Fisheries. J. Whitehouse, Florida Sea Grant, University of Florida, typed the manuscript, prepared under the auspices of NOAA Grant NA16RG2195 to the Florida Sea Grant College Program.

References

Beck, M. W., K. L. Heck, Jr., K. W. Able, D. L. Childers, D. B. Eggleston, B. M. Gillanders, B. S. Halpern, C. G. Hays, K. Hoshino, T. J. Minello, R. J. Orth, P. F. Sheridan & M. P. Weinstien, 2003. The role of nearshore ecosystems as fish and shellfish nurseries. Issues in Ecology. Number 11. Ecological Society of America: 12 pp.

Clewell, A. F., J. Rieger & J. Munro, 2000. Guidelines for developing and managing ecological restoration projects. Publications Working Group. Society for Ecological Restoration: 11 pp.

Coen, L. D. & M. W. Luckenbach, 2000. Developing success criteria and goals for evaluating oyster reef restoration: Ecological function or resource exploitation? Ecological Engineering 15: 323–343.

Deysher, L. E., T. A. Dean, R. S. Grove & A. Jahn, 2002. Design considerations for an artificial reef to grow giant kelp (*Macrocystis pyrifera*) in Southern California. ICES Journal of Marine Science. 59(Supplement 1): S201–S207.

EEA (European Environment Agency). 2003. Europe's Environment: The Third Assessment. Office for Official Publications of the European Union, Luxembourg: 61 pp.

FAO (United Nations Food and Agriculture Organization). 2002. The State of World Fisheries and Aquaculture, 2002. FAO Information Division, Rome, Italy.

Frid, C. L. J. & S. Clark, 1999. Restoring aquatic ecosystems: An overview. Aquatic Conservation: Marine and Freshwater Ecosystems 9: 1–4.

ICES (International Council for the Exploration of the Sea), 2002. Seventh International Conference on Artificial Reefs and Related Aquatic Habitats. ICES Journal of Marine Science 59(Supplement): 362 pp.

Jaap, W. C., 2000. Coral reef restoration. Ecological Engineering 15: 345–364.

Jackson, J. B. C., 2001. What was natural in the coastal ocean? Proceedings of the National Academy of Sciences 98: 5411–5418.

Jensen, A. C., K. J. Collins & A. P. M. Lockwood (eds.), 2000. Artificial Reefs in European Seas. Kluwer Academic Publishers, Dordrecht, The Netherlands.

Kennedy, V. S. & L. L. Breisch, 1981. Maryland's Oysters: Research and Management. University of Maryland Sea Grant Publication UM-SG-TS-81-04. College Park, Maryland.

Koenig, C. C., 2001. *Oculina* banks: Habitat, fish populations, restoration and enforcement. Project Report. South Atlantic Fishery Management Council, Charleston, South Carolina.

Lenihan, H. S., 1999. Physical-biological coupling on oyster reefs: How habitat structure influences individual performance. Ecological Monographs 69: 251–275.

National Coral Reef Institute, 2001. Proceedings of the International Conference on Scientific Aspects of Coral Reef Assessment, Monitoring, and Restoration. Bulletin of Marine Science 69(2).

Newell, R. I. E., R. R. Hood, E. W. Koch & R. E. Grizzle, 2003. Modeling the effects of changes in turbidity on light available for submerged aquatic vegetation. Final Report. NOAA/UNH Cooperative Institute for Coastal and Estuarine Environmental Technology. University of New Hampshire, Durham.

NOAA Restoration Center, 2004. Restoring Coastal and Marine Habitats. U. S. National Oceanic and Atmospheric Administration, Fisheries Office of Habitat Conservation. Silver Spring, Maryland: 16 pp.

OSPAR Commission, 1999. OSPAR Guidelines on Artificial Reefs in Relation to Living Marine Resources. OSPAR Commission, London. OSPAR 99/15/1-E, Annex 6.

OSPAR Commission, 2000. OSPAR Quality Status Report 2000: Region II – Greater North Sea. OSPAR Commission, London: xiii + 136 pp.

Pauly, D., V. Christensen, J. Dalsgaard, R. Froese & F. Torres Jr., 1998. Fishing down marine food webs. Science 279: 860–863.

Pitcher, T. J., R. Watson, N. Haggan, S. Guenette, R. Kennish, R. Sumaila, D. Cook, K. Wilson & A. Leung, 2001. Marine reserves and the restoration of fisheries and marine ecosystems in the South China Sea. Bulletin of Marine Science 66(3): 543–566.

Pitcher, T. J., E. A. Buchary & T. Hutton, 2002. Forecasting the benefits of no-take human-made reefs using spatial ecosystem simulation. ICES Journal of Marine Science 59(Supplement): S17–S26.

Proffitt, E., 2004. Book review: Handbook of ecological restoration. Restoration Ecology 12(1): 143–144.

Reed, D., 2002. Giant kelp. In Reed, D., S. Schroeterand & M. Page (eds) Proceedings from the Second Annual Public Workshop for the SONGS Mitigation Project. Report to the California Coastal Commission. University of California, Santa Barbara. Marine Science Institute,: 62–85.

Reed, D., S. Schroeterand & M. Page (eds), 2002. Proceedings from the Second Annual Public Workshop for the SONGS Mitigation Project. Report to the California Coastal Commission. University of California, Santa Barbara. Marine Science Institute.

SER (Society for Ecological Restoration International Science & Policy Working Group), 2004. The SER International Primer on Ecological Restoration. Tucson, Arizona.

Spalding, M. D., C. Ravilious & E. P. Green, 2001. World Atlas of Coral Reefs. University of California Press, Berkeley,: 424 pp.

Turner, R. E., N. N. Rabalais, E. M. Swenson, M. Kasprzak & T. Romaire, 2004. Summer hypoxia in the northern Gulf of Mexico and its prediction from 1978 to 1995. Marine Environmental Research 59(1): 65–77.

USACOE (U.S. Army Corps of Engineers), 1999. Central and South Florida Comprehensive Review Study Final Feasibility Report and Programmatic Environmental Impact Statement. Restoration Program Office, West Palm Beach, Florida.

United States Coral Reef Task Force, 2000. The National Action Plan to Conserve Coral Reefs. Washington, D.C.

Waters, E., 1999. Principles of estuarine habitat restoration. Report on the RAE-ERF Partnership. Estuarine Research Foundation, Port Republic, Maryland,: 24 pp.

Wilson, K. D. P., A. W. Y. Leung & R. Kennish, 2002. Restoration of Hong Kong fisheries through deployment of artificial reefs in marine protected areas. ICES Journal of Marine Science 59(Supplement): S157–S163.

Hydrobiologia (2007) 580:157–171
DOI 10.1007/s10750-006-0456-x

BIODIVERSITY IN ENCLOSED SEAS

Fish assemblages on sunken vessels and natural reefs in southeast Florida, USA

P. T. Arena · L. K. B. Jordan · R. E. Spieler

Abstract Derelict ships are commonly deployed as artificial reefs in the United States, mainly for recreational fishers and divers. Despite their popularity, few studies have rigorously examined fish assemblages on these structures and compared them to natural reefs. Six vessel-reefs off the coast of southeast Florida were censused quarterly (two ships per month) to characterize their associated fish assemblages. SCUBA divers used a non-destructive point-count method to visually assess the fish assemblages over 13- and 12-month intervals (March 2000 to March 2001 and March 2002 to February 2003). During the same intervals, fish assemblages at neighboring natural reefs were also censused. A total of 114,252 fishes of 177 species was counted on natural and vessel-reefs combined. Mean fish abundance and biomass were significantly greater on vessel-reefs in comparison to surrounding natural reef areas. Haemulidae was the most abundant family on vessel-reefs, where it represented 46% of total fish abundance. The most abundant family on natural reefs was Labridae, where it accounted for 24% of total fish abundance. Mean species richness was significantly greater on vessel-reefs than neighboring natural reefs and also differed among vessel-reefs. Both mean fish abundance and mean species richness were not significantly different between natural reefs neighboring vessel-reefs and natural reefs with no artificial structures nearby. This suggests the vessel-reefs are not, in general, attracting fish away from neighboring natural reefs in our area. Additionally, economically important fish species seem to prefer vessel-reefs, as there was a greater abundance of these species on vessel-reefs than surrounding natural reef areas. Fish assemblage structure on natural versus artificial reefs exhibited a low similarity (25.8%). Although no one species was responsible for more than 6% of the total dissimilarity, fish assemblage trophic structure differed strikingly

Guest editors: G. Relini & J. Ryland
Biodiversity in Enclosed Seas and Artificial Marine Habitats

Electronic Supplementary Material Supplementary material is available for this article at http://dx.doi.org/10.1007/s10750-006-0456-x and accessible for authorised users

P. T. Arena · L. K. B. Jordan · R. E. Spieler
Oceanographic Center (NSUOC), Dania Beach, FL, USA

P. T. Arena · R. E. Spieler
Guy Harvey Research Institute (GHRI), Dania Beach, FL, USA

P. T. Arena · L. K. B. Jordan · R. E. Spieler
National Coral Reef Institute (NCRI), 8000 North Ocean Drive, Dania Beach, FL 33004, USA

P. T. Arena (✉)
Farquhar College of Arts and Sciences,
Nova Southeastern University, Davie, FL 33314, USA
e-mail: arenap@nova.edu

between the two reef types. Planktivores dominated on vessel-reefs, accounting for 54% of the total abundance. Conversely, planktivores only made up 27% of total abundance on natural reefs. The results of this study indicate vessel-reef fish assemblages are unique and that these fishes may be utilizing food resources and habitat characteristics not accessible from or found at natural reefs in our area. Production may also be occurring at vessel-reefs as the attraction of fish species from nearby natural reefs seems to be minimal.

Keywords Artificial reef · Vessel · Coral reef · Habitat

Introduction

The popularity of recreational fishing has risen dramatically in the past 50 years and with this increase in fishers has come additional pressure on global fish stocks, the majority of which, have been classified as either fully- or over-exploited (Murray & Betz, 1994; FAO, 1997a, b). For example, the state of Florida, USA has 13 560 km of coastline, more than 800,000 registered boats, and over one million registered recreational fishing licenses (FWRI, 2004; White, 2004). Given these numbers it is clear there is mounting pressure on state resource managers to protect and sustain coastal fisheries. A popular management option currently in use is the deployment of artificial reefs, as these structures are known to harbor high abundances of fishes.

Derelict vessels have been intentionally deployed to increase fishing success since 1935 (Stone, 1985) and support for their use has come from the fishing industry (recreational and commercial), tourist industry, diving community and environmental managers (Murray & Betz, 1994; Jones & Welsford, 1997; MacDonald et al., 1999). Murray & Betz (1994) reported all groups of respondents (commercial fishermen, recreational fishermen, sport divers, and environmentalists) from a mail survey preferred artificial reefs constructed from ships and barges. In particular, sport divers have shown a preference for vessel-reefs due to high densities of fishes at these sites and the aesthetic qualities offered by the structure itself (Brock, 1994; Murray & Betz, 1994; Jones & Welsford, 1997). The demand for vessel-reefs will undoubtedly escalate as a result of the growing sport diving industry, which reported 50,000 new worldwide diving certifications each year since the early 1980s (Gilmore, 2004).

The popularity of vessel-reefs has led to legitimate questions about their effectiveness as fisheries enhancement tools (Seamen & Jensen, 2000). An understanding of fish assemblage structure on vessel-reefs is required to determine if they are achieving the goals set forth by resource managers. Yet, there have been surprisingly few studies comparing vessel-reef fish assemblages to those on adjacent natural reefs (Jones & Thompson, 1978; Markevich, 1994) and few of these have been statistically rigorous.

Broward County, Florida has a wide diversity and abundance of vessel-reefs, as well as a substantial natural reef system, which afforded us an excellent opportunity to conduct comparative surveys of fish assemblages. Our objectives were to: (1) compare the fish assemblages on six vessel-reefs to adjacent natural reefs, and (2) compare the fish assemblages among the various vessel-reefs.

Materials and methods

Study site

Broward County's reef complex is approximately 1.5 km wide and is composed of three relic coral reef terraces, each separated by sand substrate, which run parallel to the coastline in sequentially deeper water (Goldberg, 1973; Lighty, 1977; Moyer et al., 2003). The three reef terraces have been locally named the inner, middle and outer reefs (Fig. 1). These high-latitude coral communities consist of typical Caribbean fauna, however benthic community structure is highly variable and cannot be characterized by existing reef classification or zonal schemes (Goreau, 1959). Moyer et al. (2003) suggested water quality, sedimentation, and/or hurricane recurrence might determine benthic community structure in this area.

During this study, the prevailing winds were typically 10–15 knots from the southeast. Wave

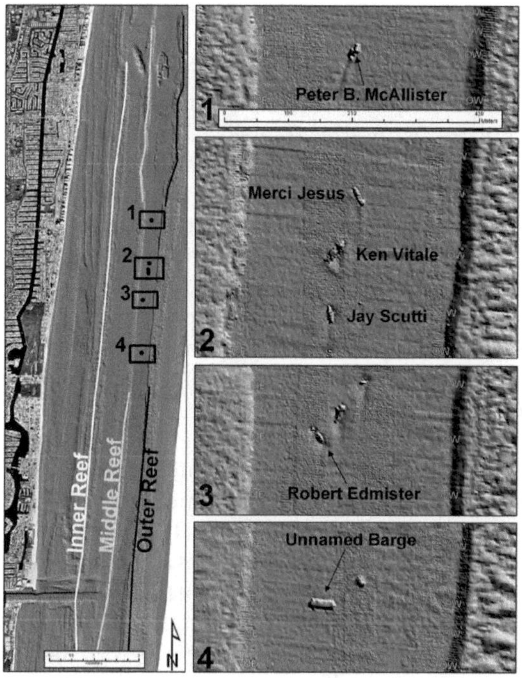

Fig. 1 Natural and vessel-reef study sites in southeast Florida, USA

action varied by season with the heaviest seas occurring during winter months (January–March). Fish surveys were conducted in currents that ranged from calm to a maximum speed estimated at 2.0 knots. Current direction usually flowed North due to the influence of the Florida Current, but the authors (unpublished), as well as Soloviev et al. (2003) have observed a reversal of this trend during late summer. Water visibility was never below a measured 7.5 m and estimated to be 10–15 m on average (maximum estimated at 33 m).

The six vessel-reefs used in this study were intentionally deployed in the sand flat that separates the middle and outer reefs at approximately 20–25 m water depth (Fig. 1). The width of this sandy substrate and subsequent distance between middle and outer reef varies from 325 m to 350 m. The six vessel-reefs varied in size, vertical relief, horizontal orientation, vessel type, deployment date, and proximity to middle and outer reef terraces (Table 1). All vessel-reefs were approximately 1.80 km from shore.

During a 13-month (March 2000 to March 2001) and 12-month period (March 2002 to February 2003) SCUBA divers used a non-destructive visual census method, commonly called a point-count, to determine species richness and abundance at vessel-reefs and nearby natural reefs (Bohnsack & Bannerot, 1986). Each vessel-reef was censused at least four times during the year, two vessel-reefs per month. The census of the adjacent natural reef occurred at irregular time intervals throughout the first 13 months but concurrent with vessel-reef censuses during the second year of the study.

Table 1 Location and physical characterization of vessel-reefs

	Unnamed Barge	*Edmister*	*Scutti*	*Tracy/Vitale*	*Merci Jesus*	*McAllister*
Latitude	26 08.520 N	26 09.193 N	26 09.520 N	26 09.573 N	26 09.635 N	26 10.185 N
Longitude	80 04.886 W	80 04.837 W	80 04.777 W	80 04.754 W	80 04.747 W	80 04.707 W
Vessel type	Barge	USCG Cutter	Tugboat	Freighter	Freighter	Tugboat
Deployment date (month/year)	c1970	12/89	09/86	03/99	08/98	06/98
Depth (m)	21.3	21.3	19.5	19.5	19.5	21.0
Distance to outer reef terrace (m)	208	223	212	185	160	193
Distance to middle reef terrace (m)	135	102	142	162	180	180
Length (m)	24	28.5[a]	29	40	27	25.5
Maximum vertical relief (m)	3.0	3.0	9.0	8.1	5.4	6.9
Estimated volume (m^3)	706	588	1208	861	556	623
Orientation	E/W[b]	N/S[b]	N/S[b]	NE/SW[b]	NW/SE[b]	NW/SE[b]

[a] Length at time of deployment

[b] Direction of bow

The census methodology used was a point-count of fishes in an imaginary 15 m diameter cylinder, extending from the substrate to the surface, providing a 176.63 m^2 footprint. The published methodology has the diver remaining in the center of the cylinder during the census (Bohnsack & Bannerot, 1986). Due to the extensive topographical relief associated with vessel-reefs, we modified this aspect of the methodology to allow the diver to swim freely within the cylinder during the census.

The diver recorded all species seen during a five-minute period. After the five-minute species count was completed, the abundance of each fish species and the minimum, maximum and mean total length were recorded to the nearest cm. A 7.5 m radius line was laid out prior to the count as an aid in estimating the cylinder boundary and the diver used a 1-m rod with a ruler attached at one end in a T-configuration to aid in length estimation.

The bow, stern, port and starboard sides were censused on five of the six vessel-reefs to obtain a mean estimate of the ship's fish assemblage per count. Two additional mid-ship counts were performed at the sixth vessel-reef (*Edmister*) due to its high complexity and extensive footprint. A total of 218 point-counts were made on vessel-reefs over the study period.

A concurrent study, also using the point-count method, counted fishes on the natural reefs of Broward County. This concurrent study inventoried the fishes on East–West running transects every 463 m along the coastline of Broward County. On each transect a point-count was made at the eastern and western edges, as well as the crest of each reef terrace (for details on methodology see Ferro et al., 2005). Ten transects were made in the vicinity of the vessel-reefs censused in this study from 2000 to 2001 (Table 1). Therefore, we have also included data from the edges of the reef terraces that border the vessel-reefs. Specifically, 10 point-counts on the eastern edge of the middle reef and 10 counts from the western edge of the outer reef are included, for a total of 32 natural reef counts during the first year (Fig. 1). During the second year of this study we performed an additional 29 counts at neighboring natural reefs. Only edge data nearest the vessel-reefs were included because of their close proximity and the fact that the edges have the most complex habitat and hold the most species and total fish of reef tract sites (Ferro et al., 2005). The assumption is, if adult fishes are moving between natural and artificial reefs, or being aggregated from natural to artificial reefs, they will most likely come from neighboring sites. Also, comparing neighboring reef areas of high topographical relief and large numbers of fishes to vessel-reefs also showing these characteristics is probably a more realistic comparison than those incorporating low relief hardbottom.

Data analysis

Prior to analysis, the estimated biomass and trophic preference of each species was determined (Froese & Pauly, 2004). Total length (TL) estimates allowed for post-census calculation of biomass using length-weight equations (Bohnsack & Harper, 1988). Fishes were classified according to their predominant trophic ecology as follows: planktivores, herbivores, piscivores, benthic carnivores, and omnivores (see Electronic Supplementary Material). The tomtate, *Haemulon aurolineatum* (Cuvier), is generally a nocturnal benthic carnivore as an adult, but both juveniles and adults commonly feed diurnally on vessel-reef planktonic prey items (personal observation). We have characterized the trophic ecology of this species as a benthic carnivore for analysis.

Fish abundance, biomass, species richness and trophic preference were examined using a mixed model analysis of variance (ANOVA) technique and a post-hoc Tukey–Kramer (TK) comparison of means per count using SAS V9.1 software (SAS Institute Inc., Cary, NC, USA). A probability value of less than 0.05 in both ANOVA and TK was accepted as a significant difference. The data that were not normally distributed and had high heteroscedasticity (i.e. abundance and biomass) were log-transformed [$\log_{10}(x + 1)$] prior to analysis (Zar, 1996)

An MDS using Bray–Curtis dissimilarity indices, an examination of similarity percentages of particular species (SIMPER) and analysis of similarity (ANOSIM) were used to examine potential differences in fish assemblage structure among sites (Field et al., 1982) using the Plymouth

Routines in Multivariate Ecological Research statistical package (PRIMER v5).

Results

Abundance

A total of 114,252 fishes was counted on natural and vessel-reefs combined (59,467 during the first sample period, 54,785 during the second). There was no statistical difference in abundance between the two sample periods, so they were pooled for subsequent analyses. With all vessel-reefs combined there were no statistical differences among months. Likewise, no differences were detected among months for natural reefs when the two edges were combined ($P > 0.05$, ANOVA). There was significantly greater mean fish abundance per count on vessel-reefs than natural reefs combined (Table 2) and the east edge of the middle terrace (Mean ± SE, 154.55 ± 22.46) was significantly greater than the west edge of the outer terrace (82.40 ± 6.30) ($P < 0.05$, ANOVA). No differences were found when comparing the abundance of individual vessel-reefs.

The abundance of individual species observed on natural and vessel-reefs is presented in Electronic Supplementary Material. The 10 most abundant species represented 79% of the total fish abundance on vessel-reefs and 64% on natural reefs. The most abundant species on all vessel-reefs combined was *H. aurolineatum*, which made up 53% of the total haemulid abundance. This species was the most abundant species on the *Edmister*, *Tracy/Vitale*, and *Scutti* (excluding *Haemulon* juveniles). The most abundant species on the *McAllister*, *Merci Jesus* and the *Unnamed Barge* were the round scad, *Decapterus punctatus* (Cuvier), bluehead wrasse, *Thalassoma bifasciatum* (Bloch), and masked goby, *Coryphopterus personatus* (Gill), respectively. The most abundant species on all natural reefs combined was the bicolor damselfish, *Stegastes partitus* (Poey).

The most abundant family on all vessel-reefs combined was the grunts (Haemulidae), which comprised 46% of total vessel-reef fish abundance. Haemulidae was the most abundant family on all vessel-reefs except the *McAllister* (where carangids were most abundant) and the *Merci Jesus* (where labrids were most abundant). There were significantly more *H. aurolineatum* on vessel-reefs than natural reefs ($P < 0.05$, ANOVA) (see Electronic Supplementary Material). In addition, there were significant differences in *H. aurolineatum* abundance among individual vessel-reefs with the *Unnamed Barge* (34.17 ± 27.80) having the lowest abundance compared to all other vessel-reefs, and the *Edmister* (170.44 ± 30.27) having a higher abundance than the *Merci Jesus* (51.25 ± 10.86) ($P < 0.05$, ANOVA, TK).

The most abundant family on natural reefs was the wrasses (Labridae), which accounted for 25% of the total fish abundance. The dominant labrid species was *T. bifasciatum*, which represented 52% of the total wrasse abundance. *T. bifasciatum* was significantly more abundant on vessel-reefs than natural reefs ($P < 0.05$, ANOVA) (see Electronic Supplementary Material). There was also a significant difference among vessel-

Table 2 Mean ± SEM per count of fish abundance, biomass and species richness between vessel- and natural reefs and among individual vessel-reefs (within a column, sites with differing subscript numbers or letters are significantly different (P 0.05, ANOVA, TK))

Site	n	Abundance	Biomass (kg)	Richness
Vessel-reefs	218	$490.80 \pm 38.70_1$	$31.71 \pm 3.04_1$	$21.51 \pm 0.28_1$
Natural reefs	61	$119.07 \pm 12.62_2$	$6.37 \pm 0.60_2$	$20.13 \pm 0.58_2$
Edmister	54	518.72 ± 98.50	39.49 ± 6.60	$22.15 \pm 0.60_A$
McAllister	32	805.53 ± 149.00	56.79 ± 10.77	$21.50 \pm 0.76_{AB}$
Merci Jesus	32	317.16 ± 28.43	27.75 ± 6.24	$20.88 \pm 0.58_{AB}$
Scutti	32	426.16 ± 48.92	15.27 ± 2.53	$22.25 \pm 0.58_{AB}$
Tracy/Vitale	32	383.09 ± 76.90	35.72 ± 9.64	$18.53 \pm 0.74_B$
Unnamed Barge	36	500.25 ± 77.35	14.60 ± 2.26	$23.14 \pm 0.58_A$

reefs. The *Edmister* (17.44 ± 1.88) had a lower *T. bifasciatum* abundance than all other vessel-reefs ($P < 0.05$, ANOVA, TK).

Twenty-six percent of the total vessel-reef abundance was classified as juveniles (mean size ≤ 5 cm TL) and 58% of the total juveniles belonged to the family Haemulidae. The mean abundance of juvenile haemulids was found to be significantly greater on vessel-reefs when compared to previous results of fish counts performed on all three local reef terraces (inshore, middle and outer) ($P < 0.05$, ANOVA, TK) (Fig. 2) (Ferro et al., 2005). Eighty percent of all juvenile haemulids were observed on two vessel-reefs, the *Edmister* (34%) and *Unnamed Barge* (46%). Natural reef fish assemblages were composed of 25% juveniles, of which 58% were bicolor damselfish.

Planktivores were most abundant on vessel-reefs, accounting for 53% of the total fish abundance and were statistically greater on vessel-reefs (60.11 ± 7.43) than natural reefs (15.01 ± 1.77) ($P < 0.05$, ANOVA), where they accounted for 27% of total abundance (Fig. 3). Planktivore abundance statistically differed among vessel-reefs with the *McAllister* (111.82 ± 28.43) ($P < 0.05$, ANOVA, TK) having a greater abundance than all vessel-reefs, except the *Unnamed Barge* (77.66 ± 17.12) ($P > 0.05$, ANOVA, TK).

Benthic carnivores accounted for 38% of the total fish abundance on both natural and vessel-reefs, but were significantly greater on vessel-reefs (23.10 ± 2.25) than on natural reef (4.46 ± 0.53) ($P < 0.05$, ANOVA) (Fig. 3). There were significantly more benthic carnivores on the *Edmister* (27.75 ± 3.64) than all other vessel-reefs and the lowest abundance was found on the *Unnamed Barge* (8.41 ± 4.48) ($P < 0.05$, ANOVA, TK).

Herbivores represented 13% of the total natural reef fish abundance and 2% of the total vessel-reef fish abundance. Natural reef herbivore abundance (4.68 ± 0.35) was significantly greater than vessel-reefs (3.29 ± 0.19) ($P < 0.05$, ANOVA) (Fig. 3). Significant differences occurred with regard to time of vessel-reef deployment with the oldest vessel-reef, the *Unnamed Barge*, harboring more herbivores than all other vessel-reefs ($P < 0.05$, ANOVA, TK). Omnivores represented 21% of total fish abundance on natural reefs and 4% on vessel-reefs; however, there was no significant difference between the two ($P > 0.05$, ANOVA) (Fig. 3).

Piscivores represented 4% of the total vessel-reef fish abundance and 1% of the total natural reef fish abundance. Vessel-reef piscivore abundance (5.94 ± 0.53) was significantly greater than natural reefs (1.45 ± 0.11) ($P < 0.05$, ANOVA) (Fig. 3). There were also significant differences among vessel-reefs with the *McAllister* (7.59 ± 1.14) having more piscivores than the *Scutti* (3.57 ± 0.41) and the *Unnamed Barge* (6.40 ± 1.56) ($P < 0.05$, ANOVA, TK).

Fig. 2 Mean abundance ± SEM per count of juvenile *Haemulon* spp. Means with differing letters are statistically different ($P < 0.05$, ANOVA, TK) (n = number of counts performed at each reef, (f = number of times *Haemulon* sp. were observed)

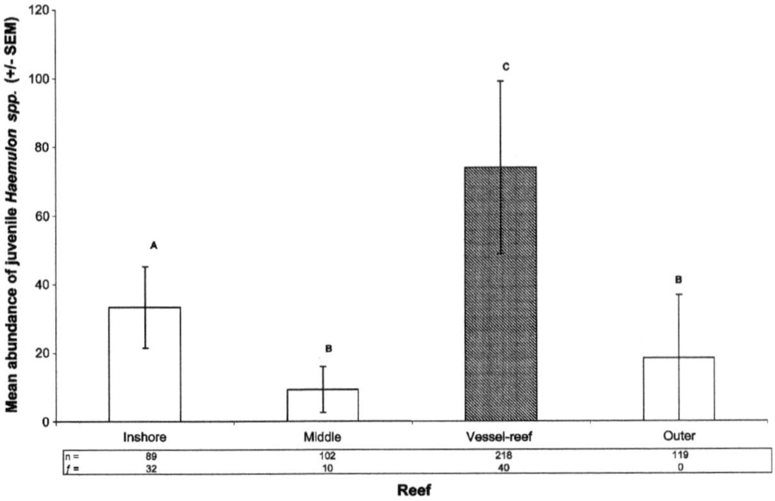

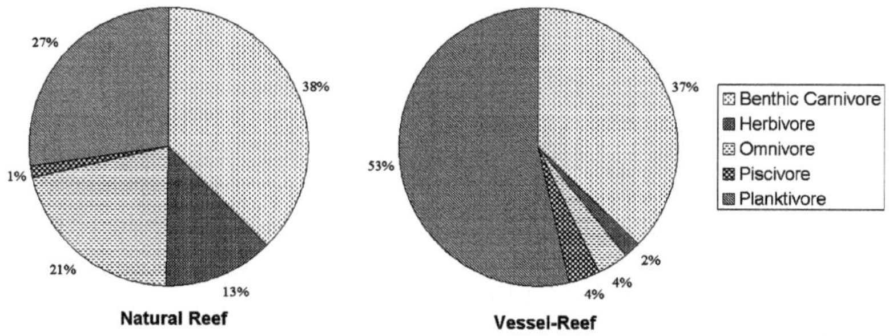

Fig. 3 Trophic composition as a percent of total fish abundance on vessel- and natural reefs

The importance of each species to fisheries (recreational and commercial) was determined by utilizing Florida's recreational and commercial regulations, as well as criteria from Bohnsack et al. (1994). The abundance of fisheries-important species was compared between natural and vessel-reefs. Vessel-reefs had a significantly greater abundance (38.90 ± 4.40) than natural reefs (6.40 ± 1.81), with both reefs having *H. aurolineatum* as the most abundant fisheries-important species ($P < 0.05$, ANOVA). Forty-eight percent of the total fish abundance on vessel-reefs was categorized as fisheries-important species. The majority of these fisheries-important species were comprised of the families Haemulidae (64%), Carangidae (25%), and Lutjanidae (9%). Sixteen percent of the total fish abundance on natural reefs was categorized as fisheries-important species. The majority of these fisheries important species were comprised of the families Haemulidae (66%), Serranidae (6%) and Labridae (5%).

The surrounding natural reefs censused in this study were compared to the results of previous research (Ferro et al., 2005), which assessed natural reefs with no artificial structures nearby. There was no significant difference in mean species richness between natural reefs surrounding vessel-reefs (20.13 ± 0.58) and natural reefs with no artificial structures nearby (20.00 ± 0.71). Additionally, mean fish abundance at natural reef sites surrounding vessel-reefs (119.07 ± 12.62), was not significantly different from natural reef areas with no artificial structures nearby (118.57 ± 8.38) (Ferro et al., 2005). Furthermore, there was no statistical difference in the abundance of fisheries-important species between these two areas.

Biomass

The mean vessel-reef biomass per count was significantly greater than natural reefs ($P < 0.05$, ANOVA) (Table 2). There were also significant differences in biomass among vessel-reefs with the *McAllister* having greater biomass than the *Scutti*, *Tracy/Vitale*, or *Unnamed Barge* ($P < 0.05$, ANOVA, TK) (Table 2). Excluding the natural reefs but with all vessel-reefs combined there was a difference among months, with February (68.76 ± 19.49) having a higher mean biomass per count than July (28.35 ± 16.48) ($P < 0.05$, ANOVA, TK). No difference in biomass was noted among months for natural reef sites.

Species richness

A total of 106,989 fishes of 159 species from 43 families was recorded from the 218 point-counts on all vessel-reefs combined. The most speciose families were groupers (Serranidae: 18 species), parrotfishes (Scaridae: 12 species) and damselfishes (Pomacentridae: 12 species). Together these three families comprised 26% of the vessel-reef species pool. There were 58 species, which were found exclusively on vessel-reefs.

On natural reefs, 7,263 fishes of 118 species were recorded in 61 point-counts. A total of 35 families was recorded on natural reefs. The most speciose families were groupers (Serranidae: 15 species), parrotfishes (Scaridae: 11 species) and damselfishes (Pomacentridae: 11 species).

Together these three families comprised 31% of the natural reef species pool. Fifty-eight species were found exclusively on natural reefs.

Although the east edge of the middle reef had significantly greater mean species richness per count (22.16 ± 0.79) than the west edge of the outer reef (18.03 ± 0.68), there were significantly more species on vessel-reefs than both natural reef areas combined ($P < 0.05$, ANOVA) (Table 2). There were also differences in species richness among vessel-reefs with the *Tracy/Vitale* having significantly lower species richness than the *Edmister* and *Unnamed Barge* (Table 2). Furthermore, there was a linear relationship between mean species richness and vessel-reef age ($R^2 = 0.06$, $P < 0.05$) (Fig. 4). No difference was noted in species richness between months on all vessel-reefs combined.

Of the 159 species recorded on vessel-reefs, 58 were not found on natural reefs and were exclusive to artificial reefs. In this study, 16 (28%) of the exclusive species on vessel-reefs were only recorded once and can be considered rare. On natural reefs 18 of the 118 species recorded were not observed on vessel-reefs. Ten (56%) of these exclusive species were observed only once (see Electronic Supplementary Material).

Assemblage structure

The results of the multidimensional scaling (MDS) showed distinct differences in fish assemblage structure between natural and vessel-reefs with little overlap (Fig. 5). The ANOSIM comparing natural and vessel-reefs produced an R-statistic of 0.718, supporting the MDS showing distinct fish assemblages between the reef types (Field et al., 1982). Further separation clearly reveals individual differences among vessel-reefs, with the *Unnamed Barge* and *Edmister* fish assemblages clearly distinct from the remaining cluster of vessel-reefs (Fig. 6). Here again, the ANOSIM R-statistic supported our findings with the highest R-values associated with comparisons between both the *Unnamed Barge* and *Edmister* with all other vessel-reefs combined. Also an R-statistic of 0.748 was produced when comparing the *Unnamed Barge* to the *Edmister*, indicating that, even though the fish assemblages at these

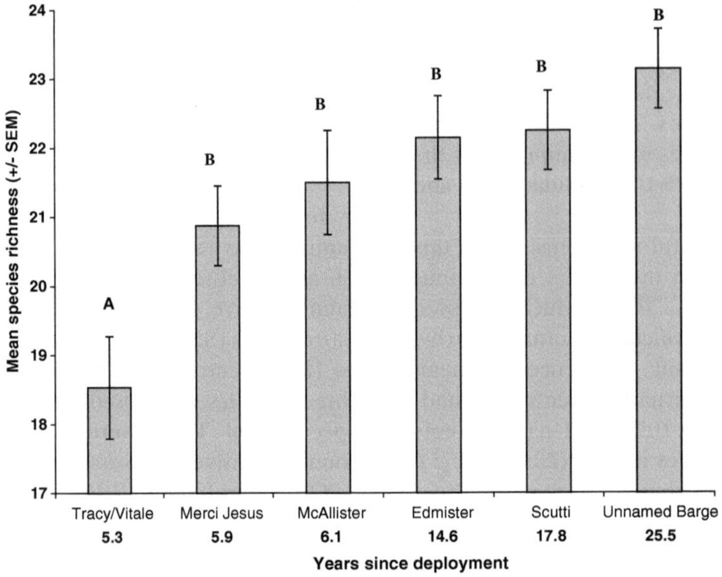

Fig. 4 Mean species richness ± SEM per count for vessel-reefs of various ages. Means with differing letters are statistically different ($P < 0.05$, ANOVA, TK)

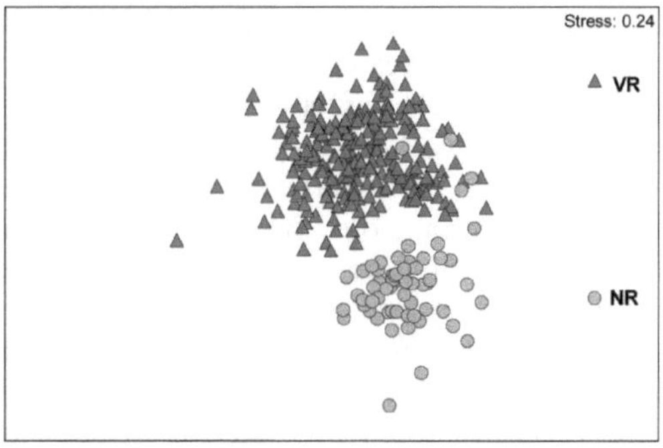

Fig. 5 MDS plot of Bray–Curtis dissimilarity indices of vessel- and natural reefs. VR = vessel-reef; NR = natural reef

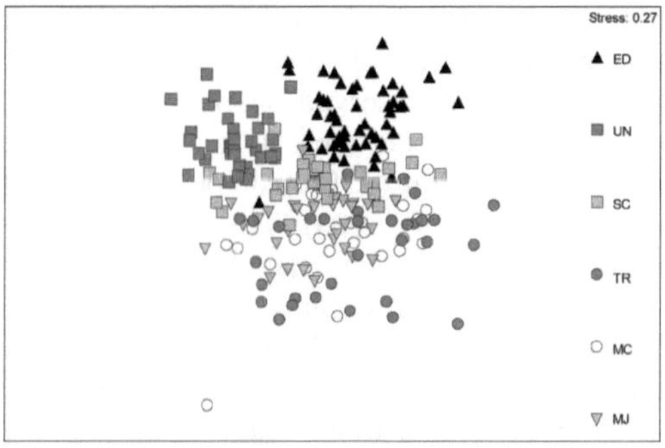

Fig. 6 MDS plot of Bray–Curtis dissimilarity indices of individual vessel-reefs. ED = *Edmister*, UN = *Unnamed Barge*, SC = *Scutti*, TR = *Tracy/Vitale*, MC = *McAllister*, and MJ = *Merci Jesus*

two vessel-reefs were quite distinct from all other vessel-reefs, they were also very different from each other.

The SIMPER analysis revealed the species contributing most to the differences indicated by the MDS plots. There was a 74% dissimilarity found between natural and vessel-reefs with *H. aurolineatum* contributing more to the dissimilarity (6.05%) than any other species (Table 3). The SIMPER analysis comparing individual vessel-reef fish assemblages to all natural reefs combined revealed the *McAllister* had the highest dissimilarity (77%) and the *Unnamed Barge* (71%) had the lowest. When individual vessel-reefs were compared to each other, the highest dissimilarity (69%) was found when comparing the oldest vessel-reef (*Unnamed Barge*) to the youngest (*Tracy/Vitale*).

Discussion

Most artificial reef research has shown artificial reefs have greater fish abundance and biomass than natural reefs with similar community structures (see Bohnsack et al., 1991). In this study there was a mean of 154.55 ± 22.46 individuals on the eastern edge of the middle terrace, 82.40 ± 6.30 on the western edge of the outer terrace and 490.80 ± 38.70 on vessel-reefs. The lower numbers on natural reefs were apparently not a function of a lower sampling frequency, which missed a period of increased abundance. A previous study in Broward County, also using point-counts, reported mean abundances of 108.00 ± 49.00 and 75.00 ± 16.00 on the eastern middle terrace and western outer terrace edges, respectively (Ferro et al., 2005). Additionally, the comparison between natural and vessel-reefs

Table 3 SIMPER percentages of the top 10 species contributing most to the differences between vessel- and natural reefs

Common name	Scientific name	Dissimilarity (%)	Cumulative dissimilarity (%)
Tomtate	*Haemulon aurolineatum*	6.05	6.05
Mask Goby	*Coryphopterus personatus*	4.18	10.23
Purple Reeffish	*Chromis scotti*	3.70	13.93
Yellowhead Wrasse	*Halichoeres garnoti*	3.13	17.06
Bicolor Damselfish	*Pomacentrus partitus*	3.05	20.11
Creole Wrasse	*Clepticus parri*	3.00	23.12
Bluehead Wrasse	*Thalassoma bifasciatum*	2.90	26.02
Grey Snapper	*Lutjanus griseus*	2.73	28.75
Tobacco Fish	*Serranus tabacarius*	2.41	31.16
Sharpnose Puffer	*Canthigaster rostrata*	2.23	33.38

provides insight into the aggregation hypothesis, which proposes that fishes on artificial reefs have been aggregated from nearby natural reefs. Because no differences in species richness or abundance were found between natural reef areas surrounding vessel-reefs and natural reefs with no artificial structures nearby (Ferro et al., 2005), it appears that vessel-reefs are not, in general, attracting fishes away from nearby natural reef areas. This conclusion is also supported by preliminary results from research studying fish colonization on a newly deployed vessel-reef in Broward County, which revealed that production, rather than strictly attraction, may be an important component contributing to vessel-reef fish assemblages (authors unpublished).

The differences in biomass, noted in this study, parallel the differences in abundance among vessel-reefs and between natural and vessel-reefs. This indicates the greater fish abundance on vessel-reefs is not due simply to large number of juveniles, as they typically weigh dramatically less than adults. This study supports the common finding of greater abundance of fishes on artificial reefs and the results of the MDS, SIMPER, and ANOSIM clearly indicate fish assemblage structure on vessel-reefs differ from nearby natural reefs.

While approximately 57% of the species recorded in this study were common to both natural and vessel-reefs, 58 species were unique to vessel-reefs. Some of these species were relatively rare (e.g., *Epinephelus itajara* (Lichtenstein), *Mycteroperca bonaci* (Poey)) noted only once or twice, and it is unclear if their presence represented a preference or simply chance occurrence (see Electronic Supplementary Material). However, some of the species unique to vessel-reefs in this study have never been recorded in natural reef surveys in Broward County (i.e. margates, *Haemulon album* (Cuvier); greater amberjack, *Seriola dumerili* (Risso); little tunny, *Euthynnus alletteratus* (Rafinesque); blackfin snapper, *Lutjanus buccanella* (Cuvier); snowy grouper, *Epinephelus niveatus* (Valenciennes); and a single 35 cm cubera snapper, *Lutjanus cyanopterus* (Cuvier)). The unique presence of the snapper and grouper fishes is particularly interesting, as they are typically deepwater species of recreational and commercial value, which appear to prefer vessel-reefs to nearby natural habitats. Arena et al. (2004) reported that juveniles of both blackfin snapper (10–26 cm) and snowy grouper (10–15 cm) were recorded only on vessel-reefs throughout Broward County and noted that previous research had also observed juveniles of these species on smaller artificial reef modules at depths both comparable to and shallower than this study. The authors suggested vessel-reefs are supplying blackfin snapper and snowy grouper with ancillary nursery/juvenile habitat that may be in short supply in deeper areas, which has been described as low-relief hardbottom.

The red grouper, *Epinephelus morio* (Valenciennes), is one species important to fisheries that was unique to natural reefs in this study. This species has been observed on vessel-reefs in the Gulf of Mexico, an area with limited natural hardbottom habitat (J. Franks, personal communication, November 2003). This suggests that *E. morio* may utilize artificial reef structures in habitat limited areas, but prefer natural reefs when they are available. Past research has shown that some reef fishes at artificial reef sites have narrower diets than those found in natural areas due to the limited availability of food resources (Sierra et al., 2001). *E. morio* may prefer natural reef areas when they are present, due to greater, species-specific food availability in those habitats.

Polovina (1991) suggested in order for artificial reefs to increase production, they need to provide habitat that can improve larval settlement, growth, and survival. The high vertical relief of vessel-reefs may increase settlement of juveniles by extending habitat into areas higher in the water column, possibly attracting larval fishes located closer to surface waters (Rilov & Benayahu, 2002). Our results indicate there were significantly more juveniles on vessel-reefs, the majority of which were *Haemulon* spp., than any natural reef terrace in our area (Ferro et al., 2005) (Fig. 2). This is an interesting result as previous research has indicated that Broward County's shallow, inshore reef was important habitat for juvenile grunts, yet our results reveal a greater mean abundance of these juveniles on vessel-reefs (Jordan et al., 2004). Another study utilizing small

1 m relief artificial reefs deployed at similar depths (21 m) and in the same sand flat between the middle and outer reefs also recorded high densities of juvenile *Haemulon* spp. and may be an indication that high vertical relief is not a requirement for all fish species (Sherman, 2000). The majority of *Haemulon* juveniles (80%) were recorded on two particular vessel-reefs, 34% on the *Edmister* and 46% on the *Unnamed Barge*. These two vessel-reefs have the lowest vertical relief (3 m) and a high amount of complexity near the seafloor, which may have increased the survival of juvenile grunts. These results suggest a possible settlement preference by *Haemulon* larvae for complex structures near the seafloor, which may enhance juvenile grunt survival and growth. This suggestion is supported by the fact that larval grunts have rarely been collected near the surface (Richards, 1981), indicating they may be epibenthic.

Although there were some species completely absent from the vessel-reefs and others absent from natural reefs, the dissimilarity in assemblage structure between natural and vessel-reefs was primarily due to differences in abundance of species common to both. *H. aurolineatum*, contributed the most to the dissimilarity between natural and vessel-reefs and was also a top contributor to differences seen among vessel-reefs. *H. aurolineatum* is one of the most abundant fish on reefs, live bottom areas and inshore habitats in the Greater Caribbean, Gulf of Mexico, and along the southeast Atlantic coast of the United States (Darcy, 1983). Juveniles are primarily diurnal planktivores; as they grow and mature they shift their feeding, in large measure, to open sand and seagrass beds, where they forage for benthic invertebrates (Darcy, 1983; Sedberry, 1985).

Due to their ubiquity and high abundance, *H. aurolineatum* may be important in transferring energy, from the sandy substrate adjacent to reef areas, to hardbottom habitat and artificial reef communities (Darcy, 1983; Meyer et al., 1983; Sedberry, 1985). Lindquist et al. (1994) speculated sandbottom benthic productivity might be more important than previously thought in supporting the nekton on the continental shelf. Additionally, *H. aurolineatum* are known to be prey for many recreationally and commercially important species, such as groupers, snappers and jacks (Darcy, 1983; Froese & Pauly, 2004). The mean size of *H. aurolineatum* on all vessel-reefs combined was 16.0 cm TL, with the maximum and minimum size recorded as 26.0 and 3.0 cm, respectively. At a mean size of 16.0 cm these fish should primarily be nocturnal feeders searching for sand dwelling invertebrates (Manooch & Barans, 1982; Sedberry, 1985), although due to the range of sizes recorded and observations of feeding behavior on vessel-reefs, these fish were also utilizing planktonic resources.

The optimal foraging theory predicts that *H. aurolineatum*, as well as most haemulids, will utilize habitats in close proximity to soft bottom feeding areas to reduce the amount of energy spent on travel between resting and foraging sites (Stephens & Krebs, 1986). *H. aurolineatum* are typically found in shelf-edge habitats, near areas of bottom relief and at the edge of rock ledges protruding into the sand (Manooch & Barans, 1982; Darcy, 1983; Sedberry, 1985). Vessel-reefs seem to be providing this species with similar habitat characteristics. The vessel-reef habitat provides shelter and resting sites within their natural foraging areas decreasing energy expenditures and risk of predation associated with travel to these sites. In addition, the ability of *H. aurolineatum* to feed diurnally in the plankton, presumably makes vessel-reefs even more advantageous to this species, which may, in turn, increase its growth rate and ultimately its fitness. *H. aurolineatum* has been shown to grow faster than many previously studied reef fishes from the South Atlantic Bight (Darcy, 1983). This exhibition of increased growth may be due to its ability to take advantage of both planktonic and benthic food resources.

Another major difference between fish assemblages on natural and vessel-reefs is the abundance of planktivores, the dominant trophic group on vessel-reefs. Of the top 10 species by mean abundance on vessel-reefs, seven were planktivores and accounted for 50.80% of the total vessel-reef fish abundance (see Electronic Supplementary Material). In contrast, only three of the top 10 species by mean abundance on natural reefs were planktivores, accounting for

22% of the total natural reef fish abundance. In addition, the SIMPER analysis showed that four of the top seven species (*C. personatus, Chromis scotti* (Emery), *Clepticus parrae* (Bloch & Snyder), *T. bifasciatum*] contributed most to the dissimilarity (13.48%) between natural and vessel-reefs were planktivores, with a fifth (*H. aurolineatum*) that feeds secondarily in the plankton (Table 3).

These results are comparable to previous studies utilizing artificial reefs with high vertical relief. Rilov & Benayahu (2000) reported the two most numerous species found on oil jetty platforms were planktivores. Linquist & Pietrafesa (1989) conducted a study that assessed fish assemblages on a tugboat with 8.3 m of vertical relief. They reported the two most abundant species on these structures were planktivores. Additionally, Stephan & Lindquist (1989) reported that planktivores dominated the fish assemblage on a dredge and FADS, with 6 and 14.7 m of vertical relief, respectively. Plankton productivity has been shown to be greatest within the top 30 m of the water column and it has been suggested that artificial reefs with high vertical relief allow the upper portion of the structure to interact with these planktonic resources closer to the surface (Rilov & Benayahu, 2000). However, artificial reefs with low vertical relief have also been shown to have a high abundance of planktivores in comparison to natural reefs (Bohnsack et al., 1994). This suggests that vertical relief may not be the only factor influencing planktivores.

Location of the artificial reef may also be an important influence on planktivore production. Diurnal planktivores depend on currents to supply food. Larger diurnal planktivores are known to move from nocturnal resting sites to diurnal feeding sites with strong currents near the shelf edge (Hobson, 1991). The vessel-reefs in this study were located between two continuous reef terraces that run in a N–S direction. The sand flat separating these reefs offers an uninterrupted current flow. Conversely, at natural reef sites, the passing currents and associated plankton are exposed to a 'wall of mouths' upstream, which can dramatically deplete plankton downstream (Hamner et al., 1988) (Fig. 1). Vessel-reefs in this study provide habitat relatively unexposed to upstream planktivores, and this may, in turn, allow resident planktivores access to more planktonic resources. Thus, planktivores, may be using the upper works of vessel-reefs to access an abundant and unexploited food resource.

Donaldson & Clavijo (1994) have suggested holozooplankton are currently underutilized as a food resource by many planktivorous fishes due to the lack of shelter from predation on open sand bottoms. Although planktonic resources or diets of planktivores were not censused here, the vessel-reefs may provide these fishes with shelter they need to access these resources that would be otherwise unavailable. Vessel-reefs not only provide planktivores with shelter from predation, but also from strong currents and passing internal waves (Grove & Sonu, 1985; Grove et al., 1991).

In addition to shelter, vessel-reef habitat may entrain planktonic resources, further increasing planktivore feeding efficiency. Lindquist & Pietrafesa (1989) suggested an upcurrent vortex reversal at vessel-reefs concentrates planktonic resources, and also reduces the amount of energy required to swim against the incoming current flow. Our experience supports this suggestion. For example, during one census at the *Unnamed Barge*, a school of brown chromis, *Chromis multilineata* (Guichenot), was observed feeding on plankton on the upcurrent side of the vessel-reef during a strong current flow. The first author swam over to their location and instantly noticed a reduction in current flow and maintained his position with little effort.

The abundance of planktivores on vessel-reefs can provide direct trophic links from open-water to coral reef communities through two main avenues. First, energy can be transferred to both demersal and pelagic piscivores, which utilize planktivores as prey. Although no studies on diet were performed, the statistically higher abundance of piscivores on vessel-reefs and repeated observations of predatory behavior by these fishes, as well as documentation of their prey items, indicate they are likely feeding on planktivores (Froese & Pauly, 2004). Additionally, demersal piscivores, such as lutjanids, at vessel-reefs may also be preying upon planktivores.

A second energy transfer is through planktivore feces, not only to detritivores, but also to other planktivorous fishes. Robertson (1982) reported that many fecal strings from planktivores are devoured by other species of fishes before they reach the bottom. The feces of some planktivores can be a valuable resource, especially when zooplankton is abundant. Hobson (1991) observed zooplankton passing through the guts of planktivores so rapidly that very little signs of digestion were observed in the feces. The majority of energy derived from planktonic resources may have not been utilized and would possibly be swept away with the current if vessel-reef habitats were unavailable (Hamner et al., 1988). Also, when planktivores return to their vessel-reef resting sites they continue to defecate and these added nutrients may enhance the production of vessel-reef benthic communities. Past research has shown corals that harbor large schools of fishes grow faster, presumably due to the constant influx of nutrients from fish feces (Meyer et al., 1983).

While distinct fish assemblages on vessel-reefs exist, the results suggest that as vessel-reef age increases they become more similar to surrounding natural reef assemblages. The SIMPER analysis comparing individual vessel-reefs to natural reefs showed the oldest vessel-reef, *Unnamed Barge*, was most similar (29%) to nearby natural reef fish assemblages. The *Unnamed Barge* had low vertical relief (3 m), possibly reducing the abundance of planktivores on that vessel-reef, thereby making it more similar to the natural reef assemblage. However, comparisons of planktivore abundance among vessel-reefs revealed the *Unnamed Barge* planktivore abundance was not significantly different from the *McAllister*, which had the greatest planktivore abundance of all vessel-reefs. Additional results revealed a strong positive linear relationship ($R^2 = 0.86$) between mean species richness and vessel-reef age (Fig. 4). These findings may be due to increased food resources and substrate complexity, provided by a richer fouling community on older vessel-reefs, and/or additional recruits that colonized vessel-reefs over time (Chandler et al., 1985; Potts & Hulbert, 1994; Tupper & Hunte, 1998).

Lastly, an important distinction between natural and vessel-reefs can be clearly seen when comparing fish species of recreational or commercial value. Vessel-reefs were found to have significantly more fisheries-important species than nearby natural reefs. The majority of vessel-reef fisheries-important species were comprised of the families Haemulidae (64%), Carangidae (25%), and Lutjanidae (9%). While the most abundant species overall was the tomtate, there were many more fisheries-important species, such as gray snapper, blackfin snapper, lane snapper, and amberjacks; all more abundant on vessel-reefs than natural reefs. These results indicate that vessel-reef habitats provide important contributions to local fishery resources.

Conclusions

This study supports the results of previous research indicating artificial reefs harbor a greater fish abundance, including fisheries-important species, and biomass than natural reef areas. Greater species richness, as well as the many exclusive species on vessel-reefs, suggests these artificial reef types are providing unique habitat characteristics, which may not be found on surrounding natural reefs. Additionally, the comparisons between natural and vessel-reef fish assemblages do not support the aggregation hypothesis and may be an indication that fish production is occurring on vessel-reefs in Broward County. However, we cannot discount the possibility that large demersal fishes, such as serranids, which have been overexploited in our study area, may be attracted to vessel-reefs in other areas. In such a situation vessel-reefs would contribute to overexploitation of some species. Acoustic telemetry, diet and growth studies, as well as estimates of fishing pressure of specific vessel-reef species, would help determine the effects vessel-reefs are having on natural reef fish assemblages.

Acknowledgements Florida Fish and Wildlife Conservation Commission, Division of Marine Fisheries funded this research through a grant to Broward County. Broward County Department of Planning and Environmental Protection (DPEP) was directly responsible for funding the study. The Guy Harvey Research Institute, South Florida Sport Fishing Classic Inc. and Patty Carr were instrumental in obtaining ancillary research funding for this project. This paper is a result of research funded by the National Oceanic and Atmospheric Administration Coastal Ocean Program under award (NA03NOS4260046 to Nova Southeastern University for the National Coral Reef Institute (NCRI). We would also like to thank Brian Walker, of the National Coral Reef Institute for his help with Fig. 1 for this paper. This project would not have been accomplished with out the help of many unpaid, yet enthusiastic volunteers from Nova Southeastern University's Oceanographic Center graduate students and Ichthyology Lab colleagues. This is NCRI contribution #83.

References

Arena, P. T., T. P. Quinn, L. K. B. Jordan, R. L. Sherman, F. M. Harttung & R. E. Spieler, 2004. Presence of juvenile blackfin snapper, Lutjanus buccanella, and snowy grouper, Epinephelus niveatus, on shallow-water artificial reefs. Proceedings of the Fifty-fifth Gulf and Caribbean Fisheries Institute 55: 700–712.

Bohnsack, J. A. & S. P. Bannerot, 1986. A stationary visual census technique for quantitatively assessing community structure of coral reef fishes. U.S. Dept. of Commerce, NOAA Technical Report NMFS 41: 1–15.

Bohnsack, J. A. & D. E. Harper, 1988. Length–weight relationship of selected marine reef fishes from southeastern United States. NOAA Technical Memorandum NMFS-SEFC. U.S. Dept. of Commerce.

Bohnsack, J. A., D. L. Johnson & R. F. Ambrose, 1991. Ecology of artificial reef habitats and fishes. In Seamen, W. & L. M. Sprague (eds), Artificial Habitats for Marine and Freshwater Fisheries. Academic Press, San Diego, 61–108.

Bohnsack, J. A., D. E. Harper, D. B. McClellan & M. Hulsbeck, 1994. Effects of reef size on colonization and assemblage structure of fishes at artificial reefs off Southeastern Florida, USA. Bulletin of Marine Science 55: 796–823.

Brock, R. E., 1994. Beyond fisheries enhancement: artificial reefs and ecotourism. Bulletin of Marine Science 55: 1181–1188.

Chandler, C. R., R. M. Sanders Jr. & A. M. Laundry Jr., 1985. Effects of three substrate variables on two artificial reef fish communities. Bulletin of Marine Science 37: 129–142.

Darcy, G. H., 1983. Synopsis of biological data on the grunts *Haemulon aurolineatum* and *H. plumieri* (Pisces: Haemulidae). FAO Fisheries Synopsis No. 133, NOAA Technical Report NMFS Circular 448. 37 pp.

Donaldson, P. L. & I. E. Clavijo, 1994. Diet of Round Scad (Decapterus punctatus) on a natural and artificial reef in Onslow Bay, North Carolina. Bulletin of Marine Science 55: 501–509.

Ferro, F. M., L. K. B. Jordan & R. E. Spieler, 2005. Spatial variability of the coral reef assemblages offshore Broward County, Florida. NOAA Technical Memorandum NMFS-SEFSC-532.

Field, J. G., K. R. Clarke & R. M. Warwick, 1982. A practical strategy for analyzing multispecies distribution patterns. Marine Ecology Progress Series 8: 37–52.

FAO (Food and Agriculture Organization), 1997a. Review of the state of the world fishery resources: marine fisheries. FAO Fisheries Circular 920, Rome, Italy.

FAO (Food and Agriculture Organization), 1997b. Review of the state of world fisheries and aquaculture: 1996. FAO, Rome, Italy .

Froese, R. & D. Pauly (eds), 2004. FishBase. World Wide Web electronic publication. www.fishbase.org, 06 April 2004.

FWRI (Fish and Wildlife Research Institute), 2004. Recreational Fisheries. Retrieved September 17, 2004 from Florida Fish and Wildlife Conservation Commission website: http://research.myfwc.com/support/view_faqs.asp?id = 25.

Gilmore, C., 2004. PADI Diver Statistics. Retrieved March 15, 2004: http://www.padi.com/english/common/padi/statistics/.

Goldberg, W. M., 1973. The ecology of the coral–octocoral communities off the southeast Florida coast: geomorphology, species composition, and zonation. Bulletin of Marine Science 23: 465–488.

Goreau, T. F., 1959. The ecology of Jamaican coral reefs. I. Species composition and zonation. Ecology 40: 67–90.

Grove, R. S. & C. J. Sonu, 1985. Fishing reef planning in Japan. In D'itri, F. M. (ed.), Artificial Reefs: Marine and Freshwater Applications. Lewis Publishers, Michigan, 187–252.

Grove, R. S., C. J. Sonu & M. Nakamura, 1991. Design and engineering of manufactured habitats for fisheries enhancement. In Seamen, W. & L. M. Sprague (eds), Artificial Habitats for Marine and Freshwater Fisheries. Academic Press, San Diego, 109–153.

Hamner, W. M., M. S. Jones, J. H. Carleton, I. R. Hauri & D. M. Williams, 1988. Zooplankton, planktivorous fish, and water currents on a windward reef face: Great Barrier Reef, Australia. Bulletin of Marine Science 42: 459–479.

Hobson, E. S., 1991. Trophic relationships of fishes specialized to feed on zooplankters above coral reefs. In Sale, P. F. (ed.), The Ecology of Fishes on Coral Reefs. Academic Press, San Diego, 69–95.

Jordan, L. K. B., P. T. Gilliam, D. S., R. L. Sherman, P. T. Arena, F. M. Harttung, R. M. Baron & R. E. Spieler, 2004. Spatial and temporal patterns of juvenile grunts (Haemulon spp.) in South Florida. Proceedings of the Fifty-Fifth Gulf and Caribbean Fisheries Institute 55: 322–336.

Jones, R. S. & M. J. Thompson, 1978. Comparison of Florida reef fish assemblages using a rapid visual technique. Bulletin of Marine Science 28: 159–172.

Jones, A. T. & R. W. Welsford, 1997. Artificial reefs in British Columbia, Canada. OCEANS 97: MTS/IEEE, Artificial Reef Society of British Columbia pp. 415–418.

Lighty, R. G., 1977. Relict shelf-edge Holocene coral reef: southeast coast of Florida. Proceedings of the Third International Coral Reef Symposium, Vol. 2, 215–221.
Lindquist, D. G. & L. J. Pietrafesa, 1989. Current vortices and fish aggregations: the current field and associated fishes around a tugboat wreck in Onslow Bay, North Carolina. Bulletin of Marine Science 44: 533–544.
Lindquist, D. G., L. B. Cahoon, I. E. Clavijo, M. H. Posey, S. K. Bolden, L. A. Pike, S. W. Burk & P. A. Cardullo, 1994. Reef fish stomach contents and prey abundances on reef and sand substrata associated with adjacent artificial and natural reefs in Onslow Bay, North Carolina. Bulletin of Marine Science 55: 308–318.
MacDonald, C. D., C. A. Mitsuyasu & E. Corbin, 1999. A planned underwater dive attractions program for Hawaii. Ocean Resources Branch, State of Hawaii Department of Business, Economic Development and Tourism. Ocean Resources Branch Contribution No. 147, pp.180–187.
Manooch, C. S. & C. S. Barans, 1982. Distribution, abundance, and age and growth of the tomtate, Haemulon aurolineatum, along the southeastern United States coast. Fishery Bulletin 80: 1–19.
Markevich, A. I., 1994. Species composition and ecological characteristics of fishes of artificial shelters in Peter the Great Bay, Sea of Japan. Russian Journal of Marine Biology 20: 169–173.
Meyer, J. L, E. T. Schultz & G. S. Helfman, 1983. Fish schools: an asset to corals. Science 220: 1047–1049.
Moyer, R. P, B. Riegel, K. Banks & R. E. Dodge, 2003. Spatial patterns and ecology of benthic communities on a high latitude South Florida (Broward County, USA) reef system. Coral Reefs 22: 447–464.
Murray, J. D. & C. J. Betz, 1994. User views of artificial reef management in the Southeastern U.S. Bulletin of Marine Science 55: 970–981.
Polovina, J. J., 1991. Fisheries applications and biological impacts of artificial reefs. In Seamen, W. & L. M. Sprague (eds), Artificial Habitats for Marine and Freshwater Fisheries. Academic Press, San Diego, 154–176.
Potts, T. A. & A. W. Hulbert, 1994. Structural influences of artificial and natural habitats on fish aggregations in Onslow Bay, North Carolina. Bulletin of Marine Science 55: 609–622.
Richards, W. J., 1981. Kinds and abundance of fish larvae in the Caribbean Sea. In Lasker, R. & K. Sherman (eds), The Early Life History of Fish: Recent Studies. ICES, Woods Hole, 240–241.
Rilov, G. & Y. Benayahu, 2000. Fish assemblage on natural versus vertical artificial reefs: the rehabilitation perspective. Marine Biology 136: 931–942.
Rilov, G. & Y. Benayahu, 2002. Rehabilitation of coral reef-fish communities: the importance of artificial-reef relief to recruitment rates. Bulletin of Marine Science 70: 185–197.
Robertson, D. R., 1982. Fish feces as fish food on a Pacific coral reef. Marine Ecology Progress Series 7: 253–265.
Seamen, W. & A. C. Jensen, 2000. Purposes and practices of artificial reef evaluation. In Seamen, W. (ed.), Artificial Reef Evaluation: With Application to Natural Marine Habitats. CRC Press, Florida, 1–20.
Sedberry, G. R, 1985. Food and feeding of the tomtate, Haemulon aurolineatum (Pisces, Haemulidae), in the South Atlantic Bight. Fishery Bulletin 83: 461–466.
Sherman, R. L., 2000. Studies on the roles of reef design and site selection in juvenile fish recruitment to small reefs. Ph.D. Dissertation. Nova Southeastern University, 173 pp.
Sierra, L. M., R. Claro & O. A. Popova, 2001. Trophic biology of the marine fishes of Cuba. In Claro, R., K. C. Lindeman & L. R. Parenti (eds), Ecology of the Marine Fishes of Cuba. Smithsonian Institute Press, London, 115–148.
Soloviev, A. V., M. E. Luther & R. H. Weisberg, 2003. Energetic baroclinic super-tidal oscillations on the southeast Florida shelf. Geophysical Research Letters 30: np.
Stephan, C. D. & D. G. Lindquist, 1989. Comparative analysis of the fish assemblages associated with old and new shipwrecks and fish aggregating devices in Onslow Bay, North Carolina. Bulletin of Marine Science 44: 698–717.
Stephens, D. W. & J. R. Krebs, 1986. Foraging Theory. Princeton University Press, NJ, 247 pp.
Stone, R. B., 1985. History of artificial reef use in the United States. In D'itri, F. M. (ed.), Artificial Reefs: Marine and Freshwater Applications. Lewis Publishers, Michigan, 3–12.
Tupper, M. & W. Hunte, 1998. Predictability of fish assemblages on artificial and natural reefs in Barbados. Bulletin of Marine Science 62: 919–935.
White, D. J., 2004. Finding common water. Retrieved September 17, 2004 from Florida Wildlife Federation website: http://www.flawildlife.org/pubs/fwn-4-01/chair.htm.
Zar, J. H., 1996. Biostatistical Analysis. Prentice Hall, N.J. 662 pp.

Hydrobiologia (2007) 580:173–180
DOI 10.1007/s10750-006-0455-y

Effect of depth and reef structure on early macrobenthic communities of the Algarve artificial reefs (southern Portugal)

A. Moura · D. Boaventura · J. Cúrdia · S. Carvalho · L. Cancela da Fonseca · F. M. Leitão · M. N. Santos · C. C. Monteiro

Abstract This study was carried out on the "Faro/Ancão" artificial reef (AR), located off Faro, deployed in May 2003. We aimed to characterise early macrobenthic community colonisation of two concrete AR groups located at different depths (16 m and 20 m depth) and to test the effect of reef structure on these communities. The non-colonial organisms were counted; barnacles and colonial species were quantified using biomass. Multivariate analyses indicated that early macrobenthic communities (6 months of immersion) were affected by depth, and that barnacles and colonial species were also affected by reef structure. Univariate analyses showed that the biomass of barnacles and colonial species was significantly different among reefs and layers of modules. Both AR groups were characterised by the species *Balanus amphitrite*, *Gregariella subclavata*, *Musculus* cf. *subpictus*, *Paleanotus* cf. *bellis* and *Syllidia armata*. *Jassa marmorata* and *Bugula neritina* were characteristic species at 16 m depth, particularly on the AR Upper layer of modules, whereas *Anomia ephippium* was particularly common at 20 m, especially on the Lower layer of modules.

Guest editors: G. Relini & J. Ryland
Biodiversity in Enclosed Seas and Artificial Marine Habitats

A. Moura (✉) · J. Cúrdia · S. Carvalho ·
F. M. Leitão · M. N. Santos · C. C. Monteiro
Instituto Nacional de Investigação Agrária e das Pescas (INIAP/IPIMAR), Centro Regional de Investigação Pesqueira do Sul (CRIPSul), Av. 5 de Outubro, 8700-305 Olhao, Portugal
e-mail: amoura@cripsul.ipimar.pt

D. Boaventura
Escola Superior de Educação João de Deus, Avenida Álvares Cabral, 69, 1269-094 Lisboa, Portugal

D. Boaventura · L. C. da Fonseca
IMAR, Laboratório Marítimo da Guia, Estrada do Guincho, 2750-374 Cascais, Portugal

L. C. da Fonseca
FCMA, DCB, Universidade do Algarve, Campus de Gambelas, 8005-139 Faro, Portugal

Keywords Macrobenthic colonisation · Depth · Benthos · Artificial reef · Portugal

Introduction

Ecological processes concerning macrobenthic communities, such as recruitment, colonisation and succession, are strongly affected by the surrounding physical and biological environment (Svane & Petersen, 2001). Factors, such as temperature, salinity, light, depth, substratum and relative position to the seafloor play an important role in the settlement and recruitment of the organisms as well as in their growth (Kocak & Zamboni, 1998; Glasby, 1999a, b; Svane & Petersen, 2001). Intra- and inter-specific interactions in the community arise in the partition of resources such as space and food, contributing to the dominance or non-presence (local extinction) of species (Kocak & Zamboni, 1998).

Artificial reefs (ARs) have become an important resource of fisheries enhancement (Bohnsack & Sutherland, 1985). Most of the commercial fish species associated with the AR systems are attracted to food habits directly linked to the macrobenthos (Ardizzone et al., 1997; Itosu et al., 1999; Steimle et al., 2002). Due to the potential role of macrobenthic communities in providing food for commercial fish fauna, studies concerning macrobenthic colonisation assume importance to explain the patterns of fish associated with AR systems. Reefs with overlapping elements are particularly effective in maximizing the surface available for the colonisation of benthic organisms. They provide different conditions, namely in light, temperature and physical–chemical parameters (Relini et al., 1994).

In Portugal, AR systems made of concrete modules were deployed along the south coast of the Algarve. The coastal area where the reefs were deployed is characterised by a water temperature between 17.4°C and 20.2°C. Salinity of the upper layers of the water column is almost constant throughout the year in the south of coast of Algarve (ranging from 36.1 to 36.3 at the surface and from 36.1 to 36.4 at 50 m depth) (see Santos, 1997 for further details). The main objective of the artificial structures was to enhance local fisheries by extending the nursery effect of estuarine–lagoon systems present along the coast through the creation of potential food resources and shelter for juveniles of commercial fish fauna (see Santos, 1997; Santos & Monteiro, 1997, 1998 for further details). Although the importance of AR systems is recognised worldwide, research regarding Portuguese ARs systems has been focused mainly on ichthyofauna, especially commercial species (Santos & Monteiro, 1997, 1998). The Algarve AR systems are composed of small (2.7 m^3) and large concrete modules (174 m^3). This study was performed in the small modules, located at 16–20 m depth. This depth range for the deployment of AR modules was determined at its lower end because of bivalve dredge fishing (an important local fishery) and at its upper and due to the characteristics of these particular reef modules (in order to maximise the reef effect at the water column).

So, the main objectives of the present study were (i) to detect the effect of depth on the macrobenthic species after 6 months of colonisation within two reef groups, and (ii) to test differences on the colonisation between two depth layers of reef modules (Upper and Lower).

Materials and methods

The present work was carried out on the "Faro/Ancão" AR located off Faro, southern Portugal, particularly on two small AR groups, deployed in May 2003, at approximately 16 m (37°00.454′ N 8°02.171′ W) and 20 m depth (37°00.062′ N 8°02.482′ W) (groups A and C, respectively). Each one of these reef groups is composed of 35 concrete module units (2.7 m^3 each unit) jumbled up, comprising roughly two different depth layers.

Macrobenthic colonisation was investigated using sample units (15 × 15 × 15 cm) made of the same concrete material of reef modules. The cubic units were suspended on the upper and lower portions (Upper layer—UL and Lower layer—LL) of each reef groups at the time of reef immersion (four units in each layer of modules *per* AR). Three replicate samples were collected by scuba diving 6 months after immersion in November of 2003. In each cubic unit, to avoid the "border effect" on macrobenthic assemblages, the central area of the surface facing outwards was scraped for posterior analysis of macrobenthic communities.

The samples were sieved through a 0.5 mm square mesh, and the retained material fixed in 4% buffered formalin. All specimens were sorted and identified to the species level whenever possible. For the calculation of diversity and richness index, all taxa were included. Non-colonial organisms were counted, and barnacles and colonial species were assessed using biomass. The wet weight of each species was obtained after a 5-min drying period on blotting paper and was measured (to 0.001 g) with a digital weighing scale.

Data analysis

Macrobenthic community was evaluated using the PRIMER v.5.0 software. The analysis was carried out using fourth-root (√√)-transformed data, which is more sensitive to changes in less abundant species. Community patterns were assessed using non-metric multidimensional scaling ordination (MDS) based on a similarity matrix using Bray–Curtis coefficient. A two-way crossed analysis of similarity (ANOSIM) was performed to determine if there were any effects of depth and reef structure on community structure. The Similarities Percentages procedure (SIMPER) of fourth-root transformed data was used to determine the contributions from individual species to the Bray–Curtis dissimilarities between depths and layers of modules.

A two-way ANOVA was used to test for differences in mean density (*D*), mean biomass (*B*), Shannon–Wiener diversity (H'; $\log_2$) and mean number of species (*S*) of each group of samples. The design included two factors: "Reef Depth" (RD) orthogonal, fixed with two levels (A and C) and "Reef Layer" (RL) orthogonal, fixed with two levels (UL and LL). Cochran's *C*-test was used to determine whether variances were heterogeneous and therefore if any data required an appropriate transformation (Underwood, 1997). Student Newman–Keuls (SNK) a posteriori comparison tests were used. ANOVA and SNK tests were carried out using GMAV5 for Windows Statistical Software (Institute of Marine Ecology, Sydney, Australia).

Results

Non-colonial organisms

A total of 78 non-colonial taxa were identified. The best represented taxonomical groups were the classes Polychaeta (31), Gastropoda (7), Bivalvia (8) and the orders Amphipoda (10), Isopoda (2) and Decapoda (7).

Data analysis for the non-colonial organisms showed similar values for *D* (20,681–25,066 individuals m^{-2}), for H' (21–27 species) and *S* (3.3–3.8) between reef and reef structure (Table 1). The two-way ANOVA showed no significant differences for the above-mentioned variables for both RD and RL factors.

Six species, *Gregariella subclavata* (Libassi), *Musculus* cf. *subpictus* (Cantraine), *Hiatella arctica* (L.), *Microdeutopus versiculatus* Chevreux, *Jassa marmorata* Holmes, and *Paleanotus* cf. *bellis* (Johnson), contributed to 67.1% of the total mean density values of the non-colonial organisms. The bivalves *G. subclavata* and *M.* cf. *subpictus* were the most abundant species. *G. subclavata* presented higher values in reef A than in reef C, and *J. marmorata* also showed a marked decrease with depth (Fig. 1).

The two-way crossed ANOSIM showed a significant difference between A and C reefs,

Table 1 Mean and standard deviation of density (*D*, number of individuals m^{-2}), number of species (*S*), Shannon–Wiener diversity (H') and biomass (*B*, ww g m^{-2}) of macrofaunal assemblages calculated for each group and layer of modules

	Reef Depth		Reef Layer			
	A	C	UL-A	UL-C	LL-A	LL-C
Non-colonial organisms						
D	23,585.2 ± 8,148.09	22,133.3 ± 5,572.99	25,066.7 ± 8,457.07	23,585.2 ± 3,393.34	22,103.7 ± 9,374.02	20,681.5 ± 7,733.50
S	25.3 ± 3.27	22.7 ± 5.35	27.7 ± 2.52	21.3 ± 7.37	23.0 ± 2.00	24.0 ± 3.46
H'	3.8 ± 0.15	3.4 ± 0.55	3.8 ± 0.21	3.3 ± 0.84	3.8 ± 0.10	3.5 ± 0.10
Barnacles and colonial organims						
B	1,309.8 ± 474.58	966.4 ± 519.45	1,716.3 ± 136.82	1,372.0 ± 374.05	903.3 ± 220.60	560.8 ± 202.81
S	7.2 ± 0.75	5.5 ± 1.38	7.3 ± 0.58	6.3 ± 1.53	7.0 ± 1.00	4.7 ± 0.58
H'	0.5 ± 0.26	0.2 ± 0.20	0.6 ± 0.26	0.2 ± 0.28	0.3 ± 0.18	0.2 ± 0.14

A = –16 m; C = –20 m; UL = Upper Layer; LL = Lower Layer

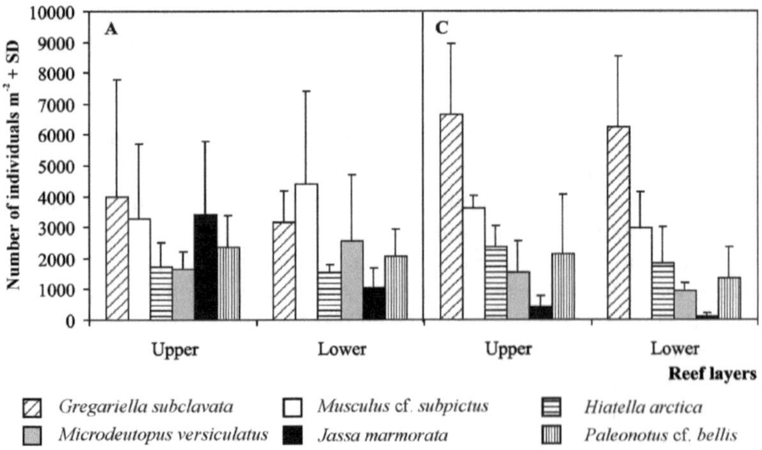

Fig. 1 Mean density values and standard deviation (SD) of the most abundant non-colonial species in the two reef groups (A and C) and layers of modules (Upper and Lower)

$R = 0.53$ ($P = 0.02$), but no dissimilarity between the Upper and the Lower layers of modules was observed ($R = 0$, $P = 0.52$). Furthermore, the MDS ordination presented an evident separation of the samples with differing depth (Fig. 2a).

As a RD effect was indicated by the ANOSIM test, a SIMPER analysis was carried out for A and C reef groups. The dominant taxa at A reef were *J. marmorata*, undetermined Turbellaria, *Syllidia armata* Quatrefages, and *Serpula vermicularis* L. (Table 2a). This reef presented exclusive taxa, such as undetermined Nudibranchia and *Nereis* cf. *zonata* Malmgren. On the other hand, *Corophium sextonae* Crawford, *Anomia* cf. *ephippium* L., *Pomatoceros triqueter* (L.) and *Pisidia* cf. *bluteli* (Risso) were particularly dominant at reef C.

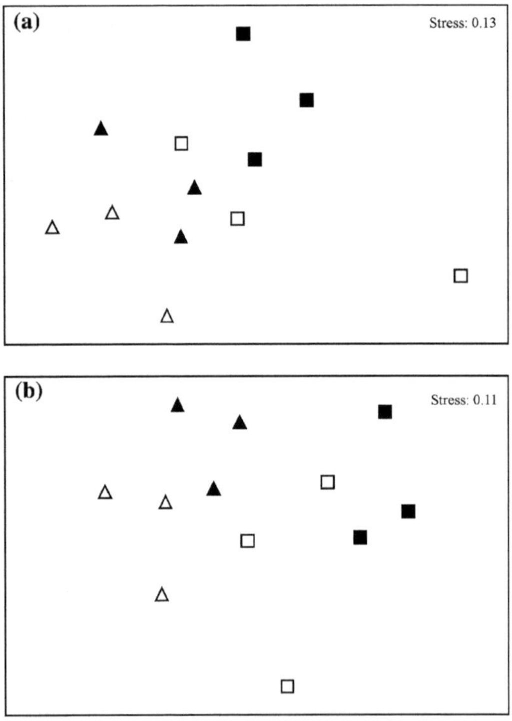

Fig. 2 MDS ordination plots for reef groups and layers of modules of (**a**) non-colonial organisms abundance and (**b**) barnacles and colonial organisms biomass (A reef: Upper layer = △; Lower layer = ▲; C reef: Upper layer = □; Lower layer = ■)

Barnacles and colonial organisms

Four species of the subclass Cirripedia were determined. Colonial organisms were mainly represented by Bryozoa (6 taxa) and Porifera (3 taxa).

Data analysis for the barnacles and colonial organisms showed higher *B* values in reef A (1,309 ww g m^{-2}) than in reef C (966 ww g m^{-2}), and in UL (1,372–1,716 ww g m^{-2}) to LL (903–560 ww g m^{-2}). *H′* and *S* values were higher in lower depths, within layers of modules *H′* and *S* values were similar (Table 1). The two-way ANOVA for *B* showed significant differences for both depth and layer factors (Table 3). B values at reef A were significantly higher than at reef C. *B* was also higher in the UL than in the LL (see SNK in Table 3). *H′* and *S* also displayed a

Table 2 Differences (< and >) in mean density (mean number of individuals m^{-2}) for non-colonial organisms and mean biomass values (ww g m^{-2}) for barnacles and colonial organisms. Contrib.% = taxa percentage contribution, from SIMPER analysis, to dissimilarities between (a) reef groups and (b) layers of modules

Taxa	A		C	Contrib.%
(a) Reef groups				
Non-colonial organisms (mean number of individual m^{-2})				
undetermined nudibranchia	385.19	>	0	4.06
Jassa marmorata	2251.85	>	266.67	3.59
Undetermined Turbellaria	414.81	>	88.89	2.73
Syllidia armata	1600	>	948.15	2.63
Corophium sextonae	266.67	<	325.93	2.6
Anomia cf. *ephippium*	177.78	<	948.15	2.59
Serpula vermicularis	385.19	>	207.41	2.51
Pomatoceros triqueter	59.26	<	237.04	2.39
Nereis cf. *zonata*	118.52	>	0	2.25
Pisidia cf. *bluteli*	177.78	<	355.56	2.19
Barnacles and colonial organisms (ww g m^{-2})				
Bugula neritina	162.45	>	0.17	22.27
Balanus perforatus	17.21	>	1.48	10.99
Scruparia chelata	6.43	>	0.03	10.57
Undetermined bryozoan sp. I	6.36	>	2.02	7.97
Megabalanus tulipiformis	6.51	<	43.69	6.92
Undetermined bryozoan sp. II	0.66	<	5.4	5.83
Balanus amphitrite	1,087.24	>	903.13	5.74
(b) Layers of modules	Upper		Lower	
Barnacles and colonial organisms (ww g m^{-2})				
Bugula neritina	145.46	>	17.16	16.26
Balanus perforatus	16.8	>	1.88	11.14
Undetermined bryozoan sp. I	0.09	<	8.29	9.21
Megabalanus tulipiformis	50.2	>	0	9.03
Balanus amphitrite	1,306.62	>	683.76	8.28
Scruparia chelata	4.35	>	2.11	7.12

A = –16 m; C = –20 m

depth effect as values were significantly higher at reef A than at reef C (Table 3).

Balanus amphitrite Darwin comprised 87.4% of the total mean biomass of the barnacles and colonial organisms. This species presented higher values in the UL of both reefs, especially at reef A. The bryozoan *Bugula neritina* (L.), ranking second in biomass values accounted for 7% of the

Table 3 Two-way ANOVA on mean biomass (*B*), number of species (*S*) and Shannon–Wiener diversity (*H′*) of barnacles and colonial organisms collected in the reef groups and layers of modules

Source of variation	df	*B*			*S*			*H′*		
		MS	*F*	*P*	MS	*F*	*P*	MS	*F*	*P*
Reef depth	1	661.41	5.69	0.04*	16.33	8.33	0.02*	0.38	5.73	0.04*
Reef layer	1	384.24	31.85	0.00***	1.33	3.00	0.12 ns	0.03	1.68	0.23 ns
Depth × layer	1	2.34	0.00	0.99 ns	48.0	1.33	0.28 ns	0.05	1.14	0.32 ns
Residual	8	106.47			19.75			0.19		
Cochran's test		*C* = 0.56 ns			*C* = 0.58 ns			*C* = 0.40 ns		
SNK tests										
Reef depth		Depth, SE = 1.02 A reef > C reef*			Depth, SE = 0.41 A reef > C reef*			Depth, SE = 0.13 A reef > C reef*		
Reef layer		Layer, SE = 1.02 Upper layer > Lower layer**								

(ns = not significant; $^*P < 0.05$; $^{**}P < 0.01$; $^{***}P < 0.001$). A = –16 m; C = –20 m

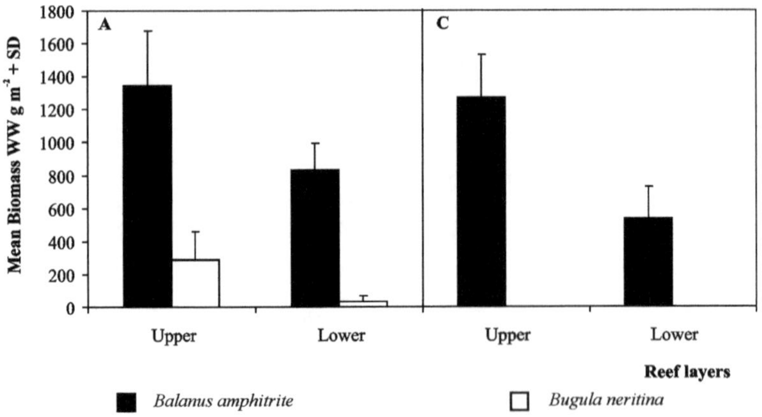

Fig. 3 Mean biomass values and standard deviation (SD) of two the most dominant barnacles and colonial species in the two reef groups (A and C) and layers of modules (Upper and Lower)

total, showing a decline with increasing depth (Fig. 3).

Multivariate analyses using biomass values showed a clear effect of depth ($R = 0.80$, $P = 0.01$), i.e. assemblages at 16 m were significantly different from those at 20 m. There were also significant differences between assemblages in UL and LL ($R = 0.54$, $P = 0.02$). The MDS ordination corroborated the differences observed in ANOSIM analyses for depth and layer factors (Fig. 2b).

SIMPER analysis showed that *Bugula neritina* and *Balanus perforatus* Brugiére have a strong contribution to the dissimilarity between depths (Table 2a), with higher B values in the reef A (Table 2a). *Scruparia chelata* (L.) and undetermined Bryozoa sp. I presented high B values in reef A, and *Megabalanus tulipiformis* (Ellis) presented highest values in reef C. Concerning the reef structure, the species *Bugula neritina* and *Balanus perforatus* presented higher values in the UL (Table 2b). Furthermore, the barnacle *M. tulipiformis* was exclusive of the UL and undetermined Bryozoa sp. I showed high B values on the LL. Despite being characteristic of every group of samples, *Balanus amphitrite* showed twice higher biomass values in UL compared to those observed in the LL.

Discussion

The biodiversity of ARs is related to different environmental and structural factors, such as morphological complexity of the reef, substratum composition, depth and distance from natural reef areas (Bohnsack et al., 1991). In the present work, the distributions of many species were depth-related, although this observation was more evident for colonial and barnacle species. The results were consistent with previous work (Relini et al., 1994; Kocak & Zamboni, 1998), with biomass and the number of sessile species decreasing with depth. Some faunal groups, like bryozoans, barnacles and molluscs occurred in different proportions according to depth. This correlation with depth was also observed in the Loano AR (Ligurian sea, Italy) (Kocak & Zamboni, 1998). Some species like *J. marmorata* and *B. neritina* were associated to lower depths, while *A.* cf. *ephippium*, *G. subclavata* and *M. tulipiformis* showed the reverse trend. Depth and reef structure alter light conditions and thus light is responsible for changes in community structure and composition (Bohnsack et al., 1991). Glasby (1999b) observed that differences in light intensity could influence (direct or indirectly) the settlement of various species, including bryozoans. The observed communities of the "Faro/Ancão" AR, after 6 months of submersion, do not reflect a mature community. The time of colonisation needed to achieve mature macrofaunal communities for the Algarve reefs is still being studied. In the present study, the macrobenthic community of "Faro/Ancão" AR corresponds to a settlement period, with serpulids, barnacles, bryozoans and molluscs dominating the community. Nevertheless, this community typically composed of pioneer taxa, showed higher biodiversity (91 macrobenthic taxa after 6 months of immersion) than the Fregene AR, Tyrrhenian Sea, Italy (10 benthic invertebrate taxa after 6 months and 135 after 4 years of immersion) (Ardizzone et al., 1989) and than the Loano AR, Ligurian sea, Italy (57 species after 12 months) (Relini et al., 1995).

The bivalves (*G. subclavata* and *M.* cf. *subpictus*) and barnacles (*B. amphitrite*) dominated the macrobenthic communities in Algarve ARs, exhibiting a similar trend to that described for the modules at the Tyrrhenian Sea (Ardizzone et al., 1989). The opposite was observed in Loano AR, where these faunal groups were never dominant (Relini et al., 1994). Contrasting with the results obtained for other European reefs (Relini et al., 1994; Collins et al., 1995; Garrido et al., 1999), macroalgae were not found colonising the reefs during the 6-month period of this study. In Portugal, as in the Adriatic (Badalamenti et al., 1992) and Tyrrhenian Sea (Ardizzone et al., 1989), the benthic communities were characterised by the absence of macroalgae, being dominated by vagile and sessile macrofauna.

Regarding the layer effect, biomass and number of barnacle species and colonial species decreased from upper to lower layers of modules. In the Alcamo Marina reef, Tumbiolo et al. (1995) observed, however, the opposite pattern, with biomass increasing from the top to the lowest layer of modules of the reef. In the Alcamo Marina reef, the amount of sediment was positively related to biomass, affecting macrobenthic communities (Tumbiolo et al., 1995). Glasby (1999b) observed differences between assemblages close to and far from the seafloor, linking them to differences in sedimentation. In the present study, however, the assemblages at the upper layer of modules in the "Faro/Ancão" AR were probably less affected by darkness and other physical factors, such as sedimentation.

As an overall conclusion, the small differences in depth (16–20 m) and the structure of the ARs influenced the development of subtidal epibiotic assemblages at the "Faro/Ancão" AR. This effect was more obvious for colonial organisms and for barnacles; and it seems that, for biomass values, the reef structure was important. The reasons for such differences are not clear, but may involve factors such as light, predation/grazing, larval behaviour and water flow at micro- or meso-scales.

Acknowledgements The authors would like to thank the INIAP/IPIMAR scientific diving team and research vessel "NI TELLINA" crew for helping in fieldwork. This work was carried out within the project "Implantação e estudo integrado de sistemas recifais", supported by the MARE programme.

References

Ardizzone, G., M. F. Gravina & A. Belluscio, 1989. Temporal development of epibenthic communities on artificial reefs in the central Mediterranean Sea. Bulletin of Marine Science 44: 592–608.

Ardizzone, G., A. Somaschini & A. Belluscio, 1997. Biodiversity of European artificial reefs. In Jensen, A. (ed.), European Artificial Reef Research. Southampton Oceanography Centre, 39–59.

Badalamenti, F., G. D'Anna, M. Gristina, M. Scalisi & L. Tumbiolo, 1992. Remarks on a method to quantify the total biomass of a benthic community on artificial substrata. Rapport Commission Internationale Mer Mediterranée 33: 377.

Bohnsack, J. A. & D. L. Sutherland, 1985. Artificial reef research: a review with recommendations for future priorities. Bulletin of Marine Science 37(1): 11–39.

Bohnsack, J. A., D. L. Johnson & R. F. Ambrose, 1991. Ecology of artificial reef habitats and fishes. In Seaman, Jr. W. & L. Sprague (eds), Artificial Habitats for Marine and Freshwater Fisheries. Academic Press, San Diego, 61–107.

Collins, K., A. Jensen & J. Mallinson, 1995. Biological development of a stabilized coal ash artificial reef, Poole Bay, UK. In Proceedings of the International Conference on Ecological System Enhancement Technology for Aquatic Environments "ECOSET-95", Tokyo, 119–124.

Garrido, M. J., R. Haroun & R. Herrera, 1999. Structure and dynamics of marine macroinvertebrate communities at Canarian artificial reefs (Central-East Atlantic Ocean). In Relini, G., G. Ferrara & E. Massaro (eds), Proceedings of the Seventh International Conference on Artificial Reefs and Related Aquatic Habitats. Erredi Grafiche Editoriali, Genova, 114–120.

Glasby, T. M., 1999a. Effects of shading on subtidal epibiotic assemblages. Journal of Experimental Marine Biology and Ecology 234: 275–290.

Glasby, T. M., 1999b. Interactive effects of shading and proximity to the seafloor on the development of subtidal epibiotic assemblages. Marine Ecology Progress Series 190: 113–124.

Itosu, C., Y. Komai & H. Sakai, 1999. Estimation of food organism production on steel-made artificial reef. In Relini, G., G. Ferrara & E. Massaro (eds), Proceedings of the Seventh International Conference on Artificial Reefs and Related Aquatic Habitats. Erredi Grafiche Editoriali, Génova, 150–157.

Kocak, F. & N. Zamboni, 1998. Settlement and seasonal changes of sessile macrobenthic communities on the panels in the Loano artificial reef (Ligurian Sea, NW Mediterranean). Oebalia 24: 17–37.

Relini, G., N. Zamboni, F. Tixi & G. Torchia, 1994. Patterns of sessile macrobenthos community development on an artificial reef in the Gulf of Genoa (northwestern Mediterranean). Bulletin of Marine Science 55: 745–771.

Relini, G., M. Relini, G. Torchia, F. Tixi & C. Nigri, 1995. Coal ash tests in Loano artificial reef. In Proceedings of the International Conference on Ecological System Enhancement Technology for Aquatic Environments "ECOSET-95", Tokyo: 107–113.

Santos, M. N., 1997. Ichthyofauna of the artificial reefs of the Algarve coast. Exploitation, strategies and management of the local fisheries. PhD Thesis, University of Algarve, 267 pp.

Santos, M. N. & C. C. Monteiro, 1997. The Olhão artificial reef system (south Portugal): fish assemblages and fishing yield. Fisheries Research 30: 33–41.

Santos, M. N. & C. C. Monteiro, 1998. Comparison of the catch and fishing yield from an artificial reef system and neighbouring areas off Faro (Algarve, south Portugal). Fisheries Research 39: 55–65.

Steimle, F., K. Foster, R. Kropp & B. Conlin, 2002. Benthic macrofauna productivity enhancement by an artificial reef in Delaware Bay, USA. ICES Journal Marine Science 59: S100–S105.

Svane, I. & J. K. Petersen, 2001. On the problems of epibioses, fouling and artificial reefs, a review. Marine Ecology 22: 169–188.

Tumbiolo, M. L., F. Badalamenti, G. D'Anna & B. Patti, 1995. Invertebrate biomass on an artificial reef in the Southern-Tyrrhenian Sea. In Proceedings of the International Conference on Ecological System Enhancement Technology for Aquatic Environments "ECOSET-95", Tokyo, 324–329.

Underwood, A. J., 1997. Experiments in Ecology. Their Logical Design and Interpretation Using Analysis of Variance. Cambridge University Press, Cambridge, 504 pp.

Hydrobiologia (2007) 580:181–191
DOI 10.1007/s10750-006-0454-z

Stakeholder perceptions regarding the environmental and socio-economic impacts of the Algarve artificial reefs

Jorge Ramos · Miguel N. Santos · David Whitmarsh · Carlos C. Monteiro

Abstract The artificial reef (AR) complex of the Algarve (Southern Portugal), deployed for the purpose of restoring and enhancing fisheries resources, is currently the largest structure of its kind in Europe, extending for over 43.5 km^2. Such a structure can be expected to have had both positive and negative impacts. To evaluate the overall perception of the effects of deployment, a survey of stakeholders' opinions was undertaken based on a set of questions addressing various dimensions (environmental, social, and economic). The survey covered 44 key-stakeholder representatives distributed in six groups: commercial fishermen associations, anglers associations and clubs, diving schools and clubs, fisheries and environmental administrators, natural and social scientists, and local council representatives in the fisheries and/or environmental sectors. The opinions of stakeholders were measured using summated rating scales. The results obtained reflect the most important issues be impacted and the possibility of using them as indicators of relative success or failure. From a total of 12 factor-sets of impacts, the results showed that in general the environmentally related were the ones having had the most positive results. The overall perception of the environmental factor-sets specified as the 'deployment area use' revealed that the artificial reefs were an incentive to users and that the structures were perceived as a satisfactory tool to support the fishery and its management. In both cases divers were the strongest supporters. A closer look at the results presented in the form of an AMOEBA plot showed that there were other factor-sets perceived as impacting positively in other dimensions. Such examples are the factor-sets 'opinion' and 'production and benefits' lying respectively in the social and economic dimensions. The latter factor-set was even the only one having the support of five out of six stakeholder-types. As expected, in general different stakeholder-types take somewhat different positions and attitudes towards AR impacts: usually scientists are the most optimistic, whereas fishermen take the most sceptic view.

Keywords Artificial reefs (ARs) · Impact analysis · Fisheries · Indicators for management

Guest editors: G. Relini & J. Ryland
Biodiversity in Enclosed Seas and Artificial Marine Habitats

J. Ramos · M. N. Santos (✉) · C. C. Monteiro
Instituto Nacional da Investigação Agrária e das Pescas/IPIMAR Centro Regional de Investigação Pesqueira do Sul (CRIPSul), Av. 5 de Outubro s/n, 8700-305 Olhão, Portugal
e-mail: mnsantos@cripsul.ipimar.pt

J. Ramos · D. Whitmarsh
CEMARE, University of Portsmouth, Boathouse no 6, College Road, HM Naval Base, Portsmouth PO1 3LJ, UK

Introduction

The Algarve (southern Portuguese coast) is a region that has been highly impacted by a multi-purpose active fleet exploiting fisheries resources (Moniz et al., 2000). These resources attain on average some of the best market prices in the country for many fish species as reported by the Fisheries and Aquaculture Directorate (DGPA, 2002), a feature which acts as an incentive to keep fishermen employed in the activity. A socio-economic characteristic of the region is that a high proportion of people depend on fishing for their livelihoods (Moniz, 1997; Moniz et al., 2000). In 2003, over 3,500 fishermen were engaged in sea-fisheries in the region (DGPA, 2004), the majority of whom fished all year round. A survey by DGPA (2000) revealed that fishing tends to be a very erratic activity, with many fishers taking advantage of other economic activities linked with tourism in order to provide an additional or substitute source of income during the summer season. Apart from commercial fishing there are many other activities in the Algarve region that are directly dependent on fish resources, notably angling and diving. The former is practised by enthusiasts from local clubs all year round, but particularly by the end of the summer and fall; whereas the latter is practised mainly during warmer months, i.e., April to September.

In recent years a number of strategies have been developed to address the misuse of fish resources that has occurred in the past. Artificial reefs (ARs) represent one such approach, and indeed have become commonly used world wide to aggregate fish species in the marine environment (Aabel et al., 1996). In the Algarve, after a successful experience with pilot ARs (Santos & Monteiro, 1997, 1998), it was decided that reefs should be deployed on a larger scale throughout the region, but particularly in the windward area (Monteiro & Santos, 2000). The main reason was that in this area there is less abundance of rocky bottoms, having instead, muddy or sandy bottoms, the latter being a prerequisite for AR deployment decision. The deployment of the main program structures started in 1996 and was concluded by summer 2003. The Algarve's ARs were created with traditional small-scale fishing use in mind (Santos & Monteiro, 2001). The initial objectives of the program were to: (a) promote bio-diversity; (b) protect juveniles of commercial species; (c) manage coastal resources; (d) reduce fishing exploitation costs; (e) recover fishing resources; (f) create fishing zones; and (g) adapt gear and fishing strategies to resources availability. The program scope was wide, encompassing both economic and social objectives, but having in mind mainly an environmental focus.

To date, scientific evidence shows that there has been an increase in abundance of economically-important fish species on the pilot ARs surveyed since their deployment in the early 1990s (Santos & Monteiro, 1997, 1998; Whitmarsh et al., 2004). Over this period, users have been accumulating greater empirical knowledge of the effects of ARs, and this has undoubtedly influenced both their behaviour (e.g. fishing patterns) and attitudes towards the reef programme as a whole. These attitudes may, of course, be negative as well as positive. Experience suggests that ARs typically give rise to a range of impacts, not all of which may be perceived as beneficial by users. For example, while CPUE and incomes may be enhanced, at least in the short term, the attraction of more vessels is likely to increase user conflict (Milon, 1989; Samples, 1989; Galvez, 1991; Murray & Betz, 1994). The aim of this paper is to investigate the local community's perceptions of the Algarve artificial reefs, and to see to what extent people regard the reef programme as having been successful. We contend that the opinions of stakeholders are crucial in this context. When there is a consensus amongst key individuals and groups over the objectives of ARs, it becomes easier to establish whether these objectives have been adequately addressed and how close they are to being reached. By contrast, lack of consensus makes it more difficult to derive a clear and unambiguous indicator for evaluating performance. Stakeholder opinions towards socio-economic as well as environmental objectives need to be considered, and this sort of data typically has to be collected via surveys (Milon et al., 2000). Such information is important to fish managers since they would like to know which impacts are acceptable and which are not. To carry out this sort of study it is important to consult properly all the local stakeholders and interest groups. Studies which have examined the impact of artificial reefs, particularly as they involve stakeholders, have commonly concerned the 'rigs-to-reefs' conversion of obsolete oil platforms to other uses (Reggio et al., 1986; McGurrin & Fedler, 1989; Reggio, 1989; Cripps & Aabel, 2002).

In this paper we have undertaken a simple analysis of perceived impacts by consulting a panel made up of different key-stakeholders involved in the AR deployment process and its use. The panel's overall perception of the effects resulting from AR deployment in the Algarve south coast may help resource managers to use achieved results in order to find out trade-offs between policy objectives.

Materials and methods

Though the reef deployment programme was only completed in 2003, the presence of artificial habitats since the early 1990s is acknowledged to have had a biological impact. While it is essential to explore the economic implications of this, particularly in respect of measurable quantities such as catches and incomes, it is important also to find out how far people regard the reef programme as successful along a wide spectrum of performance criteria. A number of techniques may potentially be used to measure stakeholder attitudes (Robson, 2002), and the particular approach adopted here is outlined below.

Questionnaire survey

As a first step in the study, three dimensions expected to be impacted by reef deployment were selected: environmental, social, and economic. For each dimension, factors likely to be affected by deployment were identified, from which an item-pool was constructed (Bell, 1987; Robson, 2002) which included all the perceived predefined impacts. The item-pool consisted of 54 ambiguous-free relevant items to be included in the survey of respondents' opinions (Table 1). The item-pool was then adapted to a specific questionnaire addressed to a range of people with

Table 1 Brief description of each of the 12 factor-sets and the number of impacts addressed to the key-stakeholders

Dimension	Factor-set	Brief description	No. items
Environmental	A. Deployment area use	To assess stakeholders' perception on the use that can be found in the area.	4
	B. Ecological impact and bio-diversity	Effects caused on the species, namely their aggregation and protection after reef deployment.	5
	C. Pollution	The contribution of the structures as a factor of pollution to the environment (water or sediment).	3
	D. Fishery and management	ARs as a management tool for fisheries (traditional fishing, off-shore aquaculture, etc).	6
Social	E. Demography and employment	Signs of changes in social aspects (people migration, employment, and social benefits).	3
	F. Enforcement and communication	The need to establish sea use rules and communication between the different players.	4
	G. Opinion	How is the AR' deployment perceived by stakeholders and the public in general.	5
	H. Conflicts	Possibilities of conflicts occurrence between the different stakeholders involved.	6
Economic	I. Production and benefits	To evaluate the chances of extra catches and returns after reef deployment.	4
	J. Costs to society	Awareness of the costs involved in the reef deployment process.	5
	K. Changes in local economy	Signs of changes in the local economy in all the sectors of activity after reef deployment.	5
	L. Safety at sea	Reefs contribution to promote safer fishing activities in their deployment area.	4

different perceptions and educational levels. The questionnaire was pre-tested and adjusted. Prior contact was established both by post or e-mail. This was accompanied by an introductory letter explaining the objectives of the proposed work, and an informative memorandum including the purposes of the ARs and their structure, organisation and location along the south-coast. The questionnaires were sent directly to each representative by hand or via post mail, and were addressed to the highest representative of each body/institution, or to the person used to work with fisheries or environmental issues. The survey was carried out approximately one year after the conclusion of the deployment phase, during a period of forty-five days (from the middle of May to the end of June 2004).

Conceptual framework and stakeholders

Though the consultation was principally a retrospective assessment of the performance of the established in situ reefs, the responses given to the questions also give an indication of the expected effects of the newly-established reefs and how far they are likely to meet the needs of stakeholders. Indeed, the attitudes of affected parties regarding the acceptability of ARs should be an element in any decision regarding future reef deployment, particularly as regards design and location. In choosing respondents to take part in the survey, individuals were pre-selected from a key-stakeholder database created from the regional yellow pages and from a fisheries events invitation list. The panel was constructed from key-stakeholders based on their agreement to take part in the survey.

The survey was addressed to representatives of: (a) fishermen associations; (b) anglers clubs and associations; (c) divers clubs; (d) environmental and fisheries administrations; (e) natural, social and economic scientists; and (f) others as borough council representatives in the environmental and/or fisheries areas. The key-stakeholders panel consulting approach was similar to the one described by McKinnon & Forster (2000), where: (i) items were kept simple, and averaged member's views encouraging a consensus within the same institution; (ii) information was collected by questionnaire, without using interviews or subject discussion; and, (iii) anonymity was guaranteed to the members of the panel.

Each one of the six stakeholder-types presented six to eight representatives. In order to understand stakeholders' involvement with the ARs, representatives were divided into two groups: those who knew the structures either by using them or by being involved since the pre-deployment process, and the others whose knowledge was solely by other means (e.g. by the media). The first three types of panel members represent the direct or potential users, whereas the other three are usually involved in the ARs process but mostly as institutional representatives.

The impact assessment validation

Key-stakeholders used 5-point Likert scales to state their positions about impacts (Murray & Betz, 1994; Cripps & Aabel, 2002; Kennish et al., 2002). Perceptions/attitudes were then measured using summated rating scales. Items were graded accordingly to the probable perturbations in the marine system caused by reef deployment as well the effects on the fishing communities nearby. After collecting all questionnaires, impacts of the AR deployment were defined according to their scores and the analyses carried out by dimension, stakeholder type, factor-sets, and the most meaningful items. To evaluate the level of the impact it was important to define a priori what constituted a 'positive impact', since this underlay the whole concept and measurement of success in policy terms.

The survey made use of an AMOEBA plot, which is a graphical device that uses a 'radar' diagram. Though the approach is simplistic it has the advantage of representing to respondents (usually managers and policy makers) the impact of an intervention in a clear and easily understandable manner (Ten Brink et al., 1991). In the current study an AMOEBA plot was used consisting of three areas: inner (negative impacts), middle (no evident impact), and outer (positive impacts). The AMOEBA reading shows that the perception on the AR complex impact assessed over 12 factor-sets is not expressed as a function of others (discrete variables). In this way we can obtain a visual impression of whether an impact on any one dimension or factor has been positive or negative.

The results obtained concerning ARs' impacts after deployment have particular significance for stakeholders. The analysis was undertaken by separating the panel of stakeholders according the group of interests. To demonstrate the differences in stakeholders' positioning, hypotheses were tested for the whole impact using a simple *t*-test (Zar, 1996). The *t*-test was carried out for the analyses on dimension, factor-set, and stakeholder-type. It was decided to work on a percentage basis where the overall score had three critical thresholds: scores over 66.7 implied that the effect was positive, those falling between 33.3 and 66.7 signified that the impacts were largely neutral, while those below 33.3 were interpreted as negative. For each item individually summated rating scales were also defined showing the top and bottom impacted ones.

Results

Key-stakeholders' characteristics

The total number of contacted stakeholders representing regional entities was 53. Of these, 9 stakeholders explicitly declined to collaborate, did not answer the calls, did not fill the questionnaire during the stipulated time, or simply filled the questionnaire in an invalid way (Table 2). The final panel consisted in 44 respondents, where 28 knew already the structures, and 16 were just somewhat familiarised with artificial reefs built off the Algarve's coast. The most familiarised group was fishermen, and the least one was the group of borough council representatives.

Table 2 The key-stakeholders contacted and its relationship with the ARs

Contacted (n = 53)				
Stakeholder type	Denied	Agreed		
		Total	AR experience	
			Heard	Known
Fishermen	2	7	0	7
Anglers	2	7	2	5
Divers	1	6	1	5
Administrators	2	8	2	6
Scientists	1	8	5	3
Others	1	8	6	2
Total	9	44	16	28

Key-stakeholders' perception

In terms of impact perception, the majority of stakeholders were positive concerning the environmental impact caused by ARs. Stakeholders who had first-hand experience of the structures were even more optimistic than those who had simply heard about them. Concerning the environmental terms, around 60% of the answers showed that reef deployment had made a positive contribution, against 20% believing the impact was negative. By contrast, for social and economic effects both types of stakeholders were more cautious in making statements about the potential impacts. The areas corresponding to the neutral position reflected in some way stakeholders' difficulty in formulating judgements. Around one third of the social and economic dimensions remain in this position (Fig. 1). For the economic and social dimensions, less than 50% of the answers were accounted as positive and more than 20% negative.

Key-stakeholders' general positioning and dimension analysis

Despite the differences found between those stakeholders who knew the reefs from first-hand experience and those that had only heard about them, it can be confirmed that the entire panel thought that the most important positive impacts belonged to the environmental dimension (Fig. 2). The hypotheses tested showed that only fishermen and anglers were not sure about the environmental overall impact of the ARs. By contrast, divers and scientists were the most optimistic (Table 3). Anglers, divers, and administrators considered that economic impacts overshadowed social impacts, whereas scientists and others claimed the opposite. In addition, fishermen representatives were the most sceptical among all concerning to the economic dimension of the reefs, contrasting with administrators who

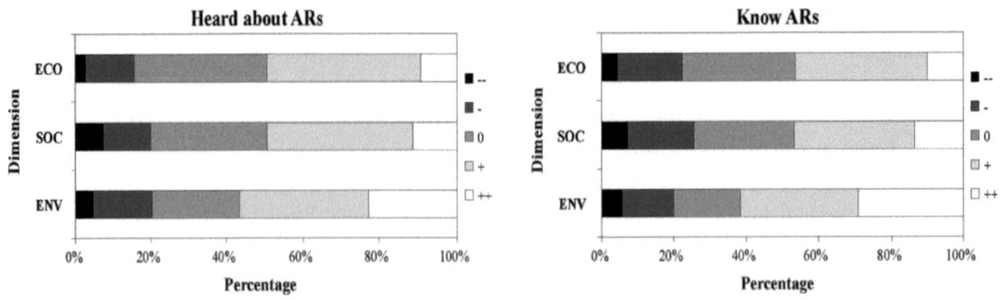

Fig. 1 Stakeholders' perception about impacts: (a) Stakeholders who only heard about ARs (n = 864 answers), (b) stakeholders that know ARs (n = 1,512 answers). "Minus" signs represent the percentage of impacts perceived as negative. "Plus" signs represent the percentage of impacts perceived as positive. Double minus/plus mean respectively high improbability/probability of occurrence. The "zero" represents answers with no clear position taken or perceived by stakeholders. Legend: ENV, Environmental; SOC, Social; ECO, Economic

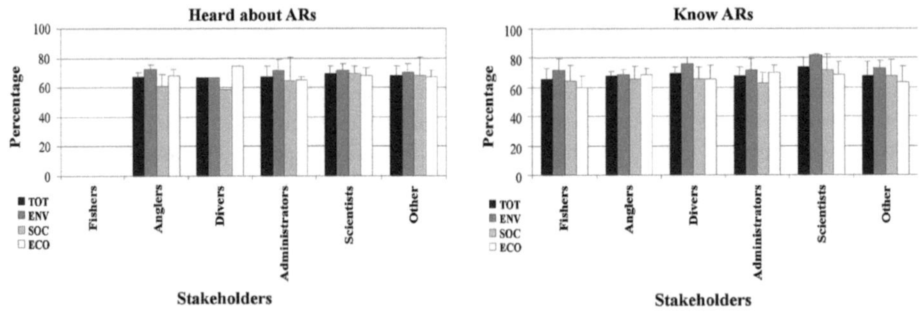

Fig. 2 Stakeholders' positioning about impacts caused in each dimension. Legend: ENV, Environmental; SOC, Social; ECO, Economic; TOT, All previous three dimensions together

strongly supported their economic role. Scientists were the most favourably inclined towards the social role of the reefs.

Factor-sets analysis

By disaggregating each dimension in their factor-sets through an AMOEBA plot, it was possible to perceive important impacts detected by the entire panel of stakeholders (Fig. 3). A refinement of the AMOEBA plot showed that among the 12 factors, only four can be considered as positively significant (Table 4). In the environmental dimension, the only factor not having a visible positive impact is related to 'pollution', whereas all the other factors are positively accepted (however, 'ecological impact and bio-diversity' was rejected by the t-test). The social dimension, by contrast, showed only a strong positive factor related with the 'opinion' demonstrating that these structures were generally welcomed. For its part, the economic dimension seemed to have factor-sets perceived sceptically in terms of some factors (for example 'costs to society' and 'safety at sea') but more favourably in terms of others (e.g. 'production and benefits').

Table 3 Statistical results using t-test for impacted dimensions. '++' for $p < 0.01$, '+' for $P < 0.05$, and the 'n.s.' for non-significant results

Stakeholder	Dimension			
	Environmental	Social	Economic	All
Fishermen	n.s.	n.s.	n.s.	n.s.
Anglers	n.s.	n.s.	n.s.	n.s.
Divers	++	n.s.	n.s.	n.s.
Administrators	+	n.s.	n.s.	n.s.
Scientists	++	n.s.	n.s.	+
Others	+	n.s.	n.s.	n.s.

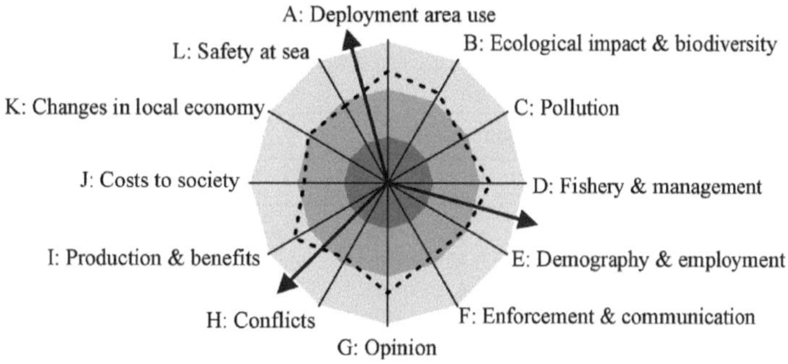

Fig. 3 The AMOEBA plot showing the 12 factor-sets of impacts

Key-stakeholders by factor-set

The disaggregation of the dimensions by factor-set and the entire stakeholder panel by stakeholder-types shows that groups are not identical in their assessment of the impacts (Table 5). After putting together all stakeholders results by factor-set it appears that fishermen were the most sceptical concerning evident positive impacts. For their part, administrators are positive about just two fundamental socio-economic aspects of the reefs: the structures' acceptance and their role as revenue generators. In contrast to the previous groups, divers are the ones believing that four out of 12 impacted factor-sets are positive.

The most positively impacted factor-set was within the environmental dimension and related to the 'deployment area use', where three of the stakeholder-types supported the suggestion that ARs will attract more users to sites. Another factor-set showing favourable results concerned the social dimension, and related to the 'opinion' about AR deployment. Here were found four out of six stakeholder-types with a confident attitude/opinion, while only fishermen and anglers remained unconvinced. The factor-set believed to have had a demonstrated positive effect was the 'production and benefits', with five out of six stakeholder types being strongly favourable to it. Fishermen alone were sceptical or did not reveal their position.

Factor-sets that do not show any significance can also give some clues about AR impact. For instance, environmentally it seems that ARs are not regarded as a source of pollution, since none of the stakeholder-types held a clear position on this aspect. The same situation was found with respect to the role of ARs as a tool to improve 'fishery and management', since apart from divers no other group revealed an attitude that was either strongly positive or negative. Socially it seems that AR deployment is not a significant contributor to 'demography and employment' in the region, a result which is consistent with the belief that ARs will not significantly impact on the local economy.

Table 4 Simple t-test statistics for the AMOEBA-approach refinement. '++' for $P < 0.01$, '+' for $P < 0.05$, and the 'n.s.' for non-significant results

Dimension	Factor-set	Statistics
Environmental	A. Deployment area use	++
	B. Ecological impact and biodiversity	n.s.
	C. Pollution	n.s.
	D. Fishery and management	+
Social	E. Demography and employment	n.s.
	F. Enforcement and communication	n.s.
	G. Opinion	++
	H. Conflicts	n.s.
Economic	I. Production and benefits	++
	J. Costs to society	n.s.
	K. Changes in local economy	n.s.
	L. Safety at sea	n.s.

Key-stakeholders agreed impacts

The survey results enable us to produce a hierarchy of items most significantly affected by the deployment of the reefs (Table 6). Ratings are indicators of stakeholders' sensitivity to impacts.

On the positive side, the overall perception is that: ARs promote a specific habitat enriched with several different species, promoting

Table 5 Simple *t*-test statistics showing stakeholder type by factor-set. '++' for $P < 0.01$, '+' for $P < 0.05$, and the 'n.s.' for non-significant results

Stakeholder	Dimension											
	Environmental				Social				Economic			
	A	B	C	D	E	F	G	H	I	J	K	L
Fishermen	n.s.	n.s.	n.s.	n.s.	n.s.	n.s.	n.s.	n.s.	n.s.	n.s.	n.s.	n.s.
Anglers	+	n.s.	n.s.	n.s.	n.s.	n.s.	n.s.	n.s.	+	n.s.	n.s.	n.s.
Divers	++	n.s.	n.s.	++	n.s.	n.s.	++	n.s.	++	n.s.	n.s.	n.s.
Administrators	n.s.	n.s.	n.s.	n.s.	n.s.	n.s.	+	n.s.	++	n.s.	n.s.	n.s.
Scientists	+	n.s.	n.s.	n.s.	n.s.	n.s.	+	n.s.	+	n.s.	n.s.	n.s.
Others	n.s.	n.s.	n.s.	n.s.	n.s.	n.s.	+	n.s.	+	n.s.	n.s.	n.s.

bio-diversity; being also able to aggregate marine fauna, and the structures are more likely to attract local fishermen than other users. The use of local fishing vessels at the reef area was considered an environmental positive impact once it is recognised as a more sustainable way of fishing, when compared with larger vessels. There are other positive impacts perceived as having the potential to augment catch rates when fishing in the reef area.

Table 6 The top and bottom impacts due the existence of the ARs. Stakeholders' rating averages are indicators of their sensitiveness to each item. Ratings vary between 1.0 (minimum), and 5.0 (maximum). Legend: CF, Commercial fishermen associations; RF, anglers associations and clubs; DV Divers clubs; AD, Administration bodies in fisheries, environment, and fisheries funds managers; SC, Natural and social scientists, and OI, Other institutions as local council representatives in the fisheries and/or environment sectors)

Rank	ARs' positive impacts	Dim	Score	Stakeholder rating averages						
Top 10				All	CF	RF	DV	AD	SC	OI
1	To promote bio-diversity	ENV	204	4.6	4.1	4.4	4.7	5.0	4.6	4.6
2	To contribute for the success in fish enhancement actions	ENV	192	4.4	4.6	3.9	4.3	4.3	4.5	4.6
3	To aggregate marine fauna	ENV	190	4.3	4.0	4.1	5.0	4.3	4.1	4.1
4	To increase the numbers of local fishing vessels in the AR area	ENV	188	4.3	4.3	4.3	4.5	4.4	4.1	4.1
5	To increase the numbers of recreational anglers in the area	ENV	187	4.2	4.0	4.3	4.2	4.4	4.4	3.9
6	To protect juveniles from inshore waters	ENV	186	4.2	4.6	4.3	3.7	4.4	4.1	4.3
7	To demonstrate to users that the program is worthy	SOC	182	4.1	4.0	4.1	4.2	4.0	4.3	4.3
8	To augment the catch in the AR area	ECO	181	4.1	4.3	4.0	4.0	4.0	4.1	4.3
9	To protect some marine life species	ENV	178	4.0	3.7	4.1	4.2	3.4	4.5	4.4
10	To increase the numbers of more divers in the area	ENV	175	4.0	4.1	4.0	4.3	3.8	4.4	3.4
10	To attract users to the near area	SOC	175	4.0	4.0	3.6	4.0	4.1	4.3	3.9
Bottom 10	ARs' negative impacts									
1	To increase the need of sea rules accomplishment	SOC	83	1.9	2.7	2.0	1.5	1.6	1.9	1.6
2	To increase fishing pressure over the AR	ENV	94	2.1	2.0	1.9	2.3	2.3	2.3	2.1
3	To augment fishing gears lost near the AR	ENV	109	2.5	1.9	2.1	1.8	2.8	2.9	3.1
4	To contaminate or pollute the water	ENV	110	2.5	3.0	2.7	2.3	2.3	2.8	2.0
5	To increase local authorities enforcement	SOC	115	2.6	2.6	2.6	2.5	2.6	2.3	3.1
6	To cause more fishing gears damages	ECO	116	2.6	2.0	3.1	2.0	2.6	2.8	3.1
7	To find out other less expansive alternatives	ECO	120	2.7	3.1	2.9	2.2	3.3	2.6	2.3
8	To make no contribution to social benefits	SOC	123	2.8	2.6	3.0	3.0	2.3	3.4	2.6
8	To generate conflicts between fishermen and anglers	SOC	123	2.8	2.6	3.0	2.3	2.6	3.3	2.9
8	To realise that there were better sites to deploy ARs	ECO	123	2.8	2.7	2.9	2.8	3.1	2.6	2.6

Negatively, the worst impact perceived is the lack of enforcement measures to keep sea use rules in the deployment area. Other negative impacts relate to the uncontrolled augmentation of fishing pressure on the reefs, associated especially with the activities of non-local boats having more powerful fishing capacity. Other perceived adverse impacts include the belief that ARs cause a loss of fishing gear which in turn entails additional costs in their replacement.

Discussion

This study shows that the deployment of the Algarve's reef structures has resulted in perceived changes. The program can be considered successful since, apart from few sceptic views, in general key-stakeholders do not oppose to it and indeed there is an overall positive attitude. For the purposes of this study key-stakeholders can be considered as licit 'judges' of the AR program by virtue of their experience, use or knowledge of these structures. Among the key-stakeholders panel it is agreed that, compared to the economic or social dimensions, the environmental dimension seems to be the one impacted most positively. The social dimension can be seen as the one that still remains relatively unaffected by reef deployment, either for better or worse. Moreover, it seems that there are no highly adverse (i.e., impacting negatively) factor-sets, whatever the dimension. The negative effects are specific and relate to the risk of losing gears, conflicts between users, and problems of enforcement. Arguably these can all be overcome through awareness campaigns on how to use the ARs. By its turn, as recommended by Murray & Betz (1994). A slightly unexpected result is that, despite the scientific evidence of increased economic abundance, some stakeholders take a cautious position regarding the economic impacts in the belief that there are no strong signs of visible positive results.

Biological findings show that AR structures are intensely colonised and attract fish assemblages. Monitoring data show that there is an average increment on catches (Santos & Monteiro, 1997, 1998). However, some potential users consider ARs' siting to be generally either unknown or of no interest. The latter situation may arise where, even if the underwater structures are detected by vessels, the quantity of harvested fish is deemed inadequate or the species composition unsuitable; consequently, the site fails to be accepted as a 'hot spot' ground amongst other fishermen. A further constraint on the effectiveness of ARs is that their use may be limited to commercial fishermen who use passive gears (as pots, traps, trammel and gillnets, etc.). Eco-tourism based around charter boats and diving is an activity that could take some advantage from reef deployment, and in particular the depth and range once they are additional features to charter boat passengers or to divers. Due the ARs' depth range, the structures can be used for several levels of divers.

ARs are also a tool to manage coastal resources insofar as they can bring about a spatial separation between inshore fishing vessels, which are attracted to the reef areas, and the more powerful commercial fishing gears (such as trawl and pelagic purse-seine) which are in effect excluded. In addition, considering that the catch can be augmented in the reef area and assuming that the access to the resources is facilitated, exploitation costs can be reduced. Economic impacts usually presuppose a change expressed by a multiplier-effect in output, revenues, and employment. However, while catches and income may well have increased as a direct consequence of deploying these structures, key-stakeholders seem unconvinced about employment effects, i.e. ARs are not believed to make a notable contribution to the number of jobs created in the nearby areas. Whether this is the case *de facto* is not clear, since even though the construction of the reefs contributed to an increase on labour for a certain period of time, there is no firm evidence for employment creation within the fisheries. This is consistent with the results of the study by Kovács (2000). Besides, a sustained increase in economic benefits arising from AR deployment depends crucially on how access is managed, and failure to restrict the number of users may result in stock depletion and a cancelling of any long-term economic gains (Whitmarsh & Pickering, 1999).

A joint collaboration between several key-stakeholders is important in order to know to what extent the impact of the ARs is perceived. A problem that is being faced is the specificity of the AR program. Many stakeholders were consulted before and during the deployment phase, and even after deployment many of the results seem to be based on expectations. This study demonstrates how the key-stakeholders perceptions of the impacts may be empirically measured. By using a summated rating scale and appropriate simple statistics it is possible to make a selection of the most important positive and negative impacts from the entire item-pool. The panel members who took part of the survey emphasised that in the future they would like to be consulted in similar surveys. This is a sign of positive interdisciplinary interest and participation in solving fisheries management problems. Finally, it can be added that these sorts of survey can give some information to fisheries managers about stakeholders' positioning, which can be used as indicators for management. It is important to get more people involved with the reefs use and awareness campaigns towards each user type in particular should be carried out in the future.

Acknowledgements The authors would like to thank the collaboration of the entities involved in answering the questionnaire. The first author is grateful to the Portuguese Foundation for Science and Technology (FCT) for providing funding for his PhD studies (PhD grant reference # SFRH/BD/6197/2001). This study was supported by the MARE program, within the project *Implantação e estudo integrado de sistemas recifais*.

References

Aabel, J. P., S. Cripps & G. Kjeilen, 1996. Oil and gas production structures as artificial reefs. European artificial reef research. In Jensen, A. C. (ed.), Proceedings of the 1st EARRN conference, Ancona, Italy, 391–404.

Bell, J., 1987. Doing Your Research Project: A Guide for First-Time Researchers in Education and Social Science. Open University Press, Buckingham.

Cripps, S. J. & J. P. Aabel, 2002. Environmental and socio-economic impact assessment of Ekoreef, a multiple platform rigs-to-reefs development. ICES Journal of Marine Science 59: S300–S308.

DGPA (Direcção Geral das Pescas e Aquicultura) 2000. As pequenas comunidades piscatórias do sul: descoberta de uma realidade... Plano de extensão pesqueira do sul - PEPE, DGPA (MADRP).

DGPA (Direcção Geral das Pescas e Aquicultura) 2002. Datapescas. 55 (Janeiro-Dezembro).

DGPA (Direcção Geral das Pescas e Aquicultura) 2004. Evolução da frota e pescadores das CPS (1998–2003).

Galvez, R. E., 1991. Some socio-economic issues in artificial reefs management: a case of Lingayen Gulf, Philippines. Tropical Coastal Area Management 6: 1–2.

Kennish, R., K. D. P. Wilson, J. Lo, S. C. Clarke & S. Laister, 2002. Selecting sites for large-scale deployment of artificial reefs in Hong Kong: constraint mapping and prioritization techniques. ICES Journal of Marine Science 59: S164–S170.

Kovács, I., 2000. Os jovens e a renovação da pesca: expectativas e aspirações em relação ao trabalho e à vida profissional. In Moniz, A. B., M. M. Godinho & I. Kovács, (eds), Pescas e pescadores: futuros para o emprego e os recursos. Celta Editora, Oeiras: 43–74.

McGurrin, J. M. & A. J. Fedler, 1989. Tenneco II artificial reef project: an evaluation of Rigs-to-Reefs fisheries development. Bulletin of Marine Science 44: 777–781.

McKinnon, A. & M. Forster 2000. Full report of the Delphi 2005 survey. European logistical and supply chain trends: 1999–2005. TRILOG consortium: TNO (Netherlands) (co-ordinators), Heriot-Watt University, Netherlands Economic Institute, Cranfield Centre for Logistics and Transportation, Chalmers Institute of Technology (Sweden), LaTTS (France).

Milon, J. W., 1989. Economic evaluation of artificial habitat for fisheries: progress and challenges. Bulletin of Marine Science 44: 831–843.

Milon, J. W., S. M. Holland & D. J. Whitmarsh, 2000. Social and economic evaluation methods. In Seaman W. Jr. (ed.): Artificial Reef Evaluation with Application to Natural Marine Habitats, CRC press LLC, Boca Raton, Florida, 165–194.

Moniz, A. B. 1997. Construção de cenários para o sistema sócio-económico das pescas: o caso português. In Monteiro, C. C. (ed.), I Encontro internacional de Vilamoura sobre pescas, Vilamoura: 97–106.

Moniz, A. B., M. M. Godinho & I. Kovács, 2000. Pescas e pescadores: futuros para o emprego e os recursos. Celta Editora, Oeiras.

Monteiro, C. C. & M. N. Santos, 2000. Portuguese artificial reefs. In Jensen, A. C., K. J. Collins & A. P. M. Lockwood, (eds.), Artificial Reefs in European Seas. Kluwer Academic Publishers, Dortrecht, 249–261.

Murray, J. D. & C. J. Betz, 1994. User views of artificial reef management in the southeastern U.S. Bulletin of Marine Science 55: 970–981.

Reggio, V., V. Van Sickle & C. Wilson, 1986. Rigs to reefs. Louisiana Conservationist 38: 4–8.

Reggio, V. C. 1989. Petroleum structures as artificial reefs: a compendium. 174. U.S. Minerals Management Service, OSC Study/MMS 89–0026.

Robson, C., 2002. Real World Research: A Resource for Social Scientists and Practitioner-Researchers, 2nd edn. Blackwell, Oxford.

Samples, K. C., 1989. Assessing recreational and commercial conflicts over artificial fishery habitat use: theory and practice. Bulletin of Marine Science 44: 844–852.

Santos, M. N. & C. C. Monteiro, 1997. The Olhão artificial reef system (south Portugal): fish assemblages and fishing yield. Fisheries Research 30: 33–41.

Santos, M. N. & C. C. Monteiro, 1998. Comparison of the catch and fishing yield from an artificial reef system and neighbouring areas off Faro (Algarve, south Portugal). Fisheries Research 39: 55–65.

Santos, M. N. & C. C. Monteiro 2001. The Portuguese experience on artificial reefs: past and future. In Coimbra, J. (ed.), NATO Science Series: Series A: Life Sciences Nieuwe Hemweg 6B Amsterdam 314: 281–294.

Ten Brink B. J. E., S. H. Hosper & F. Colijn, 1991. A quantitative method for description and assessment of ecosystems: the AMOEBA approach. Marine Pollution Bulletin 23: 265–270.

Whitmarsh, D. & H. Pickering 1999. Commercial exploitation of artificial reefs: economic opportunities and management imperatives. In Boncoeur, J. (ed.), Proceedings of the 9th annual conference of the European Association of Fisheries Economists EAFE: 66–83.

Whitmarsh, D., M. N. Santos, C. C. Monteiro & J. Ramos, 2004. Economic measures of habitat improvement in marine fisheries and their conservation. Poster presented at the World Fisheries Congress, May 2–6, Vancouver, B. C.

Zar, J. H., 1996. Biostatistical Analysis, 3rd edn. Prentice Hall International Editions, Upper Saddle River, N. J.

Hydrobiologia (2007) 580:193–217
DOI 10.1007/s10750-006-0453-0

BIODIVERSITY IN ENCLOSED SEAS

History, ecology and trends for artificial reefs of the Ligurian sea, Italy

G. Relini · M. Relini · G. Palandri · S. Merello · E. Beccornia

Abstract From 1970 to the present 10 artificial reef sites have been developed in coastal waters of the Ligurian Sea, Italy. They range from Ventimiglia, in the west, to La Spezia, in the east, with the largest and best known reef complex being located in the Gulf of Genoa at Loano and consisting of 2,745 m^3, about 5,200 t of material and covering a surface of 350 ha. Design and construction practices have advanced from an initial, unsuccessful effort that used automobile bodies (now banned) to current use of custom-designed concrete modules deployed systematically. Funding for reef construction has come since 1983. The earliest aim of reefs was as a physical barrier to protect habitats against illegal otter trawl fishing. Newer objectives include habitat restoration, enhancement of biodiversity and fishing catch, and research to test materials and designs for physical and ecological performance. Reefs also functions as environmental observation stations, with the invasive species *Caulerpa taxifolia* (Vahl) C. Agardh, being recorded on the reef at Alassio. For some Artificial Reefs (Ars), benthic organisms and fishes, settlement, biomass and development of community are recorded. In Loano AR, immersed in 1986, more than 150 algae species are recorded, more than 200 benthic animal species and 78 species (87 taxa) of fishes. Fifty-six species (61 taxa) of fishes are recorded by visual census, the others are caught only by trammel net and long line. Trammel catches at Loano are on average about 2.32 kg/100 m net. Comparisons among ARs reveal that age of the reef, location and presence of seagrass meadows are crucial for success. An indication of functional equivalence between ARs and natural rocky reefs is seen if both fish and sessile macrobenthos are compared. After 34 years of investigation a database comprising at least one hundred scientific articles based on research programs of up to 15 years, and other unpublished reports, provides information to guide future planning of reefs. On the basis of acquired experience, some management advice is suggested and the best design for the basic module in the Ligurian sea is described. The role of ARs, providing protection of coastal environment against the illegal otter trawling, nursery, microhabitat and food supply, while increasing biodiversity, biomass of benthos and fishes, and facilities for ecotourism, is outlined.

Keywords Artificial habitat · Benthos · Fishes · Visual census · Biomass

Guest editors: G. Relini & J. Ryland
Biodiversity in Enclosed Seas and Artificial Marine Habitats

G. Relini (✉) · M. Relini · G. Palandri · S. Merello · E. Beccornia
Laboratori di Biologia Marina ed Ecologia Animale, DIP.TE.RIS., Università di Genova, Corso Europa 26, 16132 Genova, Italy
e-mail: biolmar@unige.it

Introduction

Since 1970, the main aim for the deployment of Artificial Reefs (ARs) in the Gulf of Genoa was the protection of coastal habitats against disturbance and destruction by the illegal otter trawl fishery. There were subsidery aims shared with most ARs in the Mediterranean Sea:

- To protect juvenile fish, especially those of commercial interest, providing shelter from predation and from illegal fishing, reducing mortality rates;
- To provide additional food and new food webs;
- To increase feeding efficiency, which probably produces faster growth rates on ARs than in the natural environment (in particular for fish);
- To provide new habitat for settlement and shelter for eggs, larvae, juveniles, favouring recruitment;
- To protect sensitive habitats such as *Posidonia oceanica* meadows and coralligenous concretions;
- To promote biodiversity and diversification of microhabitats;
- To contribute to the recovery of quality of environment and of coastal fishing resources;
- To promote alternative and innovative forms of management, to enhance the value of the coastal zone, a rational exploitation of the fishing resources and the development of mariculture;
- To develop the potential of eco-tourism, in particular controlled SCUBA diving;
- To develop integrated studies on coastal ecosystem functioning.

The present paper is a review of the deployment of ARs in the Ligurian sea (Fig. 1) and updates previous papers, in particular those within Jensen et al. (2000). It is divided into three parts that are not only temporal, but concern the initiative and design quality as well: (1) from 1970 to 1986, (2) from 1986 to 2003, (3) ARs in progress after 2004 and those planned. The three parts recognise also an increasing interest in the ARs, and a rise in financial support from public agencies; the number of pages devoted to the three periods are quite different. Platforms, port wharfs, breakwaters, etc. (structures that can act as ARs, but that are built with

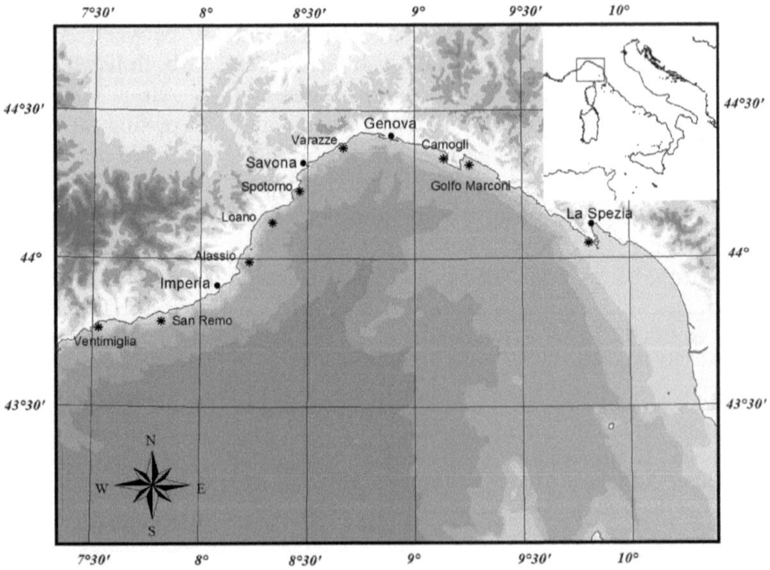

Fig. 1 Sites of deployment of artificial reefs (asterisk) along the Ligurian coast (N–W Italy) from 1970 to 2006

other aims) are not discussed here. Structures or equipment whose main role is to attract fishes, such as FADs, are not considered. A general worldwide overview on AR approaches is available in the book edited by Seaman (2000), while the ARs in Europe are described by Jensen (1997) and Jensen et al. (2000).

Artificial reefs built from 1970 to 1986

The first sizeable AR in Italy was built in the Ligurian Sea, in front of Varazze (Savona, Western Ligurian Riviera) in December 1970, using 1,300 car bodies destroyed during the flood of Genoa in October 1970. In the same year some car bodies were sunk at Portovenere (La Spezia, Eastern Riviera) near the cave of Byron. They were destroyed in a short time (less than 1 year), by wave action, and disappeared.

Sport fishermen of Varazze, following a similar initiative in U.S.A. (Carlisle et al., 1964; Turner et al., 1969) and without any scientific support (Relini & Orsi Relini, 1971), obtained permission to sink the above mentioned car bodies between 35 and 50 m depth in order to prevent trawling and to improve sport fishing in the area off Punta dell'Olmo (Fig. 2).

The Laboratori di Biologia Marina ed Ecologia Animale, Università di Genova, had the opportunity to follow settlement of organisms on different kind of materials within the car bodies, some of which contained polluting substances, such as heavy metals in paints. The preliminary survey (Relini & Orsi Relini, 1971) by SCUBA divers on the wrecks, after 11 months of immersion, and the laboratory examination of fragments, compared with non-toxic substrata, showed: (1) a reduced settlement of sessile organisms on car bodies; (2) a deterioration of the paint films, and a high rate of corrosion of metallic materials preventing settlement.

The cars, that were partially sunk in the mud bottom after some years and covered by a layer of sediment, were examined after 1, 2, 6, 10 and 12 years of immersion. The results did not come up to sport fishermen's expectations, probably for the following three main reasons: (a) unsuitable substrates; (b) heavy sedimentation; (c) placing at too great a depth in relation to the turbidity of the water column. The sediment, along with the formation of corrosion products and peeling of the paint, strongly limited the settlement of organisms, that 6 years later were still represented mainly by serpulids and bryozoans (Relini & Wurtz, 1977). In fact the community, which

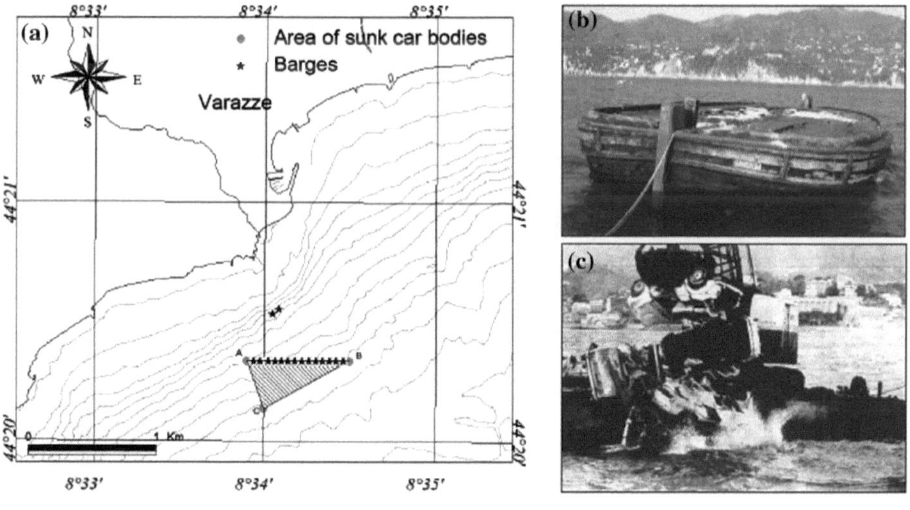

Fig. 2 Varazze AR made of car bodies, in 1970, and wooden barges, in 1979. (**a**) Map of the area with the triangle in which car bodies were sunk. The black stars represent barges. (**b**) Wooden barge. (**c**) Car bodies

lacks algae, was characterised by serpulids, especially *Serpula vermicularis* L., *Pomatoceros triqueter* (L.), *Hydroides pseudouncinatus* Zibrowius, *Spirobranchus polytrema* (Philippi) and bryozoans, especially *Schizomavella linearis* (Hassall), *Fenestrulina malusii* (Audouin), *Patinella radiata* (Audouin), *Aetea sica* (Couch), plus a few molluscs and hydroids. After 10 years (Relini, 1982b) the community was richer in species (75 taxa of invertebrate against 36 described after 6 years, see Table 1 in Relini, 1982b) because of increased settlement of sponges (*Dysidea*, *Myxilla*, *Haliclona*) and of higher number of species of bryozoans, molluscs, and serpulids.

However, 12 years after deployment, sport fishing (lines, trammels, diving) remained poor (Relini, 1982b). The results were so negative that car bodies as material for ARs were banned in Italy, particularly because of pollution concerns. The unfitness of car bodies for improving benthic and fish production by man-made reefs is proven also by the comparison between the results obtained in Varazze and those in Ancona, where concrete blocks were immersed in shallow waters (Bombace, 1977). After 1 year in Ancona (13–15 m depth) settlement of invertebrates (mainly mussels) reached 100 kg/m^2, while in Varazze (35–50 m depth), after 10 years it was still less than 0.6 kg/m^2.

The question that arose was "can the ARs in Gulf of Genoa achieve the aims previously described, after the Varazze negative results?"

To reply to the question, recovered non-toxic materials were used for small ARs in different sites (Relini, 1979). The research, funded by Programma Finalizzato "Oceanografia e Fondi Marini" by the National Council of Research, was mainly devoted to the choice of sites and materials, and studies on the settlement of sessile organisms and fish attraction.

For this purpose, large wooden barges were sunk at different depths and sites; they were obtained free of charge from the Port Authority of Genoa, which assured the necessary technical assistance for the sinking. Each barge was about 50 t of oak timber. In 1979, 16 barges were immersed between 30 and 50 m deep, close to the car bodies off Varazze (side AB of the triangle of Fig. 2) and two barges were put at 20 m depth (Relini, 1979).

In coastal waters in front of Camogli (West of the Portofino promontory), six barges, two each at 20, 30 and 40 m depth, were immersed in October 1979 (Fig. 3), but they were not studied in the following years. In December 1980, 11 barges (plus three in July 1981) were immersed at 30–40 m depth along a line crossing the eastern part of the Marconi Gulf (Fig. 4), which is quite different in character from the Camogli site, being

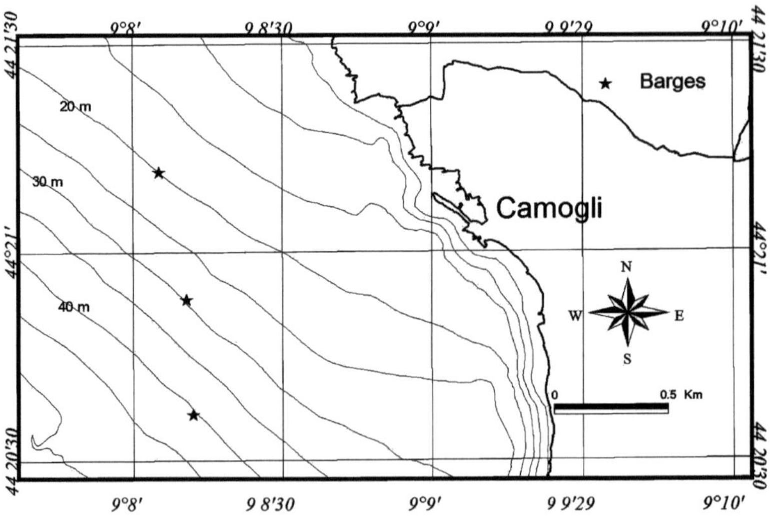

Fig. 3 Sites of barges immersion near Camogli in 1979. The barges are the same of Fig. 2b

sheltered and affected by a weak sewage outfall (Relini, 1982a), to effectively close the inner area of the Gulf to otter-trawlers (the western part is closed by a sewage conduit). One additional barge was put at 15 m depth to provide further information on settlement. In November 1983, two barges and five dock-gates of timber and iron were added, followed, 1 year later, by 10 concrete blocks (2.2 × 3.2 × 1.45 m), so the AR reached a volume of 16,185 m^3, covering a bottom area of 2.1 km^2 and protecting an area of about 6 km^2 of the Gulf. Finally, in 1985, 27,500 m^3 of gravel and coarse pebbles were discharged between the main components of the AR.

The wooden barges proved to be a more suitable substrate for settlement of sessile organisms than car bodies and were more attractive for fishes, probably related to larger size, as documented by photos and video made by divers. In all localities, adults of *Diplodus vulgaris* (Geoffr.), *D. sargus* (L.), and sometimes of *Pagellus* spp., were attracted by the barges a few months after immersion. Barges were also more effective in preventing fishing by trawlers in shallow water than car bodies because of their volume and weight. Nevertheless these structures can be destroyed by marine wood borers, and it is difficult to forecast how long they will last in the sea.

The seabed around the Marconi Gulf AR is muddy and communities belong mainly to Biocoenosis of Coastal Terrigenous Muds (CTM) (Relini et al., 1986). In general, settlement on the reef was good in terms of species richness, though algae were scarce, due to depth and turbidity, and mussels absent. On the other hand, oysters and many erect forms, such as sponges, *Sabella*

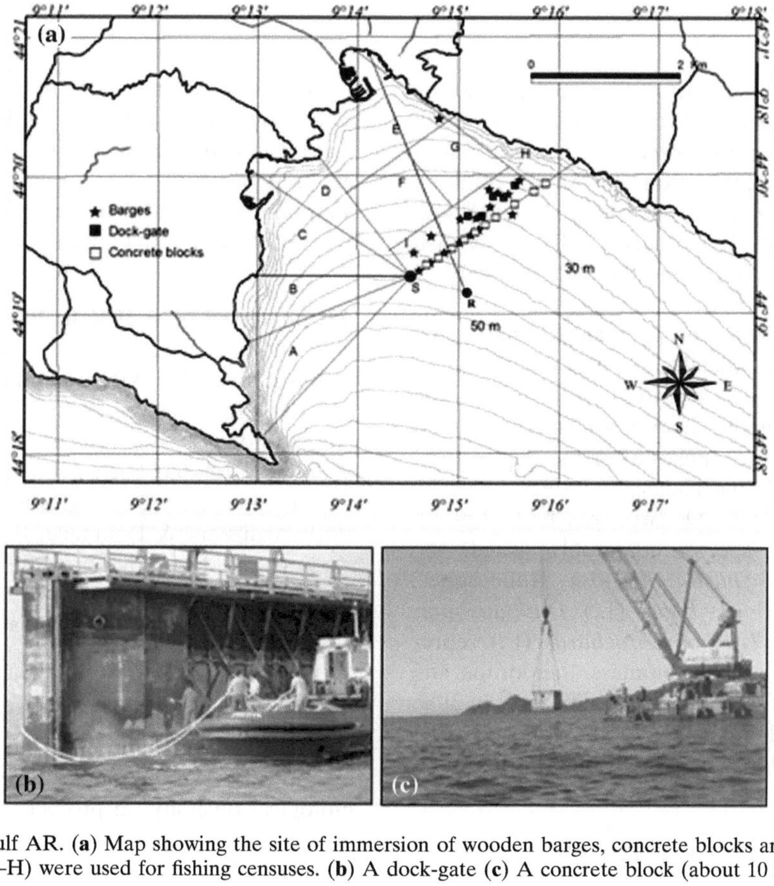

Fig. 4 Marconi Gulf AR. (**a**) Map showing the site of immersion of wooden barges, concrete blocks and dock-gates. The different zones (A–H) were used for fishing censuses. (**b**) A dock-gate (**c**) A concrete block (about 10 m^3)

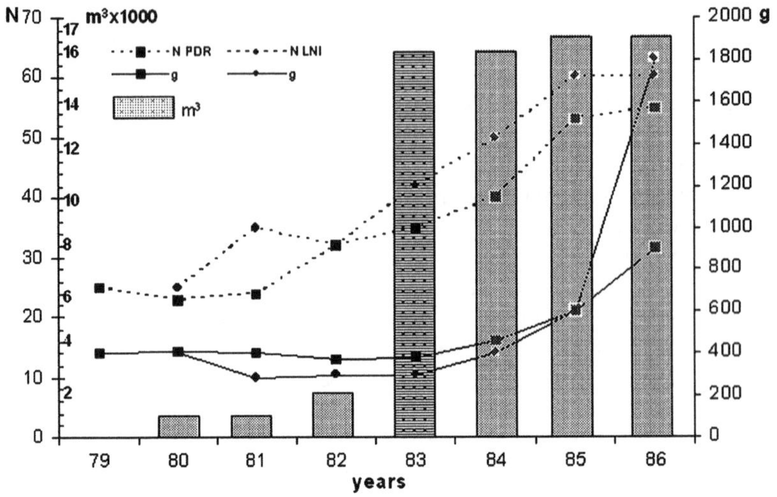

Fig. 5 Increasing volume of Marconi Gulf AR (bars; left y axis) and results of summer line competitions. Dotted lines represent the average number of sport fishermen engaged each year in summer competitions (left *y*-axis), continuous lines represent the average catch per head (weight in grams; right *y*-axis). Lines with small circles represent data of the Club Lega Navale Italiana at Rapallo, while those with squares are of the Circolo Pescatori Dilettanti at Rapallo

spallanzanii (Gmelin), *Alcyonium palmatum* Pallas, *Ascidia mentula* Müller, *Phallusia mammillata* (Cuv.), *Leptogorgia sarmentosa* (Esper), *Filograna* sp., were present. Among the vagile epifauna spiny lobster, *Palinurus elephas* (Fabricius), and *Scyllarus arctus* (L.) occurred. Thirty-six fish species were caught around the AR (see Table 5 in Relini et al., 1986).

The results (Relini et al., 1986; Relini & Orsi Relini, 1989) of official angling competitions in the Marconi Gulf, provided evidence of an increase both in the amount of activity and in the quantity of the catches after the deployment of the AR and, in particular, after 1985, when the highest volume of AR was reached (Fig. 5). Among the 25 species commonly fished, the following five, *Spicara flexuosa* Rafinesque[1], *S. maena* (L.), *Boops boops* (L.), *Diplodus annularis* (L.) and *Trachurus trachurus* (L.) represented 90% of the catch biomass. In addition, the total number of annual competitions was in 1986 four times higher than in 1980, reflecting the increased popularity of sport fishing in the Gulf.

[1] At present this species is no longer considered separated from *S. maena*.

Collecting reliable data on professional small-scale fishing was much more difficult, because the catch was sold directly to restaurants, thus avoiding registration of sales. Nevertheless, some data on trammel net catches were collected in different areas of the Marconi Gulf (see Relini et al., 1986).

AR immersed from 1986 to 2003

In this period, five ARs, Loano (1986), Spotorno and Ventimiglia (1989–1990), Lavagna (1993) and Alassio (1998–1999) were built, but most subsequent research has been done at Loano.

The experience from ARs in Italy (Bombace, 1981; Ardizzone & Bombace, 1983), from fouling studies, and from the previously described research in the Gulf of Genoa (Relini, 1983), made it possible to plan an artificial reef at Loano (LAR). The main aim was to enhance alieutic resources and to protect the *Posidonia* meadow from illegal trawling and, if possible, restore it to its past size—at least double in surface area. The meadow has an important role in maintaining biological diversity, in protecting the coast from erosion and in promoting small-scale fisheries (Relini & Moretti, 1986). This project obtained the financial support of the European Economic Community (EEC), and was realised in the Summer 1986, thanks to the strong interest of Loano Municipality and with the aid of the Engineering Contractor's Group, Rome and the University of Genoa.

The shallow bottom along the Loano coast was composed mainly of thin sand and mud partially covered by *Posidonia* and *Cymodocea* meadows. Between 3 and 12 m depth most of the area was covered by *Cymodocea nodosa* and related organisms. At a depth more than 25 m, the Biocenosis of Coastal Detritus bottoms (CD) was present and was gradually changing into CTM. The *Posidonia* bed extends from 9 to 20 m depth, but is reduced eastwards and ends off the mouth of a small river close to Loano harbour. In the past, *Posidonia* meadows extended along all the coast in the direction of Pietra Ligure. In 1984 the density of plants was not uniform; in the central part of the existing meadow there were 120–150 shoots/m^2, while near the edge no more than 50 occurred.

LAR is composed of two parts (Fig. 6): a central main group of pyramids protected all around by a grid of single blocks (1.2 × 1.2 × 1.2 m) to cover an area of about 350 ha (3,500 m on the side parallel to the coast and 1,150 m perpendicular to the coast) extending from 5 to 45 m depth; 200 blocks (total volume of 345 m^3) were positioned in the form of a grid designed to avoid any passage larger than 50 × 250 m, to prevent trawling in the area, but allowing the use of artisanal gear; for this reason no iron bars or hooks were put on blocks.

The main reef extends between 17 and 25 m depth, is composed of 30 pyramids 25 m apart (in total 150 blocks and 1,200 m^3), and it is positioned along the sides and inside the 200 × 100 m rectangle (Relini, 2000a). Each pyramid, arranged in four blocks (2 × 2 × 2 m) at the base and one at the top, rests on a base of small stones, which was initially about 50 cm high to prevent the structure from sinking into the bottom, and it has an important role for colonisation of many organisms. The top cube has a metal ring in the middle of the upper surface for hooking FADs or ropes for mariculture.

The main reef is located immediately beyond the deeper limit of the *Posidonia* meadow where the sea-grass is degraded; the *Posidonia* bed and a large wreck sited at 30 m depth are the main source of colonisers.

In summer 1989, 150 large blocks (2 m side, in total 1200 m^3) were added to reinforce the outer part of the reef because some small blocks were displaced by trawling activity. In the following year the shore area changed because of the construction of a submerged breakwater and enlargement of Loano harbour (Fig. 6).

Colonisation of the AR, both by benthos and by fishes, was followed over many years and some observations are still in progress. Benthos was studied by means of immersion of panels and collecting samples by scraping concrete blocks. Panels (concrete–asbestos 20 × 30 × 0.3 cm) were immersed at four depths (5 m, 10 m, 20 m and 36 m) and at least two exposures for different periods of time from May 1987 to May 1994. Then, some panels were immersed irregularly up to 2003.

Qualitative and quantitative aspects of sessile communities, colonisation processes and seasonal variations have been described. So far 92 taxa (61 species) of algae (47 Rhodophyta, 24 Phaeophyta, 18 Chlorophyta, two Bacillariophyta, one Cyanophyta) (Cecere et al., 1993) and 120 species of sessile animals have been identified (Relini et al., 1995a) on panels immersed at four depths. A total of 161 taxa has been recorded on the panels. The most diverse animal group is bryozoans (66 species), followed by serpulids (25 species), molluscs (24 species) and hydroids (23 species). The colonisation of the blocks by algae was studied from 1994 to 1998 by the team of Prof. Bressan of Trieste University. During seasonal observation in 1994 and 1995, 79 species (58 Rhodophyta 73.4%, 14 Phaeophyta 17.7%, 7 Chlorophyta 8.9%) have been described (Falace et al., 2002). During a further study (Falace & Bressan, 2002), on anti-grazing nets (sea urchins), 93 algae, 64 Rhodophyta, 20 Phaeophyta and nine Chlorophyta were found. The relative increase of biomass on protected surfaces seems related to the mechanical effect of the netting in reducing grazing by sea urchins (Falace & Bressan, 2002); though another experiment gave different results, probably due to low density of sea urchins, and emphasised the importance of grazing by fishes (Puccio et al., 2001).

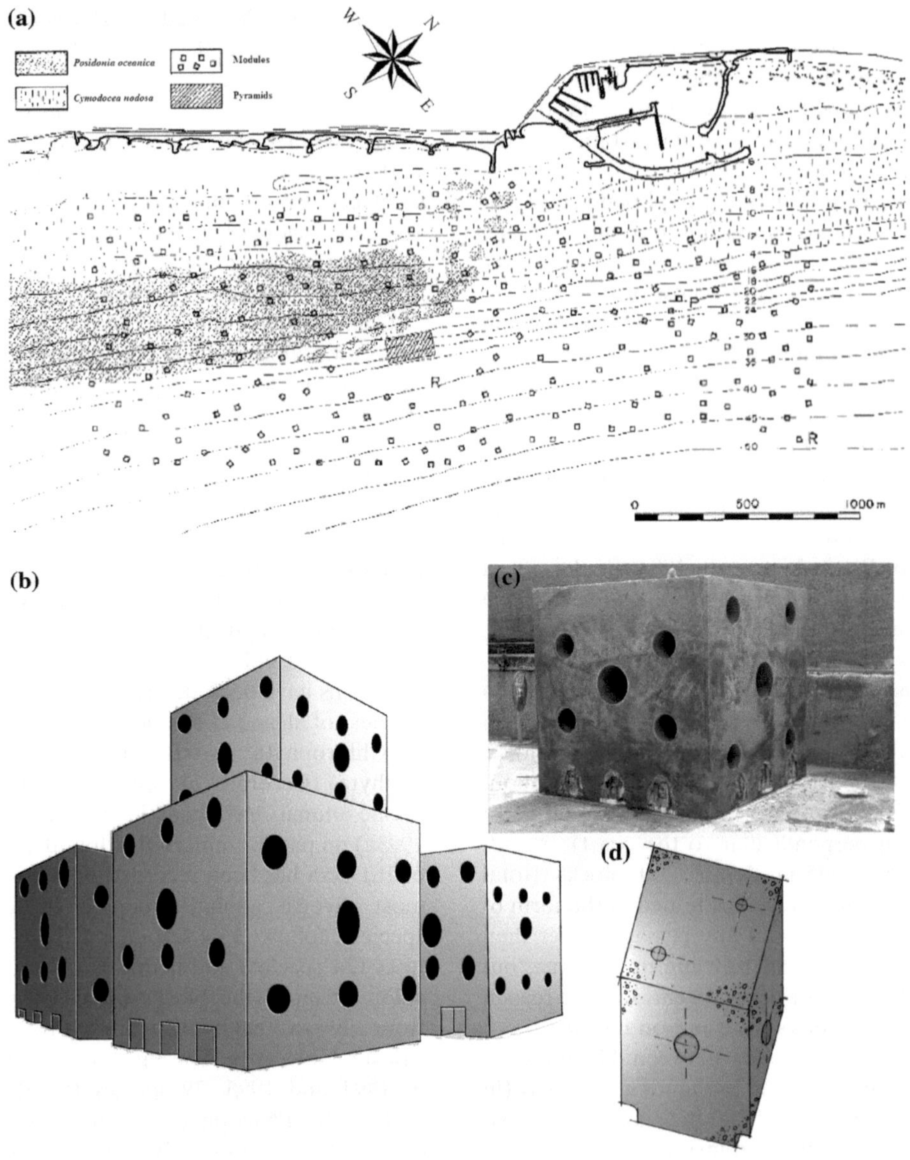

Fig. 6 Loano AR. (**a**) Map showing the distribution of *Posidonia* and *Cymodocea* meadow and of different modules. The new enlarged harbour is drawn, (**b**) Pyramid with large blocks, (**c**) Large block (2 m side), (**d**) Small block (1.2 m side)

Forty more animal species, in addition to the 161 taxa mentioned above, were found only on large concrete blocks of pyramids (see Table 6 in Relini et al., 1995a). Vagile fauna was sampled on and around blocks using an air-lift operated by a scuba diver, in relation to studies on fish diet: 82 taxa of Crustacea, mainly Amphipoda and Decapoda were identified (Maurizi, 1993; Relini et al., 1995a).

The settlement periods of the main species or groups were identified on panels immersed for 1 and 3 months (Relini & Cormagi, 1990; Relini et al., 1994a). Settlement on monthly panels was poor in term of biomass and biodiversity, and was constituted of hydroids and serpulids nearly all year, while incrusting bryozoans settled in summer, and slime films predominate from November to May. The biomass of settled organisms was evaluated as non-decalcified wet weight. The monthly values per panel at the four stations are available only for the first year and they range from 2 to 51 g/12 dm^2 (Relini & Cormagi, 1990). In general the biomass was low if compared with settlement in eutrophic environments. Differences in biomass between seasons were not as large as found in Ligurian harbours (Montanari & Relini, 1973).

The development of a sessile community on panels exposed at four depths for 1, 3, 6, 9, 12 and 24 months (Relini et al., 1990b, 1994a) and seasonal pattern of the settled community (Relini et al., 1994a; Zamboni et al., 1992) have been described. The qualitative and quantitative differences found in the colonisation between depths (four stations) were greater than the seasonal differences at the same station. That means that depth is more important than season in determining the pattern of colonisation.

On the concrete blocks of the AR pyramids the different exposure of each face to light, currents and sedimentation strongly influenced both settlement and the subsequent evolution of the community.

On the upper side, the one most exposed to light, a true ecological succession was observed, which tended towards a mature stage dominated by the algae of genera *Sargassum* and *Cystoseira.*

On the inside faces of the blocks of the pyramids typical sciophylous organisms settled, giving rise to a complex community in which sponges, coelenterates, large bryozoans, such as *Pentapora ottomulleriana* (Moll), were dominant. This community is still developing towards a "cave community" or coralligenous assemblages. The process of reaching a steady-state community at the LAR has been complicated and slowed by factos of unpredictable magnitude, such as hydrodynamism, sedimentation and predation - in particular the grazing by fishes and sea urchins, *Paracentrotus lividus* (Lamarck) and *Arbacia lixula* (L.).

In conclusion, data obtained on the Loano AR (LAR), using panels immersed for 3-year cycles and observations carried out on blocks over a period of more than 15 years, provide the most detailed and wide-ranging information available in the Mediterranean and in Europe, with regard to studies on the colonisation processes of sessile macrobenthos on ARs. A review of data on the colonisation of the main Mediterranean ARs is available in Jensen et al. (2000).

The influence of sedimentation on the settlement was also studied at Loano AR. The development of sessile benthic communities, settled on hard substrata and subject to different sedimentation rates, was studied over a period of 24 months (Relini et al., 1998a). To this aim differently placed panels were used: immersed horizontally, with and without protection from a glass sheet, 8 cm above, and vertically.

The average sedimentation rate (measured every 3 months by means of traps fixed on the structure) was 92.80 $g/m^2/day$; the maximum occurred in winter (212.98 $g/m^2/day$), the minimun in spring (10.12 $g/m^2/day$). A comparison between the surfaces protected and unprotected from sedimentation showed the importance of sedimentation rate in determining the development of benthic communities, in terms of quality (species richness) and quantity (benthic biomass). On the surfaces protected from sedimentation the communities were rich and diversified (88 taxa), characterised in their mature stages (over 24 months of observations) by oysters and many species of encrusting and erect bryozoans; biomass was high.

The equivalent surface unprotected from sedimentation showed definitely poorer communities: 69 taxa were identified, and in particular bryozoans and ascidians seemed more influenced by sediments. The whole ecological succession appeared to suffer; the cover was always lower than on the protected surface and the biomass was the lowest of all surfaces examined (including vertical panels and the upper side of horizontal ones). Although a high sedimentation rate was found on the Loano artificial reef, the influence of other factors—such as the reduction of light intensity and the different hydrodynamics on the panels protected by a glass sheet—cannot be excluded.

Settlement and macrobenthos development were also studied on different surfaces, using small cubic blocks (20 cm side) made of concrete or coal ash (70.8% fly ash, 6.8% hydrated lime and 22.4% water). No significant differences in the organism settlement were found between the two types of material (ash and concrete) of the different small blocks though higher friability of ash material influenced the amount of fouling (Relini et al., 1995b).

These data confirmed that coal ash stabilised materials can be used successfully for ARs, as shown also by tests made at Torvaldaliga Power Station (Relini, 2000b; Sampaolo & Relini, 1994).

Some observations on the settlement and growth of oysters *Ostrea edulis* L. and mussels *Mytilus galloprovincialis* Lamarck were carried out. Settlement of mussels was poor, but not the growth rate, considering the oligotrophic environment.

Mussels, removed from natural hard substrata on the shore, were suspended in cylindrical nylon nets immersed at two sites in 10 m and 18 m depth (Guidetti & Relini, 1995). From November 1992 to November 1993, the growth rate was of 2.54 mm/month of the shell length at 18 m (from 15.0 to 45.5 mm) and of 1.67 mm/month at 10 m depth (from 17.0 to 37.0 mm). These growth rates are lower than those found elsewhere in Italy, in shallower and eutrophic water (Guidetti & Relini, 1995).

Fishes, and other organisms of fishing interest, have been studied since 1989 by different methods (Table 1). A first set of data, both qualitative and quantitative (species, number of individuals, size and weight), was obtained from local fishermen who filled in a special form from 1/1/89 to 14/9/89. In a total of 191 catches, 33 of which were made using longlines and 158 trammel nets (Relini et al., 1990a), 35 fin fish species were found.

Studies on and around the LAR were carried out by divers using a direct visual census method at monthly intervals from February 1989 up to January 1994, and by limited fishing activity using long-lines and trammel nets (Relini et al., 1990a; Relini et al., 1994b; 1995b; Relini & Torchia, 1993). In addition to the above mentioned data supplied by fishermen, from May 1991 to August 1992 twelve trammel catches were obtained with the help of professional fishermen in order to improve knowledge of the catchable macrofauna present in the area. Trammels 200–400 m long and 1 m high (3.5 cm mesh) were used, the nets being immersed for 12 h from sunset to sunrise east or west of the rectangle of pyramids. Each specimen caught was measured and weighed. Information was obtained about length/weight relationships, which was then used for biomass evaluation during visual censuses. The 12 h catch per 100 m of trammel ranged from 4.61 kg to 1.01 kg, with an average of 2.32 kg. This average trammel yield was clearly higher than those obtained in ecologically comparable seas, such as in the Northern Sicily, both with regard to natural areas (0.64 kg/100 m; Toccaceli & Levi, 1990) and ARs (0.31 kg/100 m; Arculeo et al., 1990). Duclerc (1980) reported values of 0.20 kg per 12 h for approximately 100 m of trammel at the Golfe Juan reef, which is made of tyres, and 0.50 kg for neighbouring zones. In Loano the catches were similar to those obtained on ARs 1 year after installation in the highly productive environment of the Adriatic Sea (2.80–1.71 kg/100 m; Bombace et al., 1990).

Visual censuses of the fish community were carried out (almost once a month) at two pyramids (stations) in the central part (18 m depth). Each station embraced a water volume of about 400 m^3 and a volume of sunken material of 40 m^3. The technique employed involved observations made by scientific SCUBA divers from fixed points at the top of the structure and along a transect incorporating the internal passageways within the pyramid as well as along its outer margins (Harmelin-Vivien et al., 1985; Relini et al., 1995b). All the specimens encountered were identified to species (or genus) and visually sized. Three size classes, juveniles, medium and large, were defined a priori, according to the maximum recorded total length of each species in the literature. The frequencies of occurrence (percentage) of each species by size class were calculated per year and over all surveys.

Overall, 48 fish taxa had been identified up to 1993, representing 16 families (Table 1). The frequency of occurrence over the entire period indicates a nucleus of at least 16 species that were present in more than 25% of the surveys.

The biomass of visually estimated fish populations was calculated by multiplying the individual mean weight of a given size class by its population

Table 1 List of bony fish censused and caught by tramel net and long lines in Loano AR. The presence of the species in Lavagna (LA) and Alassio (AL) and in some natural rocky environments is recorded (PC: Port Cros, PT: Portofino, A: Asinara)

LOANO												
Species	Visual census (V) Loano					Tramel catches • Long line ○ Loano (T + L) 1993	V + T + L 2004	LA	AL	PC	PT	A
	1989	1989–1991	1989–1993	1989–1998	1989–2004							
Anthias anthias (Linnaeus, 1758)			•	•	•	•	•				•	•
Apogon imberbis (Linnaeus, 1758)	•	•	•	•	•	•	•	•	•	•	•	•
Arnoglossus laterna (Walbaum, 1792)						•	•					
Arnoglossus sp.						•	⊙					
Arnoglossus thori Kyle, 1913						•	•					
Balistes carolinensis (Gmelin) 1789					•		•	•	•			
Blennius ocellaris Linnaeus, 1758				•	•		•					
Blennius sp.		•	•	•	•		⊙					
Boops boops (Linnaeus, 1758)	•	•	•	•	•	•	•	•	•	•	•	•
Bothus podas (Delaroche, 1809)						•	•	•			•	
Buglossidium luteum (Risso, 1810)						•	•					
Chromis chromis (Linnaeus, 1758)	•	•	•	•	•	•	•	•	•	•	•	•
Conger conger (Linnaeus, 1758)	•	•	•	•	•	○	•	•	•	•	•	•
Coris julis (Linnaeus, 1758)	•	•	•	•	•	•	•	•	•	•	•	•
Dentex dentex (Linnaeus, 1758)	•	•	•	•	•	⊙	•	•	•	•	•	•
Dicentrarcus labrax (Linnaeus, 1758)	•	•	•	•	•	⊙	•	•	•		•	
Diplodus annularis (Linnaeus, 1758)	•	•	•	•	•	•	•	•	•	•	•	•
Diplodus puntazzo (Cetti, 1777)	•	•	•	•	•		•	•	•	•	•	•
Diplodus sargus (Linnaeus, 1758)	•	•	•	•	•	⊙	•	•	•	•	•	•
Diplodus vulgaris (E. Geoffroy St. Hilaire, 1817)	•	•	•	•	•	•	•	•	•	•	•	•
Engraulis encrasicolus (Linnaeus, 1758)			•	•	•	•	•				•	•
Epinephelus caninus (Valenciennes, 1834)				•	•		•					
Epinephelus marginatus (Lowe, 1834)			•	•	•		•	•	•		•	•
Gobius cruentatus Gmelin, 1789		•	•	•	•		•		•	•	•	•
Gobius geniporus Valenciennes, 1837									•			
Gobius niger Linnaeus, 1758			•	•	•		•		•			
Gobius sp.	•	•	•	•	•		⊙				•	
Labrus bimaculatus Linnaeus, 1758		•	•	•	•		•			•	•	•
Labrus merula Linnaeus, 1758		•	•	•	•	•	•		•	•	•	•
Labrus viridis Linnaeus, 1758			•	•	•		•	•	•	•	•	•
Lipophrys basiliscus (Valenciennes, 1836)								•				

Table 1 continued

LOANO

Species	Visual census (V) Loano					Tramel catches • Long line ○ Loano (T + L) 1993	V + T + L 2004	LA	AL	PC	PT	A
	1989	1989–1991	1989–1993	1989–1998	1989–2004							
Lepidotrigla cavillone (Lacepède, 1801)						•	•					
Lithognathus mormyrus (Linnaeus, 1785)						⊙	•	•			•	•
Lophius piscatorius Linnaeus, 1758			•	•	•		•					
Merlucius merlucius (Linnaeus, 1758)						•	•					
Monochirus hispidus Rafinesque, 1814						•	•					
Microchirus variegatus (Donovan, 1802)						•	•					
Mugil sp.	•	•	•	•	•	•	⊙			•	•	•
Mullus barbatus Linnaeus, 1758				•	•	⊙	•	•			•	
Mullus surmuletus Linnaeus, 1758	•	•	•	•	•	⊙	•	•	•	•	•	•
Muraena helena (Linnaeus, 1758)				•	•		•	•	•	•	•	•
Mycteroperca rubra (Bloch, 1793)	•	•	•	•	•		•					
Oblada melanura (Linnaeus, 1758)	•	•	•	•	•	•	•	•	•	•	•	•
Ophidion barbatum Linnaeus, 1758						•	•					
Ophidion rochei Muller, 1845						•	•					
Ophisurus serpens (Linnaeus, 1758)			•	•	•		•					
Pagellus acarne (Risso, 1826)					•	•	•	•			•	
Pagellus erythrinus (Linnaeus, 1758)		•	•	•	•	•	•				•	
Pagellus sp.	•	•	•	•	•	•	⊙					
Pagrus pagrus (Linnaeus, 1758)					•	⊙	•		•		•	
Parablennius gattorugine (Brunnich, 1768)				•	•		•	•	•	•	•	•
Parablennius incognitus (Bath, 1968)			•	•	•		•				•	•
Parablennius rouxi Cocco, 1833	•	•	•	•	•		•	•	•	•	•	•
Parablennius sp.				•	•		⊙					
Phycis phycis (Linnaeus, 1766)			•	•	•	•	•	•	•	•	•	
Sarda sarda (Bloch, 1793)						•	•				•	
Sardinella aurita Valenciennes, 1847						•	•					
Sarpa salpa (Linnaeus, 1758)				•	•	•	•	•	•	•	•	•
Sciaena umbra Linnaeus, 1758	•	•	•	•	•		•	•	•	•	•	•

Table 1 continued

LOANO											
Species	Visual census (V) Loano					Tramel catches • Long line ○ Loano (T + L) 1993	V + T + L 2004	LA	AL	PC	PT	A
	1989	1989–1991	1989–1993	1989–1998	1989–2004							
Scomber japponicus Houttuym, 1780						•	•					
Scophthalmus rhombus (Linnaeus, 1758)						•	•					
Scorpaena notata Rafinesque, 1810	•	•	•	•	•	⊙	•	•	•	•	•	•
Scorpaena porcus Linnaeus, 1758	•	•	•	•	•	⊙	•	•	•	•	•	•
Scorpaena scrofa Linnaeus, 1758	•	•	•	•	•	⊙	•	•	•	•	•	
Seriola dumerili (Risso, 1810)		•	•	•	•	⊙	•	•		•	•	•
Serranus cabrilla (Linnaeus, 1758)	•	•	•	•	•	•	•	•	•	•	•	•
Serranus hepatus (Linnaeus, 1766)	•	•	•	•	•		•		•		•	
Serranus scriba (Linnaeus, 1758)					•	⊙	•	•	•	•	•	•
Solea vulgaris Quensel, 1806						•	•					
Sparus aurata Linnaeus, 1758			•	•	•	•	•	•		•	•	•
Sphyraena sphyraena (Linnaeus, 1758)						•	•					
Spicara flexuosa Rafinesque, 1810		•	•	•	•	•	•					•
Spicara maena (Linnaeus, 1758)	•	•	•	•	•	•	•	•	•	•	•	•
Spicara smaris (Linnaeus, 1758)	•	•	•	•	•	•	•	•	•	•	•	•
Spondyliosoma cantharus (Linnaeus, 1758)	•	•	•	•	•	•	•	•	•	•	•	•
Symphodus cinereus (Bonnaterre, 1788)		•	•	•	•	•	•	•	•	•	•	•
Symphodus doderleini (Jordan), 1891				•	•		•	•	•	•	•	•
Symphodus mediterraneus (Linnaeus, 1758)	•	•	•	•	•	•	•	•	•	•	•	•
Symphodus ocellatus Forsskal, 1775								•	•			
Symphodus roissali (Risso, 1810)					•		•		•	•	•	•
Symphodus rostratus (Bloch, 1797)								•	•			
Symphodus tinca (Linnaeus, 1758)	•	•	•	•	•	•	•	•	•	•	•	•
Trachinus sp.						⊙	⊙					
Trachurus mediterraneus (Steindachner, 1863)						•	•					
Trachurus sp.						•	⊙	•				
Trachurus trachurus (Linnaeus, 1758)						•	•					
Triglidae						•	⊙					
Umbrina cirrosa (Linnaeus, 1758)		•	•	•	•	⊙	•					
Uranoscopus scaber Linnaeus, 1758						•	•					
Zeus faber Linnaeus, 1758						•	•					
Total	29	38	48	56	61		87	44	43	38	52	42

density using census data from February 1989 to July 1993. The values of study area (280 m^2, 1,026 m^3 of water volume, with 80 m^3 of the artificial reef) are a rough estimate of biomass because of the methodology used. The biomass ranged between 117.9 g m^{-2}, 33 g m^{-3} (considering water volume), 424.44 g m^{-3} (based on the volume of the two pyramids) for February '92 and 9.59 g m^{-2}, 2.69 g m^{-3}, 34.52 g m^{-3} for February '93. The average of the 62 censuses was 31.68, 8.89 g m^{-3}, 114.06 g m^{-3}. These are high biomass values compared to both natural areas and other studies carried out on Mediterranean ARs (see Table 4 in Relini et al. 1995b) and, in particular, compared to muddy-sandy bottom areas around the LAR, where only low numbers and biomass were observed (3–4 species and 5–10 g m^{-3}).

From 1994 only visual censuses were undertaken (Relini et al., 1997b, 2002b). The role of a FAD attached to a reef pyramid was also studied (Relini et al., 1995a). While the species diversity proved to be more or less the same around the pyramid with or without a FAD, the densities registered around the FAD were three times higher than those obtained at the control station. In particular, there were four species, distributed throughout the column of water, which were notably more abundant close to the FAD: *Chromis chromis, Oblada melanura, Spicara smaris* and, especially, *Spicara maena*, which registered density levels 17 times greater on the FAD station than on the control station. Among the nektobenthonic species, only the white bream *Diplodus sargus* experienced an increase on the pyramid with the FAD, which can be explained by the presence of fouling exploited by the white bream as a source of food.

The 87 taxa (78 at species level) recorded at LAR from 1989 to 2004, of which 61 taxa (56 at species level) were monitored by visual census, are listed in Table 1. Up to February 2005 (Palandri et al., 2006) 61 taxa were recorded by censuses.

In addition, five cephalopods, *Illex coindetii* (Venary), *Loligo vulgaris* Lamarck, *Octopus vulgaris* Cuvier, *Sepia officinalis* Linnaeus, *Todarodes sagittatus* (Lamarck) and four crustacean decapods of commercial interest, *Homarus gammarus* (L.), *Palinurus elephas* (Fabricius), *Scyllarus arctus* (L.), *Squilla mantis* (L.) have been recorded.

There was an obvious seasonal change in the fish community on the LAR, shown by the fact that some species were present only during certain periods of the year, and by the regular seasonal variations in the number of individuals of other species (Relini et al., 1994b). Two different populations, one Summer–Autumn, the other Winter–Spring, seemed to occur. These variations were often linked to the arrival of juveniles, an event which occurred in mid-Summer and in Autumn for the majority of the species, indicating that the AR acted as a nursery.

On the whole, changes in the composition of the assemblages led to a minimum number of individuals in June and a maximum in late Summer and Autumn, as described on the artificial reef near Marseilles, in the western Mediterranean (Bregliano & Ody, 1985). Here too, the authors obtained low values of total abundance in May and June, and a maximum from September to December.

The development of the fish community was closely linked to changes in the macrobenthic settlement on the blocks of the LAR. In 1990, there was a good development of algae (*Dyctyota* sp., *Sargassum vulgare*) covering the horizontal surfaces of the pyramids. In relation to this change, making the environment more complex and richer, the number of fish species, and in particular of young Labridae, which lived on the reef, increased considerably.

One year later, the disappearance of the large *Sargassum* and decrease of other algae, probably due to severe deposition of mud and increased water column turbidity, led to a notable reduction in species richness censused. The recovery was slow in the following years. More in general a change of fish community over time occured (Figs. 7, 8).

A study (Relini et al., 1997a) was devoted to recruitment of juveniles at two depths (18 m, pyramids; 10 m small pyramids and heaps of small blocks). In total the young of the year of 18 fish species were recorded. These were divided into three different groups:

(a) Four taxa (*Symphodus* spp., *Chromis chromis, Apogon imberbis, Coris julis*) which lived on the AR from their larval stages;

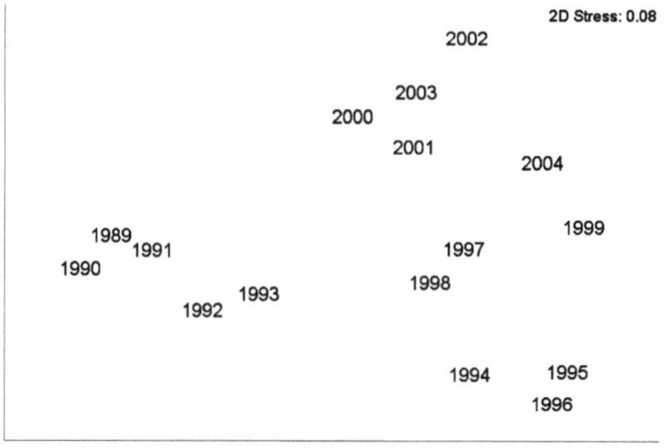

Fig. 7 MDS of fish species abundance (based on Bray-Curtis Similarities) for 16-years census on LAR

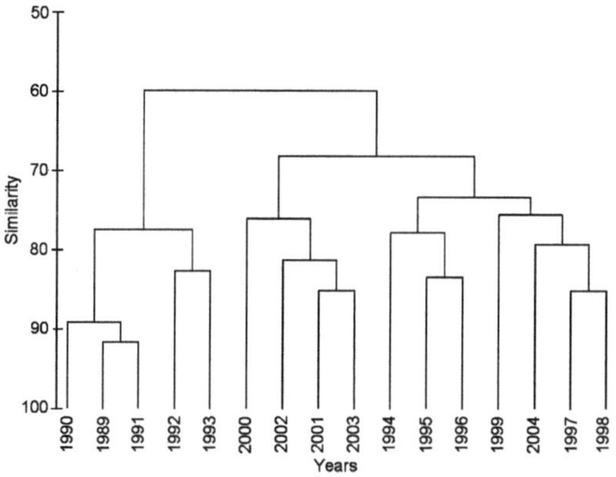

Fig. 8 Dendrogram of similarity of fish species richness from 1989 to 2004

(b) Eight species (*Mullus barbatus, M. surmuletus, Oblada melanura, Phycis phycis, Scorpaena notata, S. porcus, S. scrofa, Serranus cabrilla*) recorded on the AR some months after hatching. It is likely they spend the first days of their lifespan in other environments and then come to the AR;

(c) Six species (Diplodus annularis, D. sargus, D. vulgaris, Pagellus acarne, Spondyliosoma cantharus, Seriola dumerili) which came to the AR from shallow water when they were 4–12 months old.

The shallow water sites (10 m depth) were characterised by higher percentage of juveniles (up to the 80–90% of the total) than at 18 m depth, where the number of juveniles reached 40–50% of the total at the peak of recruitment.

At this depth there were also differences between reef structures: the pyramids consisting of five blocks, one meter side each, proved to be more effective (number of species and individuals per m^3 of material) than units made up of small blocks. Data of different structures were tested with the Wilcoxon test and showed very significant results ($p < 0.001$). The spatial distribution of fish and their behaviour in sharing spaces in the pyramids of LAR has also been described (Relini & Torchia, 1993).

Some species were tagged during 1994 and 1995 (grouper *Epinephelus marginatus*, common spiny lobster *Palinurus elephas* and the comber *Serranus cabrilla*). All tagged *S. cabrilla* registered during the tagging phase of the study were recaptured or observed on the same pyramid where they were released, suggesting that individuals stay on same pyramid for long time (Davies, 1995). One *P. elephas*, among 16 (70–82 mm CL) tagged with two T-bars and released in the LAR in May 1995, was recaptured in Sori (Eastern Ligurian Sea) after 18 months, having covered a distance more than 70 km (Relini & Torchia, 1998). The increment of the carapace length was 15 mm in 18 months.

Trophic relationships between fishes and the LAR were investigated (Relini et al., 2002a) analysing the feeding habits of four resident fish species (*Diplodus annularis, Scorpaena porcus, S. notata* and *Serranus cabrilla*) caught either by spear fishing or by trammel net.

Scorpaena porcus and *S. notata* caught with trammel nets had fed very little on reef species (only 5% and 6%, respectively, of individuals belonging to species common to stomach contents and the blocks). The guts of specimens of *S. notata* and *S. cabrilla*, caught by spear fishing on the reef blocks, contained 58% and 78%, respectively, of individuals belonging to species common to both. *D. annularis* feeds on a wider spectrum of species, almost all of which are present on the reef blocks (91%).

In conclusion, the fish population at the LAR showed a high species richness, high biomass and good catch rates with trammels; 33.3% of species censused by the visual method can be considered as belonging to the resident population.

By comparing the Loano list with that of fish censused in the natural rocky areas of marine parks of Port-Cros (Marseilles, France: Harmelin, 1987), Portofino (Gulf of Genoa, Italy: Tunesi & Vacchi, 1993; Tunesi & Molinari, 2005), Asinara (NW Sardinia: Tunesi et al., 2001), it can be seen that the three natural environments have certain fish characteristics in common with Loano. Thirty-eight species out of 47 recorded at Port Cros were common to both biotopes and *Chromis chromis* was numerically dominant in both communities. The most numerous families were Sparidae and Labridae, followed by Serranidae. Fifty-two species out of 85 censused at Portofino on rocky substrate were in common, and 20 species out of 33 non-listed in Loano belong to cryptic species mainly Blennidae and Gobidae. The same was true for Asinara, which shared with Loano 42 out of 58 common species.

Many of the species censused at Loano were the same as those caught or observed in *Posidonia* meadows (Bell & Harmelin-Vivien, 1982; Harmelin-Vivien, 1982; 1983). These authors also stated that the majority of species that frequent *Posidonia* meadows were not exclusive to that biocoenosis.

Spotorno and Ventimiglia

In 1989–1990, an artificial reef was built at Spotorno (Fig. 9a) and a similar one off Ventim-

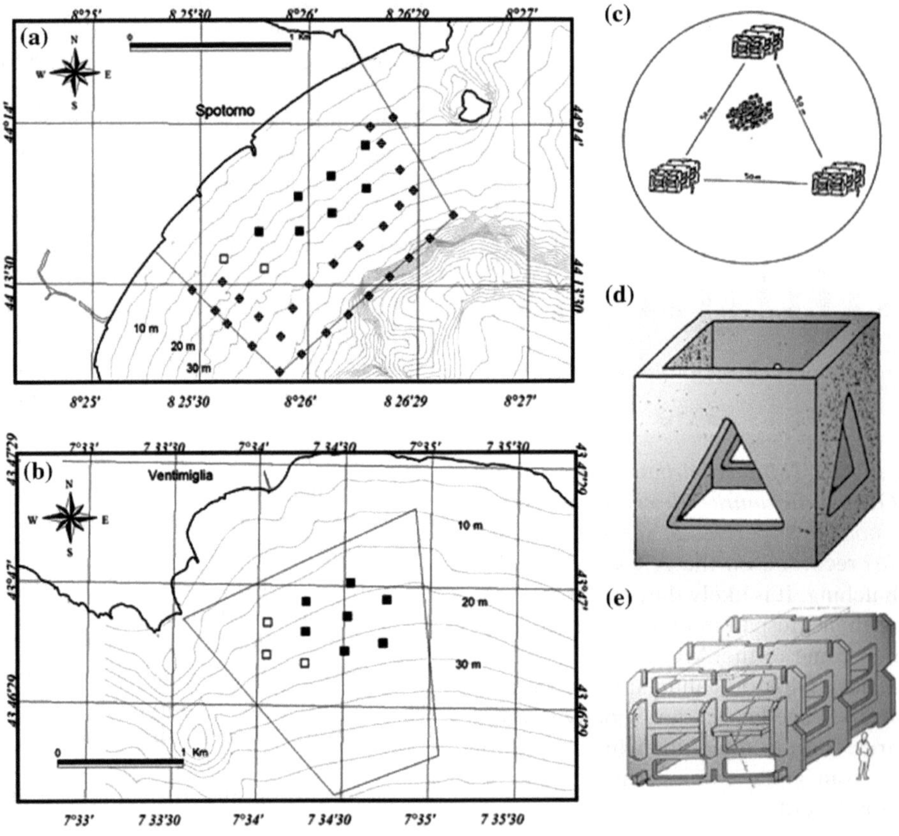

Fig. 9 The Spotorno and Ventimiglia ARs. (**a**) Map of the Spotorno AR sited between the town of Spotorno and Bergeggi island. The black spots represent the 7 units, the two white ones are planned and the crossed circles are the piles. (**b**) Map of Ventimiglia AR. (**c**) A unit is composed of three big modules positioned at the vertices of a triangle (50 m side) inside which 14 small modules are sited. (**d**) A small module, an open cube of 1.25 m side. (**e**) A big module of about 158.4 m^3, open parts included

iglia. The construction was financed by the EEC, planned and followed by Studio Volta of Savona and from the scientific point of view, by GIS-Posidonie, Marseille which monitored the environment before the immersion of the structures.

The AR off Spotorno extends to a depth of 48 m covering a surface of 167 ha. The area is protected from the illegal trawl fishing by a double row of 24 steel piles (400 mm in diameter and 6 m in length) driven into the sand and by seven concrete blocks with piles placed on rocky substrates. Inside this area there are seven "groups of modules" or units (Fig. 9) set in a double row, at the distance of "biological interaction" of 200 m. Each group of modules is composed of three "big modules", aimed to attract the pelagic species (Fig. 9), and of 84 "small modules" for benthic species.

The "big module" (Bonna type) is a structure composed of nine pre-fabricated cement panels each, of 6 × 6 × 4.4 m with a total volume of 158.4 m^3 (open parts included), 172.8 m^2 of surface. The weight of the big module is 26 t. The small module for benthic species is an empty cube of 1.25 m side with a triangular hole in each lateral face.

There are no published data on these two ARs and so it is impossible to know the results obtained.

The same design was followed to plan ARs off Varazze, Pietra Ligure, Diano Marina by Studio Volta, but fund raising was not successful.

Lavagna

The AR of Lavagna, made of eight pyramids (Fig. 10), was deployed on the seabed in 17 m of water in October 1993. The Ansaldo Company wished to evaluate small concrete blocks (15 cm side) made of stabilised muddy materials, dredged from harbours, with the aim of using this recycled material in marine constructions, including ARs (Relini et al., 1995c). The rack with its small stabilised mud and concrete (as control) blocks, was in the middle of an area protected against illegal otter trawling by eight pyramids 10 m apart, each made of five perforated concrete blocks of 2 m side, similar to those off Loano. The pyramids were lying directly on the seabed without the mattress of stones, used as a foundation layer, described in LAR. The absence of this bed allowed the blocks to sink partially into the seabed and destabilised the structure so that the pyramids fell over in storm conditions.

The seabed, influenced by the River Entella, is muddy and sandy with some patches of sea grass *Cymodocea nodosa*. Settlement of benthic organ-

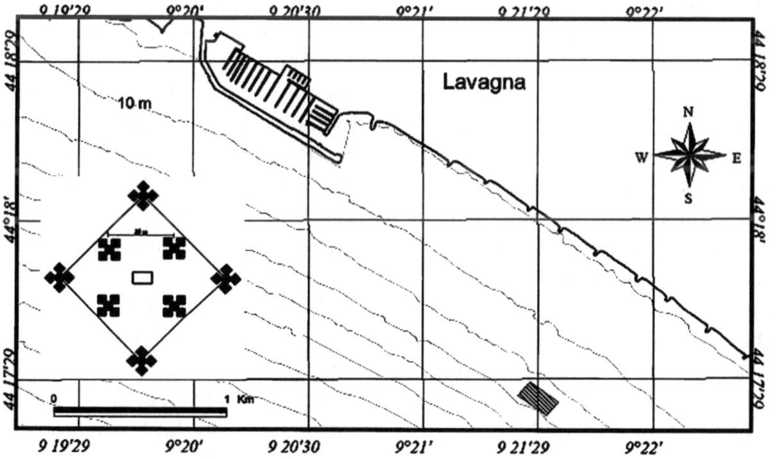

Fig. 10 Lavagna AR. Map of the site with the small AR between 15 m and 20 m depth (R). On the left: the eight pyramids protecting the rack sited in the centre

isms (Relini et al., 1995c; 1998b) and fishes (Relini et al., 1996; Relini et al., 1997a, b) were studied immediately after the immersion of blocks. The total number of taxa (identified mostly to species level) that settled on small blocks (both concrete and stabilised mud) was similar to that of Loano where, during the same period (1993–1994), small concrete and coal ash blocks were investigated (Relini et al., 1995b, c).

Fishes were studied by means of two visual censuses per season on two pyramids (about 80 m^3) from November 1993 (one week after the immersion of the AR) to May 1995. During this 18 months period, 25 species—23 of them of commercial interest—were recorded (Relini et al., 1996). Table 1, eighth column, shows the list of fish species monitored by visual census at Lavagna AR up to spring 2004.

Comparison between Loano and Lavagna

Comparing the settlement on small concrete blocks (using only the series of increasing time of exposure, 3, 6, 9, 12 months) immersed in Loano and in Lavagna for 1 year, 53 taxa were recorded, 31% of which were common to both sites, 13 taxa were exclusive to Lavagna and 9 to Loano (Relini et al., 1998b).

A comparison between the 13 surveys, carried out at Lavagna and Loano between November 1993 and July 1995, showed that 14 fish species were in common (Relini et al., 1997b). Twenty-four species were seen at Loano and 21 at Lavagna. Each survey was carried out in the two areas with large pyramids (blocks of 2 m side) covering a volume of water of about 400 m^3 and a volume of sunken material of 40 m^3. The lower number of species recorded at Lavagna and Loano was caused by the use of only one pyramid and not two as in previous surveys, so some rare species have not been censused during this comparative study.

At the LAR there was a greater number of fish species with higher percentage of occurrence: eight in the 1st class of frequency of occurrence (>75% presence in the census) and 2 in the 2nd one (50–74.9%). At Lavagna there were only two species in the 1st class and 5 in the 2nd. Thus, at Loano the resident nucleus was more numerous and more stable. At Lavagna there were many species visiting the reef for short periods only.

Table 2 Space occupation classes of Mediterranean littoral rocky fish species (after Harmelin, 1987)

Spatial category	Description
Cat. I	Highly mobile gregarious, pelagic erratic species
Cat. II	Planktophagous and relatively sedentary species, living throughout the water column
Cat. III	Demersal mesophagous species, with medium-amplitude vertical movements and relatively broad horizontal movements
Cat. IV	Demersal species, with slight vertical and high lateral movements
Cat. V	Sedentary demersal mesophagous species
Cat. VI	Highly sedentary cryptic benthic species

Fifty-four percent of the species present at Loano were sedentary (spatial categories II, V and VI, as defined by Harmelin, 1987) (Table 2) and 46% were active with medium to high mobility (spatial categories I, III and IV). At Lavagna active fish dominated, representing 66% of the species, while the remaining 33% were more sedentary.

Seven species were exclusive to Lavagna: 5 active (spatial categories I and III) and 2 sedentary (spatial category V). Ten species were exclusive to Loano: two were active (spatial category III) and 8 were sedentary (spatial categories II, V and VI).

At Lavagna the seasonal variations were much greater. In the first 7 months the population increased quickly, reaching a very high density (almost six specimens m^{-3} in comparison with LAR), subsequently falling and then picking up again later. These fluctuations were in part due to the high number of pelagic species that periodically visited the reef in large schools (50–300 individuals).

Differences in the fish populations at Lavagna and Loano can be attributed to the greater age of the LAR, where a stable fish community had become established, as opposed to the developing community on the Lavagna reef. Indicators of the immaturity of the Lavagna community are: the very few but increasing numbers of sedentary species, the large qualitative and quantitative variations between surveys, and the presence of a high number of juveniles and sub-adults (see Table 3 in Relini et al., 1997b).

A comparison between Lavagna and Loano suggests that the main variables include: age of the two reefs (8–9 years for the large pyramids at Loano, 1–19 months for Lavagna); the geographical area (eastern and western Ligurian Riviera) and the presence of a sparse *Posidonia* meadow at Loano, absent at Lavagna, where only *C. nodosa* meadow occurs.

Alassio

The Alassio AR, planned by Studio Gaggero, IDRA and University of Genoa, was immersed in 1998–1999 with the financial support of Local Municipality that wished to restore the degraded *Posidonia* bed and so to contribute to sandy shore protection and enhancement of fish population. The *Posidonia* meadow was distributed as a narrow strip (120–250 m) parallel to the shoreline between 12 and 20 m depth. The AR lies outside the remains of the *Posidonia* meadow at about 20 m depth and is composed of three main parts (Fig. 11): two recovering areas (REA) at two extremities (C, D in Fig. 11a) and one protected area (PRO) in between. The REA is composed by three rings of eight pyramids (five concrete blocks of 2 m side, Fig. 11b, e), similar to those of Loano, giving a volume of 960 m^3, an area of 7,850 m^2. One REA is located near the small wharf (A) in the middle of the Gulf of Alassio and the other one eastward, near S. Croce Cape (B in Fig. 11a).

The PRO covers a surface of 70,000 m^2 with 80 blocks of 1 m side (Fig. 11d) and 93 tetrapods (2 m high, 3.2 m^3; Fig. 11c) in total 378 m^3 and joints the two REA.

Before the deployment of AR the biocoenoses living in the area between Laigueglia and Gallinara Island down to 30 m depth, in particular *Posidonia oceanica* and *Cymodocea nodosa* meadows, were mapped to facilitate the choice of reef site. Some trammel net catches (500 m long, 1.5 m high, 8 cm mesh, exposure during the night from sunset to sunrise) and visual censuses were undertaken. Only eight species (0.02 individuals/m^2) were recorded by visual census (3 transects 50 m each in summer and winter) and most of them near the small wharf. The average trammel net catch was poor, 0.67 kg/100 m of net, in comparison with LAR, but 22 species were caught. It is significant that fishes characteristic of *Posidonia* meadows and hard bottoms were absent before the deployment of AR.

During six visual census surveys, 24 fish species were recorded at wharf REA and 28 at S. Croce REA, 15 were common to both sites. Some fishes typical of hard bottoms occurred; 32 species were caught by trammel net, 28 near the pyramids of eastern REA and 19 on the tetrapods, while the average catch rates were respectively 1031 ± 489 g/100 m and 776 ± 226 g/100 m (unpublished data).

Fifty-one animal sessile macrobenthic taxa settled on panels immersed during 1, 3 and 6 months. The pioneer settlement on blocks of pyramids was monitored by photos taken inside frames fixed to some blocks. Algae were important colonisers. Finally the spreading of the non-indigenous invasive alga *Caulerpa taxifolia* was monitored in the area of the AR and its surround.

Further visual censuses showed an increase in the fish species richness; up to spring 2004, 43 bony fish species were recorded (Table 1, ninth column).

ARs in progress during 2004

Sanremo

The AR off Sanremo, financed by the Municipality and planned by Studio Gaggero, IDRA and University of Genoa, is under construction: a first part was immersed during 2004, between 20 and 30 m depth. It comprises two main parts, one off Portosole harbour (Fig. 12a, point A), the second in front of Pian di Poma (point B in Fig. 12a). At present (Summer 2004), in zone A only nine pyramidal blocks have been immersed, while in zone B there are two REAs (Restoring Area) joined by 20 pyramidal blocks (Fig. 12). A REA is composed of four pyramids, each of five perforated blocks. Both the small pyramidal blocks and pyramids of five cubic blocks are placed on a mattress of small stones.

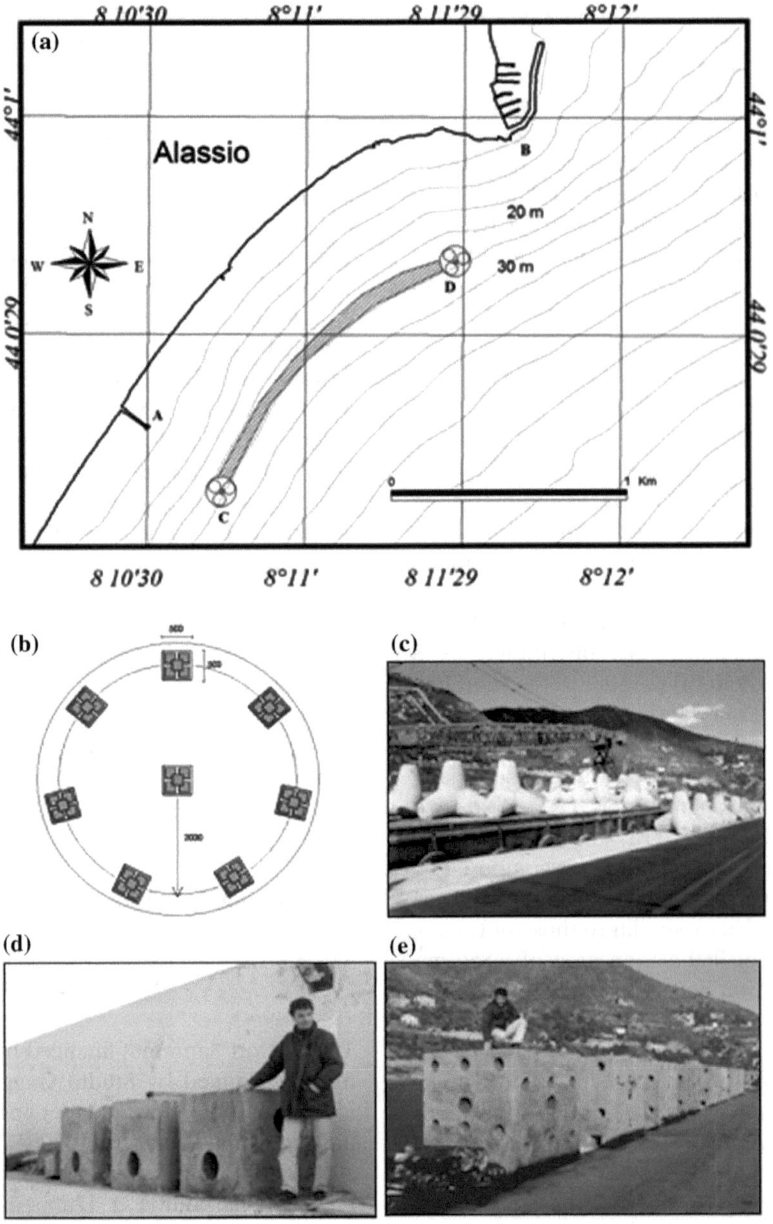

Fig. 11 Alassio AR. (**a**) Map showing the area covered by the AR between wharf Bestoso (A) and S. Croce Cape (B). (**b**) One of the three rings of REA (Restoration area: C, D) composed of 7 pyramids at the border and one in the middle of a circle of 45 m in diameter. (**c**) The tetrapods. (**d**) Small blocks (1 m side) immersed with tetrapods and making up the PRO (Protection area sited between C and D). (**e**) The large blocks (2 m side) for pyramids

The aim of this AR is the same as that for Alassio and Loano. New types of modules are being tested, in particular the cubic block containing a lot of tubes of different diameters to increase structural complexity and to offer refuges of different sizes as suggested by previous experience.

Monitoring of fishes and benthos respectively is planned through visual census by SCUBA

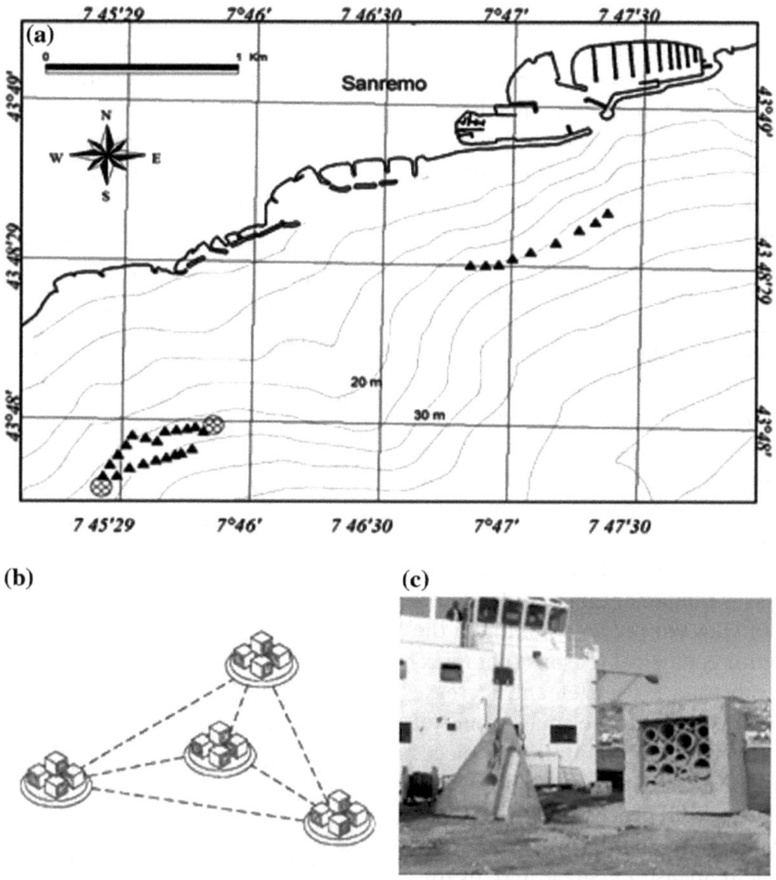

Fig. 12 The AR of Sanremo (**a**) Map of the AR placed in two zones off Portosole (A) and near Pian di Poma (B), (**b**) One of the two groups of pyramids in the site B. Each group is constituted of four pyramids, laying on a base of small stones, with five cubic blocks 2 m side. (**c**) A pyramidal block and a cubic one before immersion

divers, photo shots taken inside a frame fixed to some blocks and epibiota samples taken by scraping of settled organisms.

There are many other proposals for ARs along the Ligurian Coast, but there is no overall plan because there is a lack of financial support.

Conclusion

The experience acquired and the results obtained during about 30 years of studies carried out in the Gulf of Genoa on ARs and published in about one hundred papers, confirm that the main aims of reef deployment (in particular, protection against otter trawling, and increase of species richness and biomass) have been achieved, and experience allows workers to suggest the best approach and design for future ARs to be built in particular in the Ligurian Sea.

The concrete blocks of ARs act in the same way as a natural rocky bottom, if both fish and sessile macrobenthos are compared. Most of the fishes censused at Loano AR are in common with those present in protected natural rocky areas (Port Cros, Portofino, Asinara) and a high percentage have been recorded in other Italian and French ARs.

The numerical predominance of species belonging to Sparidae and Labridae, sometimes followed by Serranidae, seems to be one of general characteristics of rocky bottoms in the N–W Mediterranean (Harmelin, 1987).

Ligurian ARs show no affinities at all with those of the western Adriatic sea (Bombace, 1981, 1989; Bombace et al., 1994) and low affinity with the Fregene AR laid in the mid-Tyrrhenian sea (Ardizzone et al., 1989, 2000), probably because of the high siltation rates and nutrient loads in these (eutrophic) waters. Similarities are shown with Sicilian reefs (D'Anna et al., 2000) and those of Provençal coasts (Charbonnel, 1990) and Balearic Islands (Moreno, 2000), all laid in oligotrophic waters.

The more complex and diversified the structure and the shape of the module, the more microhabitat, shelters, and substrate for settlement are available, and so in turn the richer in number and biomass will be the species. Holes and crevices at different levels are very important as refuges for eggs and young fish.

The presence of a bed of small stones under the pyramids of blocks is fundamental not only from a structural point of view (to prevent sinking in the soft bottom), but for ecological reasons (complexity, new type of substrata for settlement, increase of species richness). Among these small stones of the Alassio AR, the natural implantation of pieces of *Posidonia* broken by waves occurred—a very important and unexpected process, described also in Sicily by D'Anna et al. (2000).

The management measures suggested by the results obtained in the Gulf of Genoa are immediate and clear: in order to increase the specific richness and the abundance of individual fish, it is necessary to diversify the environment by increasing the attraction systems used in the water column; but to avoid the construction of ARs on sites with heavy sedimentation, as this interferes with colonisation by macrobenthos and consequently affects the halieutic biomass.

Another important recommendation deals with the type of module to be used in the Ligurian Sea and, more generally, in the Mediterranean Sea to achieve the main aims described, and in particular protection against trawlers and the supply of new microhabitats.

The best module must have the following characteristics:

(a) to last for a long time (>50 years) without leaching pollutants;
(b) to be strong and heavy to avoid displacement or damage by otter trawl activity;
(c) to have the maximum possible number of holes and crevices;
(d) to present sufficient surfaces that are protected from silting and so available for settlement of sessile organisms.

Concrete seems to be the most suitable material to build modules with these characteristics. The large module can be assembled in pyramids or other configurations, with the addition of small structures to increase the complexity of the shape.

It is confirmed that ARs could be used as a successful tool in the management of coastal waters, providing protection of the natural environment, increasing biodiversity and biomass of fishes and benthos, providing shelter and nursery areas for young fishes, and producing food to be used locally or to be exported. The utilisation of AR or parts of them for tourism (diving, fish watching, photography) is an interesting and promising development for ecotourism.

Finally, it is encouraging to report that people and Public Administration are so strongly convinced of the successful role of ARs along the Ligurian coast, that the two last ARs (Alassio and Sanremo) were built mainly or completely thanks to local (Municipality) financial support.

References

Arculeo, M., G. Bombace, G. D'Anna & S. Riggio, 1990. Evaluation of fishing yields in a protected and an unprotected coastal area in N/W Sicily. FAO Fisheries Report 428: 70–83.

Ardizzone, G. D. & G. Bombace, 1983. Artificial Reef experiments along a Tyrrhenian sea coast. Journée de Etudes sur le Récif artificielle et Mariculture suspendus. CIESM. Cannes, 45–51.

Ardizzone, G. D., C. Chimenz & A. Belluscio, 1989. The development of epibenthic communities on artificial reefs in the central Mediterranean Sea. Bulletin of Marine Science 44(2): 592–608.

Ardizzone, G. D., A. Somaschini & A. Belluscio, 2000. Prediction of Benthic and Fish Colonization on the Fregene and other Mediterranean Artificial Reefs. In Jensen A. C., K. J. Collins & A. P. M. Lockwood (eds.), Artificial Reefs in European Seas. Kluwer Academic Publishers, The Netherlands, 113–128.

Bell, J. D. & M. L. Harmelin-Vivien, 1982. Fish fauna of French Mediterranean *Posidonia oceanica* seagrass meadows. I. Community structure. Téthys 10(4): 337–347.

Bombace, G., 1977. Aspetti teorici e sperimentali concernenti le barriere artificiali. Atti IX congresso S.I.B.M., Ischia, 29–42.

Bombace, G., 1981. Note on experiments in artificial reefs in Italy. In management of living resources in the Mediterranean coastal area. Studies and reviews. General Fisheries Commission for the Mediterranean 58: 309–324.

Bombace, G., 1989. Artificial reefs in the Mediterranean Sea. Bulletin of Marine Science 44(2): 1023–1032.

Bombace, G., G. Fabi & L. Fiorentini, 1990. Preliminary analysis of catch data on artificial reefs in Central Adriatic. FAO Fisheries Report 428: 86–98.

Bombace, G., G. Fabi, L. Fiorentini & S. Speranza, 1994. Analysis of the efficacy of artificial reefs located in five different areas of the Adriatic Sea. Bulletin of Marine Science 55: 559–580.

Bregliano, P. & D. Ody, 1985. Structure du peuplement icthyologique de substrat dur à travers le suivi des récifs artificiels et d'une zone naturelle témoin. Quatrième Colloque Pluridisciplinaire Franco-Japonais, Marseille 16–21 Septembre 1985, 6: 101–112.

Carlisle, J. C. jr., C. H. Turner & E. E. Ebert, 1964. Artificial habitat in the marine environment. California Department Fish Game Fish Bulletin 124: 1–90.

Cecere, E., A. Petrocelli, G. Relini & O. Saracino, 1993. Phytobenthic communities on fouling panels at Loano (Savona, Ligurian Sea). Oebalia, 19(suppl.): 163–172.

Charbonnel, E., 1990. Les peuplements ichtyologiques des récifes artificiels dans le département des Alpes-Maritimes (France). Bulletin Société Zoologie Françãis, 115(1): 123–136.

D'Anna, G., F. Badalamenti & S. Riggio, 2000. Artificial Reefs in North-West Sicily: Comparisons and Conclusions. In Jensen A. C., K. J. Collins & A. P. M. Lockwood (eds) Artificial Reefs in European Seas. Kluwer Academic Publishers, The Netherlands, 97–112.

Davies, C.R., 1995. Movement of a small serranid *Serranus cabrilla* within the Loano artificial reef in the Ligurian sea, Italy. Biologia Marina Mediterranea 2(1): 91–94.

Duclerc, J., 1980. Rèserve marine de Golfe-Juan. Etude de point zéro. Comité Local des Pêches et ISTPM. Centre de Recherches de Sète. 15 pp.

Falace, A. & G. Bressan, 2002. A qualitative and quantitative analysis of evolution of macroalgae vegetation on an artificial reef with antigrazing nets (Loano–Ligurian Sea). ICES Journal of Marine Science 59, S150–S156.

Falace, A., A. Zuliani & G. Bressan, 2002. Macrophytobenthic colonization of an artificial reef in the Ligurian Sea (Loano–Savona). Biologia Marina Mediterranea 9(2): 20–28.

Guidetti, P. & G. Relini, 1995. Accrescimento di *M. galloprovincialis* nella barriera di Loano. Biologia Marina Mediterranea 2(1): 187–202.

Harmelin, J. C., 1987. Structure et variabilité de l'ichtyofaune d'une zone rocheuse protégée en Méditerranée (Parc -national de Port-Cros, France). P.S.Z.N.I: Marine Ecology 8(3): 263–284.

Harmelin-Vivien, M. L., 1982. Ichtyofaune des herberies de Posidonies du parc national de Port-Cros. I: Composition et variation spatio-temporelles. Travaux Scientifiques du Parc National de Port-Cros 8: 69–92.

Harmelin-Vivien, M. L., 1983. Ichtyofaune des herberies de Posidonies des côtes provencales françaises.–Rapport Commission Internationale Mer Mèditerranée. CIESM. Monaco 28 (3): 161–163.

Harmelin-Vivien, M. L., J. C. Harmelin, C. Chauvet, C. Duval, R. Galzin, P. Lejeune, G. Barnabe, F. Blanc, R. Chevalier, J. Duclerc & G. Lasserre, 1985. Evaluation des peuplements et populations de poissons. Méthodes et problèmes. Revue d'Ecologie. (La Terre et la Vie) 40: 467–539.

Jensen, A. C. (ed.), 1997. European Artificial Reef Research. Proceedings of the first EARRN Conference, March 1996 Ancona, Italy. Southampton Oceanography Centre. 449 pp.

Jensen, A. C., K. J. Collins & A. P. M. Lockwood, 2000. Artificial Reefs in European Seas. Kluwer Academic Publishers, 508 pp.

Maurizi, A., 1993. Rapporti trofici nella barriera artificiale di Loano. Tesi di laurea in Ecologia Animale, Università degli Studi di Genova, 218 pp.

Montanari, M. & G. Relini, 1973. Variazioni stagionali del fouling di pannelli immersi a diverse profondità nell'Avamporto di Genova. Atti V Congresso Società Italiana Biologia Marina. Ed. Salentina, Nardò, 305–326.

Moreno, I., 2000. Artificial Reef Programme in the Balearic Islands: Western Mediterranean Sea. In Jensen A. C., Collins K. J. & Lockwood A. P. M. (eds.) Artificial Reefs in European Seas. Kluwer Academic Publishers, The Netherlands: 219–233.

Palandri G., E. Beccornia, M. Relini & G. Relini, 2006. Sedici anni di osservazioni sui pesci della barriera artificiale di Loano. Biologia Marina Mediterranea 13(1): 132–140.

Puccio, V., G. Torchia, M. Relini & G. Bressan, 2001. Valutazione del ruolo dei ricci nello sviluppo del fitobenthos dei substrati artificiali della barriera di Loano (SV), Mar Ligure. Biologia Marina Mediterranea 8(1): 617–622.

Relini, G., 1979. Ricerche in corso in Liguria sulle barriere artificiali. *(Artificial reef investigation in progress in the Ligurian Sea)*. Atti 8° Simposio Nazionale sulla Conservazione della Natura, Bari, Aprile 1979: 79–87.

Relini, G., 1982a. La barriera artificiale nel Golfo Marconi (Mar Ligure). *(The artificial reef in Marconi Gulf (Ligurian Sea)*. Naturalista Siciliano, S IV,VI (suppl.) 3: 593–599.

Relini, G., 1982b. Esperienze di barriere artificiali in Liguria. *(Investigation on artificial reefs in the Ligurian Sea)*. Atti del Convegno delle Unità Operative afferenti ai sottoprogetti Risorse Biologiche e Inquinamento Marino. (Roma, 10–11 Nov. 1981): 155–164 .

Relini, G., 1983. Twelve years of experiments on artificial reefs in the Gulf of Genoa (Italy). Journée Etudes Recife artificielle et Mariculture suspendus. CIESM. Cannes 1982: 73–75.

Relini, G., 2000a. The Loano artificial reef. In Jensen A. C., K. J. Collins & A. P. M. Lockwood (eds.) Artificial Reefs in European Seas. Kluwer Academic Publishers, The Netherlands, 129–149.

Relini, G., 2000b. Coal ash for artificial habitats. In Jensen A. C., K. J. Collins & A. P. M. Lockwood (eds.) Artificial Reefs in European Seas. Kluwer Academic Publishers, The Netherlands: 343–364.

Relini, G., & P. Cormagi, 1990. Colonisation patterns of hard substrata in the Loano artificial reef (Western Ligurian Sea). FAO Fisheries Report, 428: 108–113.

Relini, G., S. E. Merello & G. Torchia, 1998a. Influenza della sedimentazione sulla colonizzazione e successione ecologica di substrati duri. Biologia Marina Mediterranea (1998), 5(1): 154–163.

Relini, G. & S. Moretti, 1986. Artificial reef and Posidonia bed protection off Loano (Western Ligurian Riviera). FAO Fisheries Report, 357: 104–108.

Relini, G. & L. Orsi Relini, (1971) 1971–1972. Affondamento in mare di carcasse di automobili ed inquinamento. *(Car bodies in the sea and pollution)*. Quaderni della Civica Stazione di Idrobiololgia Milano, 3/4: 31–43.

Relini, G. & L. Orsi Relini, 1989. The artificial reefs in the Ligurian Sea (N–W Mediterranean): aims and results. Proceedings of fourth International Conference on Artificial Habitats for Fisheries. Bulletin of Marine Science 44(2): 743–751.

Relini, G., A. Peirano, L. Tunesi & L. Orsi Relini, 1986. The artificial reef in the Marconi Gulf (Eastern Ligurian Riviera). FAO Fisheries Report (357): 95–103.

Relini, G., M. Relini & G. Torchia, 1990a. Fishes of the Loano artificial reef (Western Ligurian Sea). FAO Fisheries Report, 428: 120–127.

Relini, G., M. Relini & G. Torchia, 1995a. La barriera artificiale di Loano. Biologia Marina Mediterranea 2(1): 21–64.

Relini, G., M. Relini & G. Torchia, 1997a. Reclutamento delle specie ittiche su alcune Barriere Artificiali della Liguria. Biologia Marina Mediterranea 4(1): 269–276.

Relini, G., M. Relini & G. Torchia, 1997b. Fish assemblages in the Ligurian Artificial Reefs (N–W Mediterranean). In Hawkins, L. E., S. Hutchinson, A. C. Jensen, J.A. Williams & M. Sheader, (eds), The Responses of Marine Organisms to their Environment, 362 pp.: 337–343.

Relini, G., M. Relini, G. Torchia & G. De Angelis, 2002a. Trophic relationships between fishes and an artificial reef. ICES Journal of Marine Science, 59: S36–S42.

Relini, G., M. Relini, G. Torchia & G. Palandri, 2002b. Ten years of censuses of fish fauna on the Loano artificial reef. ICES Journal of Marine Science, 59: S132–S137.

Relini, G., M. Relini, G. Torchia, F. Tixi & C. Nigri, 1995b. Coal ash tests in the Loano artificial reef. Proceedings ECOSET '95. Japan International Marine Science and Technology Federation, 1: 107–113.

Relini, G., F. Tixi, M. Relini & G. Torchia, 1998b. Settlement on concrete blocks immersed in the Ligurian Sea (N–W Mediterranean). International Biodeterioration and Biodegradation, 41(1): 57–65.

Relini, G., N. Trentalance, M. Relini, G. Torchia, & F. Tixi, 1995c. Stabilized harbour muds for artificial reef blocks. Proceedings ECOSET '95. Japan International Marine Science and Technology Federation, 1: 114–118.

Relini, G. & M. Wurtz, 1977. La scogliera artificiale di Varazze (Mar Ligure) a sei anni dall'immersione. (The artificial reef at Varazze (Ligurian Sea) after six year immersion). Atti IX Congresso Società Italiana Biologia Mediterranea, Ischia, Maggio 1977: 363–371.

Relini, G., N. Zamboni & F. Sönmezer, 1990b. Development of macrobenthos community in the Loano artificial reef. Rapport Commission Internationale Mer Mèditerranée. CIESM. Monaco. 32(1): 27.

Relini, G., N. Zamboni, F. Tixi & G. Torchia, 1994a. Patterns of sessile macrobenthos community development in an artificial reef in the Gulf of Genoa (NW-Mediterranean). Bulletin of Marine Science, 55(2–3): 745–771.

Relini, M., G. Relini & G. Torchia, 1994b. Seasonal variation of fish assemblages in the Loano artificial reef (Ligurian Sea NW-Mediterranean). Bulletin of Marine Science, 55(2–3): 401–417.

Relini, M. & G. Torchia, 1993. Distribuzione spaziale dei pesci sulla barriera artificiale di Loano (SV). Biologia Marina, Supplemento Notiziario SIBM, 1: 195–200.

Relini, M. & G. Torchia, 1998. Inaspettato percorso migratorio di una aragosta in Mar Ligure. Biologia Marina Mediterranea 5(1): 641–643.

Relini, M., G. Torchia & G. Relini, 1995a. The role of a fad in the variation of fish assemblages on the Loano Artificial Reef (Ligurian Sea NW-Mediterranean). Proceedings ECOSET '95. Japan International Marine Science and Technology Federation, 1: 1–5.

Relini, M., G. Torchia & G. Relini, 1995b. Fish population patterns in a coastal artificial habitat in the northwestern Mediterranean. Biology and Ecology of Shallow Coastal Waters. In Eleftherion, A.D. Ansell & C.J. Smith (eds), Proc. 28th European Marine Biology Symposium. (Crete). Olsen and Olsen, Fredensbourg, DK, 359–368.

Relini, M., G. Torchia & G. Relini, 1996. Evoluzione del popolamento ittico sulla barriera artificiale di Lavagna. Biologia Marina Mediterranea 3(1): 230–236.

Sampaolo, A. & G. Relini, 1994. Coal ash for artificial habitats in Italy. Bulletin of Marine Science, 55(2–3): 1277–1294.

Seaman, W. Jr., 2000. Artificial Reef Evaluation, with application to Natural Marine Environment. CRC Press, Boca Raton, Florida, 246 pp.

Toccaceli, M. & D. Levi, 1990. Preliminary data on an experimental trammel net survey designed to estimate the potential of a planned artificial reef near Mazara del Vallo (Italy). FAO Fisheries Report 428: 154–158.

Tunesi, L. & A. Molinari, 2005. Species richness and biogeographic outlines of the fish assemblages of Portofino marine protected area (Ligurian Sea). Biologia Marina Mediterranea 12(1): 116–123.

Tunesi, L., E. Salvati, G. Lauriano & L. Mariani, 2001. Study of the coastal fish fauna of the Asinara island to support the zoning of the marine protected area. Atti A.I.O.L., 14: 189–198.

Turner, H. C., E. E. Ebert & R. R. Given, 1969. Man-made reef ecology. California Department Fish Game Fish Bulletin 1–46: 79–128.

Zamboni, N., F. Sönmezer & G. Relini, 1992. Variazioni stagionali delle comunità pioniere nella barriera artificiale di Loano. *(Seasonal variations of pioneer communities in the artificial reef of Loano).* Oebalia, suppl. 17: 435–437.

Hydrobiologia (2007) 580:219–224
DOI 10.1007/s10750-006-0452-1

Settlement and early survival of red coral on artificial substrates in different geographic areas: some clues for demography and restoration

L. Bramanti · S. Rossi · G. Tsounis · J. M. Gili · G. Santangelo

Abstract The red coral *Corallium rubrum* (L 1758) is a long-lived, slow-growing gorgonian, endemic to Mediterranean rocky bottoms. Because of its high economic value, red coral has long been harvested, and most populations have been depleted. In the present study, 54 marble tiles were placed in June 2003 within red coral populations over 3 different geographic areas (Calafuria–Livorno and Elba MPA in Italy and Medes Islets MPA, in Spain), on vertical cliffs between 25 and 35 m. In each area 2 different sites were randomly selected. Tiles were subsequently sampled photographically. Between July and August 2003 red coral recruits settled on tiles in all the geographic areas and sites, exhibiting wide variability in their density. On the basis of a 2-factors nested ANOVA a significant variability between different sites at a few hundred metres distance occurred, indicating high variations in the recruitment process within the same red coral population. Mortality, measured in June 2004, widely varied between different geographic areas.

Guest editors: G. Relini & J. Ryland
Biodiversity in Enclosed Seas and Artificial Marine Habitats

L. Bramanti · G. Santangelo (✉)
Dip. to Biologia, Ecol, Evol, Università di Pisa, Via Volta 6, I-56126 Pisa, Italy
e-mail: gsantangelo@biologia.unipi.it

S. Rossi · G. Tsounis · J. M. Gili
Institut de Ciencias del Mar CSIC, Barcelona, Spain

Keywords Artificial substrates · *Corallium rubrum* · Octocorals · Recruitment · Settlement · Western Mediterranean

Introduction

Data on recruitment and early growth rates are fundamental to population dynamics studies (Ebert, 1999). Direct field study of these demographic parameters for populations of slow-growing, deep-dwelling, epibenthic suspension feeders is particularly difficult, and few direct measurements of such populations are available (Garrabou & Harmelin, 2002). Long-term settlement plates of suitable materials could provide reliable measures of these basic life-history traits in these populations (Mundy, 2000).

Corallium rubrum (the precious red coral) is a long-lived, slow-growing gorgonian, endemic to Mediterranean and neighbouring Atlantic rocky bottoms between 20 and 200 m depth (Zibrowius et al., 1984). Red coral is one of the components of the Mediterranean *coralligenous* community (Sarà, 1973). Because of their high economic value, red coral colonies have long been harvested, and most populations have been greatly depleted. Thus, strategies to foster conservation and management of this species must be set out (Cicogna et al., 1999). Such strategies need to be based on a sound knowledge of population demographic features and life-history traits (Beissinger & McCullough, 2002; Santangelo et al., 2004; Santangelo et al., 2006). Long-lived species may require demographic measurements over many years (Bos et al., 2006) and several populations in order to sample the wide range of environmental and demographic variability (Fisk & Harriot, 1991; Caley et al., 1996; Todd, 1998; Beissinger & McCullough, 2002). The main objective of this study was thus to check red coral recruitment and mortality rates in different geographic areas via suitable, artificial substrates.

On the basis of a previous four-year study on red coral recruitment and early survival on marble tiles (Bramanti et al., 2005) carried out at Calafuria (Ligurian Sea, Italy), the net recruitment rate (recruitment minus mortality) showed a positive trend, indicating the tendency of red coral to permanently colonise the tiles. Moreover, out of the 388 colonies that settled on these tiles over 4 years, none was affected by boring sponges, which affected more than 50% of the colonies living in the same area and are one of the main causes of red coral mortality (Corriero et al., 1997). These findings, together with the early sexual maturity of this species (Santangelo et al., 2003), suggest that red coral re-colonisation of depleted areas could be fostered by transferring and re-fixing colonised marble tiles, rather than single colonies as in Cerrano et al. (1999).

In the research presented here we tried to determine the recruitment rate and variability in different red coral populations, in 3 different geographic areas in the north-western Mediterranean Sea, located at hundred kilometer distances.

Materials and methods

Three different geographic areas (Fig. 1) where red coral populations thrive were chosen for this

(a)

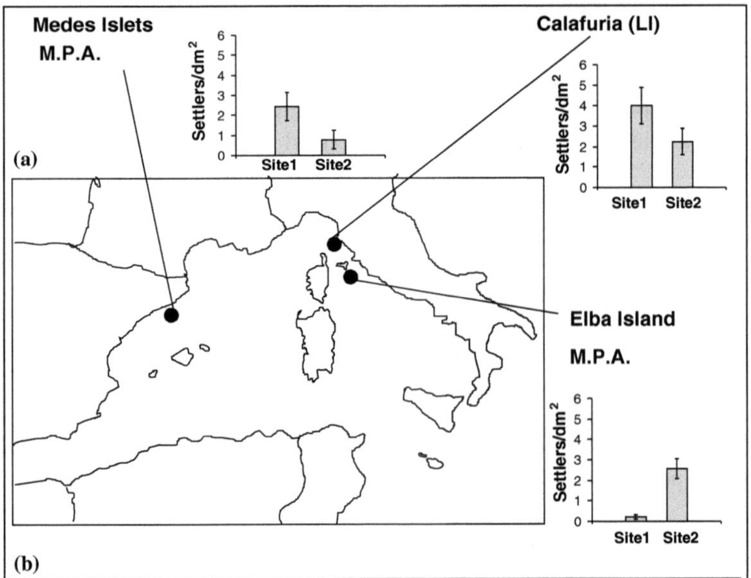

(b)

Source	SS	DF	MS	F	P	F versus
Area	1.59	2	0.7974	0.49	N.S.	Site(Area)
Site (Area)	4.86	3	1.6231	5.40	***	RES
RES	14.43	48				
TOT	20.89					

Fig. 1 (**a**) Map of recruitment density in the different areas and sites. (**b**) ANOVA of recruit density in the different areas and sites. The only significant factor was Site (2 levels, nested in area), indicating that the highest variability occurs between sites

experiment: the Medes Islets (Gerona, Spain, 42°03′ N, 3°13′ E); Elba Island (Tuscan Archipelago, Italy, 42°44′ N, 10°17′ E) and Calafuria (Livorno, Italy 43°30′ N, 10°20′ E). While the first two populations dwell in MPAs, the third is located in an unprotected coastal suburban area (Santangelo et al., 2003).

Previous genetic analyses of the Calafuria and Elba red coral populations showed that, even if located at about 100 km apart, they are discrete distinct genetic units (Abbiati et al., 1993).

In early June 2003, 54 white marble tiles (9 × 12 cm wide) were fixed via a central Fisher's screw (Bramanti et al., 2003) to the *coralligenous* substrate (*sensu* Sarà, 1973) onto the vaults of crevices in which red coral colonies dwell. All the marble tile method is patented (Patent Number PI2006A000126). In order to test if significant differences in red coral settlement exist within and between different geographic areas and neighbouring sites, the marble tiles were placed following a nested ANOVA model with 2 factors: 1) area, fixed with 3 levels: Calafuria, and Elba MPA (Italy) and Medes MPA (Spain); 2) site (nested in area), with 2 levels (2 sites randomly chosen in each area a few hundred metres apart). The factor site was thus random and nested in area. In each site, 9 tiles (replicates) were located.

Mortality variability was examined following a different, simplified one-way ANOVA model which factor was area (3 levels) with 9 replicates for each level. To perform a balanced experiment, as only 9 tiles were colonised at Elba and Medes, 9 tiles too were randomly selected from those colonised at Calafuria.

A small numbered buoy was placed near each tile. Tiles were sampled by photograph (Nikonos V 35 mm lens, macro 1:3 extension tube and TTL strobe) in October 2003 and June 2004. Mortality was determined as the percentage of settlers that died between October 2003 and June 2004.

ANOVAs were processed with the GMAV5 program for Windows (Sage, 2002). In order to test variance homogeneity, the Cochran test was applied, and as this resulted significant, a square root data transformation was performed (Underwood, 1997).

Results

Between July and August 2003 red coral planulae settled on the tiles in all the geographic areas and sites. In October 2003 all settlers (0–9 settlers on each tile) were easily recognised, and no further settlement was ever observed in the following months. Overall 66.6% of the tiles were colonised at Calafuria and 50% at Elba Island and Medes Islets (Table 1). Wide variability in the percentage of colonised tiles was also found between different sites of the same area; in one site of Elba Island only one tile was colonised (Table 1).

Settler density also showed wide variability (Calafuria 2.77 ± 3.04; Medes 1.6 ± 1.96; Elba 1.1 ± 1.4 recruits dm^{-2}; average ± SD; Fig. 1a), being higher at Calafuria, where the average density is comparable with that measured previously in the same area on the same kind of tiles (5.38 ± 4.09 col dm^2; Bramanti et al., 2005). Settlement was slightly higher at the Medes than at

Table 1 Percentage of tiles colonised by red coral and post-settlement mortality (October 2003 and June 2004) in the different sites and geographic areas

		Colonised tiles (%)	Settlers (October 2003)	Survived (July 2004)	Mortality (%)
Calafuria	Site 1	77.7	36	31	13.9
	Site 2	55.5	14	12	14.3
	Whole area	66.6	50	43	14
Elba	Site 1	11.0	1	1	0
	Site 2	88.8	19	17	10.5
	Whole area	50.0	20	18	10
Medes	Site 1	66.6	22	7	68.2
	Site 2	33.3	7	1	85.7
	Whole area	50.0	29	8	72.4

Elba and failed almost completely in one of the 2 sites at Elba, where only one settler was found.

As the results of the nested ANOVA revealed (Fig. 1b), site was the only significant factor, representing the main variance component. This finding indicates that the highest variability occurred between neighbour sites located in the same geographic area, rather than between different areas.

Mortality affected only 14 and 10% at Calafuria and Elba, respectively, and 72.4% of the settlers at Medes Islets (Table 1). Some minor variability in mortality was also found between different sites in the same area. According to one-way ANOVA, mortality varied significantly ($P < 0.0001$) between different areas, being higher at Medas and similarly lower at Calafuria and Elba (SNK test; Underwood 1997).

Discussion

Red coral colonised the marble tiles in all the sites, confirming marble provide a highly suitable substrate for red coral settlement in the different geographic areas. The whole."marble-tile settlement method" is now patented.

Our data indicate high variability in settlement, with the lowest levels at Elba Island. In one site of this area only one settler was found. As no data on density or recruitment of the Elba population has ever been collected before the present study, no further comparisons were possible.

Medes Islet too showed low settlement rates, confirming the reduced red coral settlement suggested previously for this area (Linares et al., 2002; Giannini et al., 2003). This finding could be due to the wide difference in adult colony density between the Medes Islets and Calafuria (26 vs. 0.2 col. dm^2; Santangelo & Abbiati, 2001; Rossi & Gili, 2003). The reproduction rates of the Calafuria and Medes populations also differ greatly, both in terms of fertility (number of fertile polyps per colony) and fecundity (number of planulae produced by each fertile polyp), the values at Calafuria being higher (Santangelo et al., 2003; Tsounis et al., 2006). Such spatial and sexual interpopulation differences may well lead to the different settlement rates found in the different populations.

The settlement levels found at Calafuria are the highest we found; they are comparable with those measured during a previous 4-year study on settlement and recruitment on marble tiles in the same area (Bramanti et al., 2005), confirming the high reproduction rate of this population. Overall, the percentage of colonised tiles at Calafuria is lower in this experiment (66 vs. 96% in Bramanti et al., 2005) but the following yearly mortality is also lower (14 vs. 24.35%) indicating some weaving of these demographic parameters within the same population.

A low mortality was also found at Elba while the highest mortality found at Medes suggests that several tiles could not be permanently colonised at this station.

A recent study by Garrabou & Harmelin (2002) on red coral recruitment on artificial limestone substrates in a cave near Marseille indicates low settlement and mortality rates: 0.11 settlers dm^2 and a 21-year mortality of –39.5% (equal to –1.9% year^{-1}). Such low recruitment and mortality values could be due to the peculiar environment in which the Marseille population settled (a cave with a 25-m horizontal extension), in which larval and plankton supply could be limiting factors for red coral settlement (Gili & Coma, 1998), reducing also density-dependent mortality of red coral settlers.

The high recruitment density and the low mortality recorded on many tiles implanted at Calafuria, on some of those implanted at Elba and at Medas, suggests that these tiles, if transferred to areas where the red coral populations have been depleted, could be a useful tool to foster their recovery (Oren & Benayahu, 1997). The whole method is now patented.

Lastly, these findings suggest that if the red coral populations of either Elba or Medes MPAs (both showing low recruitment values and, in the case of Medes, also a high post-settlement mortality) were affected by repeated mass mortality events, similar to that which struck the red coral in another Mediterranean area only a few years ago (Garrabou et al., 2001), they would have a limited capacity for recovery. On the contrary, the dense and highly reproductive Calafuria population (even if not dwelling in an MPA) would likely exhibit high resilience in the years following an anomalous mortality event or a negative net recruitment rate over more than one year (Bramanti et al., 2005). In the other two geographic areas, a reduction in settlement (Bassim & Sammarco, 2003) or some increase in mortality could greatly reduce the population resilience.

Acknowledgements This research was supported by a CSIC/ICM (Spain) - CNR (Italy) common research project: Population dynamics of the red coral. We thank A. Cafazzo for his revision of the English text.

References

Abbiati, M., G. Santangelo & S. Novelli, 1993. Genetic variation within and between two Tyrrhenian populations of the Mediterranean alcyonarian *Corallium rubrum*. Marine Ecology Progression Series 95: 245–250.

Bassim, K. M. & P. W. Sammarco, 2003. Effects of temperature and ammonium on larval development and survivorship in a scleractinian coral (*Diplora strigosa*). Marine Biology 142: 241–252.

Bramanti, L., G. Magagnini & G. Santangelo, 2003. Settlement and recruitment: the first stages in the life cycle of two epibenthic suspension feeders. The Italian Journal of Zoology 70: 175–178.

Bramanti, L, G. Magagnini, L. DeMaio & G. Santangelo, 2005. Recruitment, early survival and growth of the Mediterranean red coral *Corallium rubrum* (L 1758), a four-year study. Journal of Experimental Marine Biology and Ecology 314: 69–78.

Beissinge, S. R. & D. R. McCullough, 2002. Population viability analysis. The University of Chicago Press, Chicago.

Bos, O. G., C. J. M. Philippart, G. C. Cadée & J. Van Der Meer, 2006. Recruitment variation in Macoma balthica: a laboratory examination of the match/mismatch hypotesis. Marine Ecology Progress Series 320: 207–214.

Caley, M. J., M. H. Carr, M. A. Hixon, T. P. Huges, G. P. Jones & B. A. Menge, 1996. Recruitment and the local dynamics of open marine populations. Annual Review of Ecology and Systematics 27: 477–500.

Cerrano, C., G. Bavestrello, F. Cicogna & R. Cattaneo-Vietti, 1999. New experiences on transplantation and red coral harvesting effects in the Ligurian Sea. In Cicogna F. & R. Cattaneo-Vietti (eds), Red coral and other Mediterranean octocorals, biology and protection. Ministero Risorse Agricole, Alimentari, Forestali, Roma, 62–67.

Cicogna, F., G. Bavestrello & R. Cattaneo-Vietti (various authors), 1999. Red coral and other Mediterranean octocorals, biology and protection. Ministero Risorse Agricole, Alimentari, Forestali, Roma.

Corriero, G, M. Abbiati, & G. Santangelo, 1997. The sponge complex inhabiting a Mediterranean red coral population. PSZN Marine Ecology 18: 147–155.

Ebert, T. A., 1999. Plant and animal populations. Methods in demography. Ac. Press, S. Diego CA.

Fisk, D. A. & J. A. Harriot, 1991. Spatial and temporal variation in coral recruitment on the Great Barrier Reef: implications for dispersal hypotheses. Marine Biology 107: 485–490.

Garrabou, J., T. Perez, S. Santoretto & J. G. Harmelin, 2001. Mass mortality event in red coral *Corallium rubrum* populations in the Provence region (France, NW Mediterranean). Marine Ecology Progress Series 217: 263–272.

Garrabou, J. & J. G. Harmelin, 2002. A 20-year study on life-history traits of a harvested long-lived temperate coral in NW Mediterranean: insights into conservation and management needs. The Journal of Animal Ecology 71: 966–968.

Giannini, F., J. M. Gili & G. Santangelo, 2003. Relationships between the spatial distribution of red coral *Corallium rubrum* and coexisting suspension feeders at Medes Islets Marine Protected Areas (Spain). The Italian Journal of Zoology 70: 233–239.

Gili, J. M. & R. Coma, 1998. Benthic suspension feeders in marine food webs. Trends in Ecology and Evolution 13: 297–337.

Linares, C., B. Hereu & M. Zabala, 2002. Avaluació de la Població de corall vermell (*Corallium rubrum*) de les Illes Medes. Exercici 2002. In Zabala, M., J. D. Ros, R. Coma & J. Romeiro (eds), Seguiment temporal de les Illes Medes. Informe tècnic per al Departament de Medi Ambient, Generalitat de Catalunya. Barcelona: 45–62. (In Catalan).

Mundy, C. N., 2000. An appraisal of methods used in coral recruitment studies. Coral Reefs 19: 124–131.

Oren, U. & Y. Benayahu, 1997. Transplantation of juveniles corals: a new approach for enhancing colonization of artificial reefs. Marine Biology 127: 499–505.

Rossi, S. & J. M. Gili, 2003. Estudio y seguimiento del estado de las poblaciones de coral rojo Corallium rubrum en el litoral catalán, Noviembre 2001–Noviembre 2003. Informe final de proyecto, Generalitat de Catalunya PCC 30103, Barcelona, (In Catalan).

Sage, M., 2002. GMAV for Windows. Instruction manual. Centre for research on Ecological Impacts of Coastal Cites. University of Sydney, Sydney.

Santangelo, G. & M. Abbiati, 2001. Red coral: conservation and management of an overexploited Mediterranean species. Aquatic Conservation: Marine Freshwater Ecosystems 11: 253–259.

Santangelo, G., E. Carletti, E. Maggi & L. Bramanti, 2003. Reproduction and population sexual structure of the overexploited Mediterranean red coral *Corallium rubrum*. Marine Ecology Progress Series 248: 99–108.

Santangelo, G., E. Maggi, L. Bramanti & L. Bongiorni, 2004. Demography of the overexploited Mediterranean red coral. Scientia Marina 68: 199–204.

Santangelo, G., L. Bramanti, M. Iannelli, 2006. Population dynamics and conservation biology of the overexploited Mediterranean Red Coral. Journal of Theoretical Biology 245 (published online).

Sarà, M., 1973. Research on coralligenous formations: problems and perspectives. Publicazioni della Stazione Zoologica di Napoli 37: 174–179.

Todd, C. D., 1998. Larval supply and recruitment of benthic invertebrates: do larvae always disperse as much as we believe? Hydrobiologia 375/376: 1–21.

Tsounis, G., S. Rossi, M. Aranguren, J. M. Gili & W. Arnz, 2006. Effects of spatial variability and colony size on the reproductive output and gonadal cycle of the Mediterranean Red Coral (*Corallium rubrum* L 1758). Marine Biology 148: 513–527.

Underwood, J. A., 1997. Experiments in ecology: their logical interpretation and design. Cambridge University Press, Cambridge.

Zibrowius, H., M. Montero & M. Grashoff, 1984. La ripartition du *Corallium rubrum* dans l'Atlantique. Thetis 11: 163–170.

Hydrobiologia (2007) 580:225–231
DOI 10.1007/s10750-006-0451-2

BIODIVERSITY IN ENCLOSED SEAS

A fourteen-year overview of the fish assemblages and yield of the two oldest Algarve artificial reefs (southern Portugal)

Miguel Neves Santos · Carlos Costa Monteiro

Abstract Artificial reefs have been deployed worldwide for the last three decades in response to problems concerning coastal resources, ecosystems and fisheries. In many countries they have became important elements of integrated fisheries management plans. In Portugal two artificial reef systems (ARSs) were deployed by the Portuguese Institute of Marine Research (IPIMAR) in 1990, in the southern coast (Algarve). They were located off Faro and Olhão, over different sea-bottom types and located at different distances from the coastline. To analyse the effect of ARSs deployment on local fish assemblages and to evaluate their effectiveness in terms of mean fishing yields and mean number of species caught, fishing surveys have been conducted over 14 years (256 net sets) using a gillnet to sample the ARS of Faro and Olhão and respective control sites. The fishing yields from the ARs continually exceeded those from the control sites, in both the mean number of species caught and the mean CPUE in weight (1.8–2.6 times); both were higher at Faro. Moreover, the comparison between fish assemblages from the ARS and respective control sites showed that the deployment of the man-made structures did not change the composition of the fish assemblages caught by the gill nets, or the equilibrium of the community, since the relative proportion of the different functional groups of fish remained stable. The ARSs are a useful management tool on the Algarve coast, enhancing and diversifying the catches, thus contributing to improved local artisanal fisheries, which play a major role in this region.

Keywords Artificial reefs · Fish assemblages · Fishing yield · Fisheries management

Guest editors: G. Relini & J. Ryland
Biodiversity in Enclosed Seas and Artificial Marine Habitats

Electronic supplementary material Supplementary material is available for this article at <http://dx.doi.org/10.1007/s10750-006-0451-2> and accessible for authorised users

M. N. Santos (✉) · C. C. Monteiro
IPIMAR, Centro Regional de Investigação Pesqueira do Sul (CRIPSul), Av. 5 de Outubro s/n, 8700-305 Olhao, Portugal
e-mail: mnsantos@cripsul.ipimar.pt

Introduction

Artificial reefs as they have been developed in the last few decades are a response to problems concerning coastal resources, ecosystems and fisheries. Currently, they are important elements of integrated management plans in several countries (Seaman & Hoover, 2001; Anon., 2003; Wilson et al., 2003). Surpassed, in great measure, its initial and exclusive use as fish aggregation structures—with the single objective of increasing fishing revenue—artificial reefs have had broader applications, namely at the ecological level, contributing to biological production, to promote biodiversity and juvenile protection and to revitalise ecosystems, among others (Santos & Monteiro, 1997, 1998; Pondela et al., 2002; Stephens & Pondela, 2002). Currently, many countries are no longer reluctant to spend money on research and management. Subsequently, they are focusing their fisheries policies more towards habitat manipulation, by investing in artificial reefs development and deployment. Many artificial reefs have been established in the western Atlantic (Haroun & Herrera, 2000; Jensen et al., 2000; Monteiro & Santos, 2000) and in the Mediterranean (Allemand et al., 2000; Barnabé et al., 2000; Bombace et al., 2000; D'Anna et al., 2000; Moreno, 2000; Relini, 2000; Revenga et al., 2000; Spanier, 2000). The amount of money spent in such countries is considered a good investment due to potential economic returns in the long-run. A pilot project consisting of two artificial reef systems (ARS) was developed by the Portuguese Institute of Marine Research (IPIMAR) in 1990 and monitored since then (Santos et al., 1996, 2002; Santos, 1997; Santos & Monteiro, 1997, 1998). The objectives of the present study were (1) to analyse the effect of ARS deployment on local fish assemblages and (2) to evaluate their effectiveness in terms of mean fishing yields and mean number of species caught. This was done by comparing the gill net catches from the artificial reefs with those from the control sites.

Materials and methods

The artificial reef systems

Each artificial reef system consists of a protection reef (PR) and an exploitation reef (ER). The PR consists of 735 concrete small cubic units (2.7 m^3 each unit), distributed in 21 groups, occupying a total area of approximately 39 ha, at depths that range from 16 m to 22 m. The ER consists of 20 large blocks, of two different shapes (130 m^3 and 174 m^3), distributed in 5 groups, occupying a total area of 21 ha, at depths that range from 25 m to 40 m.

Study sites

The Faro (F) and Olhão (O) artificial reef systems (ARS) were deployed in 1990 off a highly productive ecosystem—Ria Formosa—that acts as a nursery supplying the most important fish stocks of the Algarve (southern Portugal) coastal waters (Monteiro et al., 1987, 1990; Monteiro 1989). The FARS was deployed on a flat sandy bottom, 2.6–4.8 km off Faro. The OARS was deployed 2.0–3.0 km offshore Olhão on a sandy-muddy bottom. The control sites (C) were located 1.8–4.6 km East and West of each AR, at the same range of depths and similar distance from the coast, which helped avoid or reduce the possibility that it might be affected by the presence of the man-made reefs. Natural reefs as well as a few scattered patches of bedrock were present on the sandy bottom at control site of Faro protection reef.

Data collection

Both the ARS and the control sites were sampled using bottom gill net, from October 1990 to January 2004. No data were recorded for the period 1996–1997. The gill nets were deployed following Santos et al. (1996). The standard net was 750 m long and 2.8 m wide, with a 60 mm mesh (stretched). The 60 mm mesh size was chosen because it is the legal size. Noteworthy that for Martins et al. (1992), according to the characteristics of the gillnet fishery in the Algarve (multi-species) and the biology of the species caught, this mesh size may be the most appropriate from a management perspective. The fishing operations were conducted seasonally at each site, corresponding to a total of 256 net sets. The ARS and the control sites were fished simultaneously. The nets were set at night (2–3 h before the sunrise) and retrieved 1 h after the sunrise, for a total of 3–4 h fishing duration. This duration of sample time corresponds to the general procedure of local fishermen. The fish nomenclature adopted was that of Whitehead et al. (1984/1986).

Data analysis

To determine how differently fish responded to the ARS, three functional groups were distinguished according to Fabi & Fiorentini (1994): pelagic (P), nekto-benthic (NB) and benthic (B). The species were assigned to the different functional groups based on their behaviour towards the ARS after in situ observations Santos (1997).

The fish assemblages from the different sites were compared by means of the relative species richness coefficient, following Monteiro (1989). It was expressed as:

$$S_i/S_t$$

where S_i is the number of species found at each site and S_t is the total number of species caught in all sites combined within the same area (Faro and Olhão).

Statistical comparisons between the ARs and respective control sites were performed using the Wilcoxon's matched pairs test (Scherrer, 1984).

Results

Catch structure

A total of 68 species belonging to 32 families were caught (see Electronic Supplementary Material) during this study. Sixty-five species were caught in the Faro area of which 13 were pelagic, 28 nekto-benthic and 24 benthic species, while 49 were caught in the Olhão area (7 P, 21 NB and 21 B). Nineteen species were caught only in the Faro area (6 P, 9 NB and 4 B) while 3 were caught only in the Olhão area (2 NB and 1 B). Forty-six species were common to both areas.

At the Faro ARS 53 species were caught (10 P, 22 NB and 21 B), of which four were found only at Faro (all B species). At the respective control sites, total of 60 species were recorded (13 P, 27 NB and 20 B), among these 11 were exclusive species (3 P, 5 NB and 3 B). At the Olhão ARS, all 49 species were recorded (7 P, 21 NB and 21 B), of which 14 were exclusive species (2 P, 7 NB and 5 B). At the respective control sites were caught 35, being 5 pelagic, 14 nekto-benthic and 16 benthic species.

Relative species richness ranged between 0.631–0.815 and 0.612–0.898, at the Faro and Olhão areas, respectively (see Electronic Supplementary Material). In the Faro area, the highest value was recorded at control site of the protection reef, followed by the exploitation reef and respective control site. In contrast, at Olhão the highest values were observed at the ARs (0.898 and 0.837 at the exploitation and protection reefs, respectively) and the lowest at their respective control sites (0.612 and 0.633, respectively).

Benthic species comprised the majority of fish caught, followed by the nekto-benthic and the pelagic fishes, respectively. This pattern was consistent at all sites with the exception at the control of Faro protection reef. Benthic fishes consisted chiefly of species that inhabited the original sandy-muddy bottom, such as species from the Soleidae and Triglidae families. Nekto-benthic fishes consisted chiefly of sparids and were more abundant at the Faro area. Pelagic fish were represented mostly by transient species and the mean number of pelagic species caught was slightly lower in the Olhão area (see Electronic Supplementary Material).

Mean number of species caught

The mean number of species caught per standard catch was continuously higher at the artificial reefs than at the control sites (Fig. 1). At the Faro area the highest value was recorded at the exploitation reef, followed by the protection reef

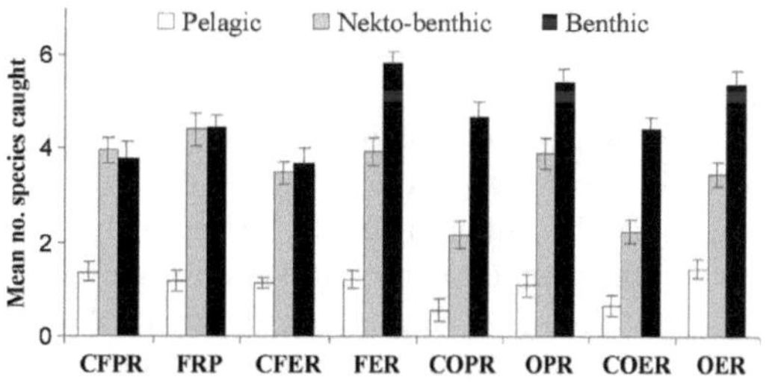

Fig. 1 Mean number of species caught (±SE) per group of fish at Faro (F) and Olhão (O) artificial reefs (PR—protection reef; ER—exploitation reef) and at respective control sites (CFPR, CFER, COPR and COER)

Table 1 Results (valid *n* in parentheses) of Wilcoxon's matched pairs test conducted on the mean number of species caught and mean fishing yield, for the all catch and on the different groups of species caught. Faro (F) and Olhão (O) artificial reefs (PR—protection reef; ER—exploitation reef) and respective control sites (CFPR, CFER, COPR and COER)

Pair	Overall	Benthic	Nekto-benthic	Pelagic
	Mean number of species			
COPR versus OPR	+	ns	+	ns
COER versus OER	+	+	+	ns
OPR versus OER	ns	ns	ns	ns
COPR versus COER	ns	ns	ns	ns
CFPR versus FPR	ns	+	ns	ns
CFER versus FER	+	+	ns	ns
FPR versus FER	+	+	ns	ns
CFPR versus CFER	+	ns	ns	ns
	Mean CPUE in weight			
COPR versus OPR	+	+	+	+
COER versus OER	+	+	+	ns
OPR versus OER	ns	ns	ns	ns
COPR versus COER	+	ns	+	ns
CFPR versus FPR	+	+	+	ns
CFER versus FER	+	+	+	ns
FPR versus FER	ns	ns	ns	ns
CFPR versus CFER	ns	ns	ns	ns

and their respective control sites, respectively. Significant differences were found between the artificial structures, the control sites, and the exploitation reef and respective control site (Table 1). In the Olhão area, the highest value was found at the protection reef, followed respectively by exploitation reef and their respective control sites. Significant differences were found between the artificial structures and respective control sites (Table 1). For the benthic species significant differences were found between the ARs and control sites, while for the nekto-benthic group differences were only recorded in the Olhão ARs. For the pelagic species no differences were observed (Table 1).

Fishing yield

The mean fishing yield per standard catch was always higher at the artificial reefs than at the control sites. In the Faro area, the highest value was recorded at the exploitation reef, followed by the protection reef and respective control sites (Fig. 2). In the Olhão area the highest value was found at the protection reef, followed respectively by exploitation reef and respective control sites. For both areas significant differences were found between the artificial structures and respective control sites (Table 1).

The benthic species were chiefly represented by species that typified the original soft-bottom community, such as species of the Soleidae and Triglidae families. Species with the highest biomass were *Microchirus azevia*, *Scorpaena notata*, *Dicologoglossa cuneata*, *Mullus surmuletus* and *Trachinus draco*. The nekto-benthic fish captured included mainly sparids. The most important species with regards to abundance were *Diplodus bellottii*, *Pagellus acarne*, *Merluccius merluccius* and *P. erythrinus*. The pelagic species were generally transients, chiefly *Scomber japonicus*, *S. scombrus* and *T. trachurus*.

Benthic fishes were the dominant group by weight, nekto-benthic fishes were the next most dominant by weight in the Olhão area, while the opposite was observed in the Faro area (Fig. 2). Significant differences were found between the

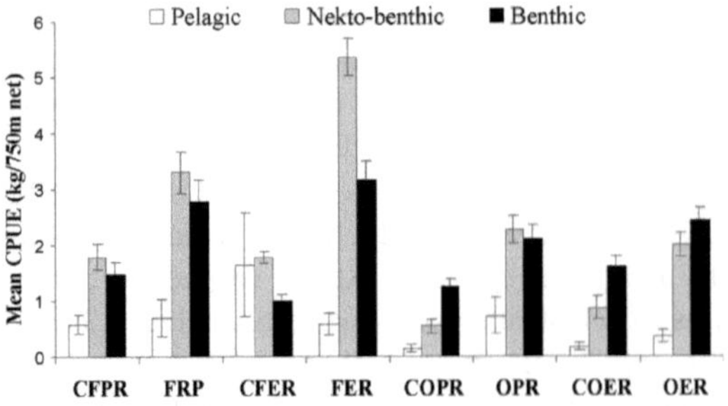

Fig. 2 Mean CPUE in terms of weight (±SE) per group of fish at Faro (F) and Olhão (O) artificial reefs (PR—protection reef; ER—exploitation reef) and at respective control sites (CFPR, CFER, COPR and COER)

man-made structures and control sites, both for the benthic and nekto-benthic groups and at both areas. Standard fishing yield for pelagic fish was low (<0.8 kg/3 h/15 panels for most sites) and no significant differences were found between most of the ARs and respective control sites (Table 1).

Discussion

Fourteen years after the deployment of the artificial reefs off the south coast of the Algarve, the global species richness is still increasing at Faro artificial reef system. This statement is based on the fact that the total number of species found at Faro protection reef was fewer than at its respective control site. This was probably a consequence of the presence of a natural reef and some scattered patches of bedrock in the area, which provided a heterogeneous environment favouring the occurrence of a high number of species characteristic of soft and hard bottoms. In contrast, in the Olhão area artificial reef deployment resulted in an enhancement of species availability. According to Bohnsack and Sutherland (1985) equilibrium community structure is usually achieved within 1 to a maximum of 5 years. Both artificial reef systems showed to be habitats able to support a diverse ichthyofaunal assemblage, comprising benthic, nekto-benthic and pelagic species. Noteworthy is that attractiveness of the artificial reefs was achieved keeping the relative proportion of the different functional groups and thus, the balance between them. Most exclusive species found at the different sites correspond to rare species, which did not allow us to explain the differences recorded. According to Amanieu & Lassèrre (1982), the higher species richness found at such a complex habitat as the artificial reefs indicates a high capacity of reception by the habitat. An increase of this ecological index was previously reported by the authors (Santos & Monteiro, 1997, 1998) but also by several other authors (D'Anna et al., 1994; Bortone et al., 1994; Fabi & Fiorentini, 1994).

Although AR deployment has not resulted in an increase in the total number of species caught in the Faro area, it has significantly increased the mean number of species at both ARS. This was especially note among benthic species, while this was only true for the OARS for the nekto-benthic species. The presence of some natural reef in the Faro area is responsible for this difference between the two areas. The natural reef provides a heterogeneous environment favouring the presences of species characteristic of soft and hard bottoms.

Catch yields at the artificial reef systems were always higher than at control sites, as consequence of AR deployment. Bohnsack & Sutherland (1985) reported that many authors found higher catch rates at artificial reefs than at control sites. More recently several studies have reported the same results (D'Anna et al., 1994; Fabi & Fiorentini, 1994; Johnson et al., 1994; Bombace et al., 2000; Zalmon et al., 2002). These differences were due chiefly to both nekto-benthic and benthic species, especially *D. bellottii*, *P. acarne*, *P. erythrinus*, *S. notata*, *D. cuneata* and *M. surmuletus*. However, no species links were found between these species and the two different types of modules (protection and exploitation). The low catch rates of pelagic species were a result of their behaviour towards the reefs (transient and occasional species in some cases) and/or due to the fishing gear selectivity, which did not allow us to analyse this group properly. For these purposes, other sampling techniques should be used. These could include destructive (e.g., purse seine) or non-destructive techniques (e.g., acoustic and visual census). Compared to those previously reported by Santos & Monteiro (1997, 1998), the increase in the catch associated with the reefs was of the same level at Faro artificial reef system (1.77–2.06) but greater now at Olhão artificial reef system (1.80–2.63). Although, an increase in catch was observed at most sites. Compared to those reported by the later authors, these were on the order of 20% and 40% (on average) at the control and reef sites, respectively. The difference in terms of standard catch within the artificial structures (0.35 kg) was of the same level as those within the control sites (0.71 kg) at Olhão, contrasting to what was observed at Faro (0.59 and 2.31 kg, within control and reef sites, respectively). The nekto-benthic species (*Diplodus bellottii* and *Pagellus acarne*) were the responsible for this difference, as the Faro exploitation reef particularly attracted them.

Overall both the protection and exploitation modules of the Faro and Olhão artificial reef systems enhanced fishing yield and diversity from the accessible fish community. Oppositely, these effects did not change the balance (or the equilibrium) of the fish assemblages, since the relative proportion of the different functional groups of fish caught by the gill nets over the 14-year period remained stable. Thus, we can say that it seems these artificial reefs are an interesting tool for management proposes for the Algarve coast, especially for the gillnet fishery, which is one of the most active in the region.

Acknowledgements We thank the staff of IPIMAR/CRIPSul (E. Brás, T. Simões, J.L. Sofia, R. Machado and L. Oliveira) and the crew of "NI DONAX" for catching and collecting these data. This work was supported by the Portuguese-EU MARE Programme—project *Implantação e estudo integrado de sistemas recifais*.

References

Anon., 2003. State of Florida artificial reef strategic plan. Florida Fish and Wildlife Commission. Division of Marine Fisheries, 15 pp.

Allemand, D., E. Debernardi, & W. Seaman Jr., 2000. Artificial reefs in the Principality of Monaco: protection and enhancement of coastal zones. In Jensen, A. C., K. J. Collins & A. P. M. Lockwood (eds), Artificial Reefs in European Seas. Kluwer Academic Publishers, Chap. 9: 151–166.

Amanieu, M. & G. Lassèrre, 1982. Organisation et évolution des peuplements lagunaires. Oceanologica Acta, Special Number, Sept., 201–213.

Barnabé, G., E. Charbonnel, J. Y. Marinaro, D. Odi & P. Francour, 2000. Artificial reefs in France: analysis, assesments and prospects. In Jensen, A. C., K. J. Collins & A. P. M. Lockwood (eds), Artificial Reefs in European Seas. Kluwer Academic Publishers, Chap. 10: 167–184.

Bohnsack, J. A. & D. L. Sutherland, 1985. Artificial reef research: a review with recommendations for future priorities. Bulletin of Marine Science 37(1): 11–39.

Bombace, G., G. Fabi & L. Fiorentini, 2000. Artificial reefs in the Adriatic Sea. In Jensen, A. C., K. J. Collins & A. P. M. Lockwood (eds), Artificial Reefs in European Seas. Kluwer Academic Publishers, Chap. 3: 31–64.

Bortone, S. A., J. van Tassell, A. Brito, J. M. Falcón, J. Mena & C. M. Brundrick, 1994. Enhancement of the near-shore fish assemblage in the Canary Islands with artificial habitats. Bulletin of Marine Science 55(2–3): 602–608.

D'Anna, G., F. Badalamenti & S. Riggio, 2000. Artificial reefs in North West Sicily: comparisons and conclusions. In Jensen, A. C., K. J. Collins & A. P. M. Lockwood (eds), Artificial Reefs in European Seas. Kluwer Academic Publishers, Chap. 6: 97–112.

D'Anna, G., F. Badalamenti, M. Gristina & C. Pipitone, 1994. Influence of artificial reefs on coastal nekton assemblages of the Gulf of Castellammare (Northwest Sicily). Bulletin of Marine Science 55(2–3): 418–433.

Fabi, G. & L. Fiorentini, 1994. Comparison between an artificial reef and a control site in the Adriatic Sea: analysis of four years of monitoring. Bulletin of Marine Science 55(2–3): 538–558.

Haroun, R. & R. Herrera, 2000. Artificial reefs of the Canary Islands. In Jensen, A. C., K. J. Collins & A. P. M. Lockwood (eds), Artificial Reefs in European Seas. Kluwer Academic Publishers, Chap. 14: 235–248.

Jensen, A., K. Collins & P. Smith, 2000. The Pool Bay artificial reef project. In Jensen, A. C., K. J. Collins & A. P. M. Lockwood (eds), Artificial Reefs in European Seas. Kluwer Academic Publishers, Chap. 16: 263–288.

Johnson, T. D., A. M. Barnett, E. E. DeMartini, L. L. Craft, R. F. Ambrose & A. J. Purcell, 1994. Fish production and habitat utilization on a southern California artificial reef. Bulletin of Marine Science 55(2–3): 709–723.

Martins, R., M. N. Santos, C. C. Monteiro & M. L. P. Franca, 1992. Contribuição para o Estudo da Selectividade das Redes de Emalhar de Um Pano Fundeadas na Costa Portuguesa no Biénio 1990–1991. Relatório Técnico Científico INIP, n° 62: 26 pp.

Monteiro, C. C., 1989. La Faune Ichtyologique de la Lagune Ria Formosa (Sud Portugal). Repartition et Organisation Spatio-Temporelle des Communatés: Aplication à l'Aménagement des Ressources. Thèse Doctorat Univ. Scien. Techn. Languedoc, Montpellier, 219 pp.

Monteiro, C. C. & M. N. Santos, 2000. In Jensen, A. C., K. J. Collins & A. P. M. Lockwood (eds), Artificial Reefs in European Seas. Kluwer Academic Publishers, Chap. 15: 249–262.

Monteiro, C., T. Lam Moi & G. Lasserre, 1987. Distribution chronologique des poissons dans deux station de la lagune Ria Formosa (Portugal). Oceanologica Acta 10(3): 359–371.

Monteiro, C. C., G. Lasserre & T. Lam Hoi, 1990. Organisation spatiale des communautés ichtyologiques de la Lagune Ria Formosa (Portugal). Oceanologica Acta 13(1): 79–96.

Moreno, I., 2000. Artificial reef programme in the Balearic Islands: Western Mediterranean Sea. In Jensen, A. C., K. J. Collins & A. P. M. Lockwood (eds), Artificial Reefs in European Seas. Kluwer Academic Publishers, Chap. 13: 219–234.

Pondela II, D. J., J. S. Stephens Jr. & M. T. Craig, 2002. Fish production of a temperate artificial reef based on the density of embiotocids (Teleotei: Perciformes). ICES Journal of Marine Science 59: S88–S93.

Relini, G., 2000. The Loano artificial reef. In Jensen, A. C., K. J. Collins & A. P. M. Lockwood (eds), Artificial Reefs in European Seas. Kluwer Academic Publishers, Chap. 8: 129–150.

Revenga, S., F. Fernandéz, J. L. González & E. Santaella, 2000. Artificial reefs in Spain: the regulatory framework. In Jensen, A. C., K. J. Collins & A. P. M. Lockwood (eds), Artificial Reefs in European Seas. Kluwer Academic Publishers, Chap. 11: 185–194.

Santos, M. N., 1997. Ichthyofauna of the artificial reefs of the Algarve coast. Exploitation strategies and management of local fisheries. Ph.D. Thesys, Universidade do Algarve, 268 pp.

Santos, M. N. & C. C. Monteiro, 1997. Olhão artificial reef system (south Portugal): fish assemblages and fishing yield. Fisheries Research 30: 33–41.

Santos, M. N. & C. C. Monteiro, 1998. Comparison of the catch and fishing yield from an artificial reef system and neighbouring areas off Faro (south Portugal). Fisheries Research 39: 55–65.

Santos, M. N., C. C. Monteiro & M. B. Gaspar, 2002. Diurnal variations in the fish assemblage at an artificial reef. ICES Journal of Marine Science 59: S32–S35.

Santos, M. N., C. C. Monteiro & G. Lassèrre, 1996. Faune ichthyologique comparée de deux récifs artificiels du littoral de la Ria Formosa (lagune Portugal): résultats préliminaires. Oceanologica Acta 19(1): 89–97.

Seaman, W. & A. Hoover, 2001. Artificial reefs: the Florida Sea Grant connection - science serving Florida's coast. Florida Sea Grant, SGEF-144: 4 pp.

Scherrer, B., 1984. Biostatistic. Gaëtan Morin Ed., Canada, Québec, Chicoutimi, 850 pp.

Spanier, E., 2000. Artificial reefs off the Mediterranean coast of Israel. In Jensen, A. C., K. J. Collins & A. P. M. Lockwood (eds), Artificial Reefs in European Seas. Kluwer Academic Publishers, Chap. 1: 1–20.

Stephens J. Jr., & D. Pondela II, 2002. Larval productivity of a mature artificial reef: the ichthyoplankton of King Harbor, California. ICES Journal of Marine Science 59: S51–S58.

Whitehead, P. J. P., M. -L. Bauchot, J. -C. Hureau, J. Nielsen & E. Tortonese, 1986. Fishes of the North-eastern Atlantic and the Mediterranean. UNESCO (I,II,III), Paris: 1473 pp.

Wilson, K. D. P., A. W. Y. Leung & R. Kennish, 2003. Restoration of Hong Kong fisheries through deployment of artificial reefs in marine protected areas. ICES Journal of Marine Science 59: S157–S163.

Zalmon, I. R., R. Novelli, M. P. Gomes & V. V. Faria, 2002. Experimental results of an artificial reef programme on the Brazilian coast north of Rio de Janeiro. ICES Journal of Marine Science 59: S83–S87.

Hydrobiologia (2007) 580:233–240
DOI 10.1007/s10750-006-0450-3

Long-term changes in a benthic assemblage associated with artificial reefs

L. Nicoletti · S. Marzialetti · D. Paganelli · G. D. Ardizzone

Abstract The aim of the study was to evaluate the long-term development of a hard bottom benthic assemblage over a period of 20 years in an area off the mouth of a large river. The artificial reef of Fregene was selected because benthic assemblage data were available for the period 1981–1992. This artificial reef is located in the mid Tyrrhenian Sea, 5 nautical miles north of the two mouths of the Tevere River (Latium, Italy) and 1.5 nautical miles offshore from Fregene (Rome, Italy). The artificial reef was deployed in March 1981 for fisheries enhancement in 10–14 m of water on a sandy-silty seabed. The Tevere River carries suspended materials and a heavy load of organics since it transports Rome's effluent, resulting in the eutrophic state of area waters. Benthic sampling was conducted in 2001 by SCUBA diving; two standard surfaces of 400 cm^2 were scraped from the vertical walls of the same uppermost block in four different periods. All organisms were identified and counted. The methodology used is the same as that adopted in the previous periods, so that the 2001 data could be compared with past collected data. The benthic assemblage was analysed by cluster analysis using the Bray-Curtis index and clustered using the group average clustering algorithm. The SIMPER procedure was used to identify those taxa that characterize each station group identified by cluster analysis. Changes in benthic assemblages and hydrological trends of the Tevere River were investigated using the cumulative sum series method. The 20-year development of the benthic community, starting from the new substratum, is composed of different phases characterised by different benthic assemblages. In particular five different phases were distinguished: 1. Pioneer species recruitment (May 1981–June 1981); 2. *Mytilus galloprovincialis* (mussel) dominance (August 1981–November 1983); 3. *M. galloprovincialis* regression (July 1984–October 1985); 4. *M. galloprovincialis* absence (91–92); 5. Bryozoans bioconstruction dominance (2001). The dynamic succession of the observed benthic assemblages exhibited a good relation with the Tevere River flow. The Tevere River flow, and the subsequent sedimentation process, seems to have strongly influenced the benthic assemblage succession of the Fregene artificial reef.

Guest editors: G. Relini & J. Ryland
Biodiversity in Enclosed Seas and Artificial Marine Habitats

L. Nicoletti (✉) · S. Marzialetti · D. Paganelli
ICRAM, Via di Casalotti, 300-00100 Rome, Italy
e-mail: l.nicoletti@icram.org

G. D. Ardizzone
University "La Sapienza", Rome, Italy

Keywords Temporal evolution · Benthic assemblage · Artificial reef · Tyrrhenian Sea · River flow · Long-term changes

Introduction

Knowledge of long-term benthic assemblages development, and in particular assemblages associated with hard bottom, is poor. Many research projects have been undertaken that describe benthic succession using different experimental approaches involving artificial reefs or defaunated natural substrata (Bourget et al., 1994; Brown & Swearingen, 1998; Kocak & Zamboni, 1998; Glasby & Connell, 2001). Most of these studies were concerned with benthic primary colonisation patterns and therefore were short-term studies. On the other hand, long-term investigations that follow later development phases of benthic assemblages have rarely been conducted. A long-term study, including origination on a newly deployed substratum, could offer insight into understanding the most important changes that occur in the benthic assemblage structure during and after the colonisation process (Wiens, 1997). Due to their sessile habit, benthic fauna facilitates the evaluation of environmental quality, as modification of such communities may be directly traceable to natural or anthropogenic environmental variations (Pearson & Rosenberg, 1978; Gray, 1981; Gray et al., 1990; Warwick & Clarke, 1991). A long-term investigation can help to elucidate the response that benthic communities manifest to environmental changes in terms of sedimentation and turbidity. The objective of this study is to evaluate the development of a hard-bottom benthic assemblage over a period of 20 years in an area off the mouth of a large river. We chose to evaluate the benthic assemblage associated with an artificial substratum which, due to its structural simplicity, facilitated quantification relative to area and time (Riggio, 1995; Glasby & Connell, 2001; Smith & Rule, 2002). In particular, this study analysed the benthic assemblage of the Fregene artificial reef, placed in an area where the Tevere River strongly influences the water conditions through a significant organic and inorganic load. The Tevere River is the main river of the central Tyrrhenian Sea and has an effect on the eutrophication process, in particular on the north coast. In fact, the principal direction of the current, and thus of suspended sediment, is northwest along the coast (Bellotti & Tortora, 1985; La Monica & Raffi, 1996). In the last few decades, coastal erosion along the Latium coast of Italy has been particularly evident and different factors have contributed to this process. One of these was the decrease in the transportation of solids due to dam construction, sand extraction from riverbeds, etc. The river's solid loads vary with river flow which is also influenced by climatic events that affect the drainage basin (Bellotti & Tortora, 1996; Bencivenga & Ranieri, 1997). The changes in sedimentation and turbidity in the deltaic areas off large rivers can lead to alteration in associated benthic community. An extended time series of data is necessary to evaluate changes in both benthic communities and river flows.

Materials and methods

The artificial reef off Fregene is located in the mid Tyrrhenian Sea, 5 nautical miles North of the two mouths of the Tevere River (Latium, Italy) and 1.5 nautical miles offshore Fregene (Rome, Italy) (Fig. 1). The artificial reef was placed at 10–14 m depth on a sandy-silty seabed. The reef is composed of 280 concrete cube-shaped units (2 m on each side) arranged in a pyramid consisting of five units to a set, four at the base and one on the top. A total of 60 sets were deployed as a reef group that covered 6 ha. Each cube contains hollows and cavities moulded into the shape in order to increase the surface/volume ratio (Bombace, 1977). The Tevere River carries suspended materials and a significant organic load, resulting in eutrophic conditions. Benthic sampling was performed in 2001 (February, June, September and December) using SCUBA. During each sampling event, two standard surfaces of 400 cm^2 were scraped from the vertical wall of the same top block. Different parts of the vertical wall were chosen in order to avoid a re-sampling on surfaces previously scraped. The methodology used is the same as that adopted in the previous periods, so that the 2001 data could be compared with past collected data (Ardizzone et al., 1989; Somaschini et al., 1997). Samples were fixed in buffered formaldehyde (10%). All organisms were sorted under a stereomicroscope. Polychaetes, molluscs, crustaceans and echinoderms were identified to

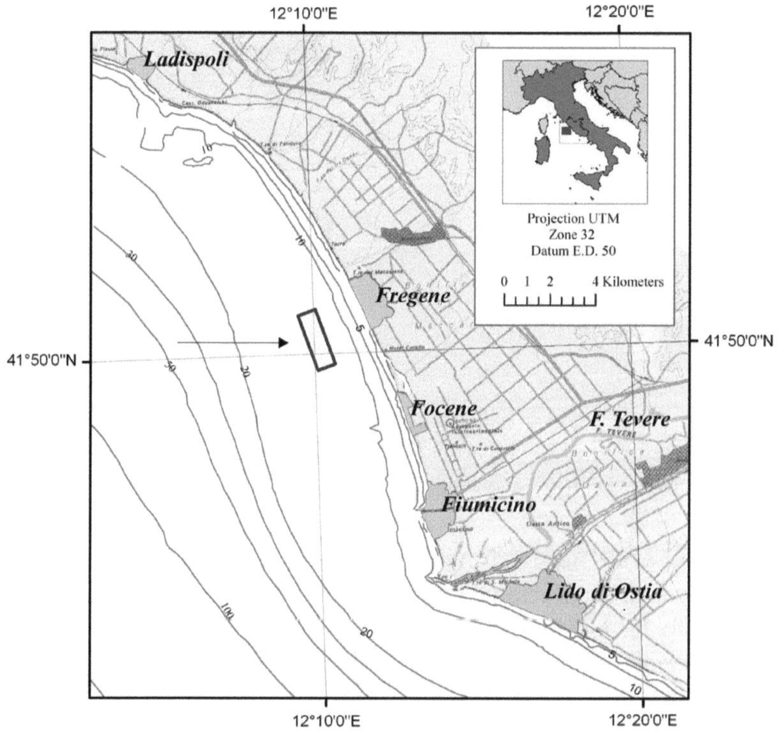

Fig. 1 Map showing the study area and location of the artificial reef at Fregene

species level and counted. The Bray-Curtis index was used as faunal similarity coefficients calculated using species/abundance matrix. This index was clustered using the group average clustering algorithm. The SIMPER procedure (similarity percentage analysis) was used to identify those taxa that characterize each station group identified by cluster analysis (Clarke, 1993). Hydrological trends of the Tevere River were investigated using the cumulative sum series method (Ibanez et al., 1993; Salen-Picard et al., 2003).

Results

Composition of benthic assemblage in 2001

In 2001 a total of 15,306 benthic invertebrate individuals and a total of 172 taxa were collected. Most benthic species were polychaetes (109 species and 6,918 individuals), followed by crustaceans (22 species and 7,257 individuals) and molluscs (24 species and 1,148 individuals). Among the polychaetes the most abundant species were sessile and suspension-feeder species such as *Sabellaria spinulosa*, *Hydroides pseudouncinatus pseudouncinatus*, *Serpula concharum*, and soft bottom deposit-feeder species such as *Aphelocheta marioni*. *Balanus perforatus* was the most abundant crustacean species. The molluscs *Striarca lactea* and *Gastrochaena dubia*, characteristic sessile species of hard-bottoms, were very abundant. The benthic assemblage is also characterized by the presence of colonies of two bryozoans, *Schizoporella errata* and *Turbicellepora magnicostata*, which cover the whole reef surface by three-dimensional colonies.

Temporal trend in assemblage structure and composition

The number of species showed a progressive increase during benthic assemblages' temporal succession (Fig. 2). The number of individuals

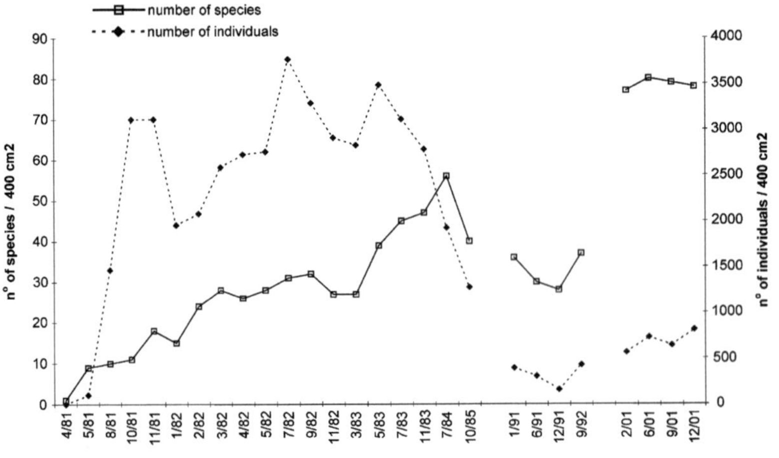

Fig. 2 Temporal trend of species richness (S) and abundance (N)

increased from 1981 to May 1983 and then decreased until December 1991. Subsequently, the number of individuals increased again until 2001. Cluster analysis indicates that the year is a chief factor in faunal similarity (Fig. 3). The benthic assemblage sampled in May 1981, 2 months after the artificial reef deployment, when the community was composed of very few pioneer species, is least similar to all other assemblages. A second cluster of assemblages (similarity value 20%) is distinct from the assemblages collected from the 1980s to the ones of the 1990s and 2001. The faunal assemblage collected from the 1980s indicates that the assemblage collected in June 1981 was less similar (30%) to the others. This period also represents a time when the assemblage on the reef was still represented by pioneer species and few individuals. The benthic faunal samples collected between 1981 and 1985 were similar to each other at level of 60%. At the 65% level of similarity the assemblages formed three distinct clusters relative to the periods August 1981–February 1982, March 1982–November 1983 and July 1984–October 1985. The assemblages from 1991–1992 and 2001 were different from the previous assemblages; their similarity was 50% and their respective stations appeared homogeneous. Table 1 shows the results of SIMPER analysis for the identified assemblages clusters.

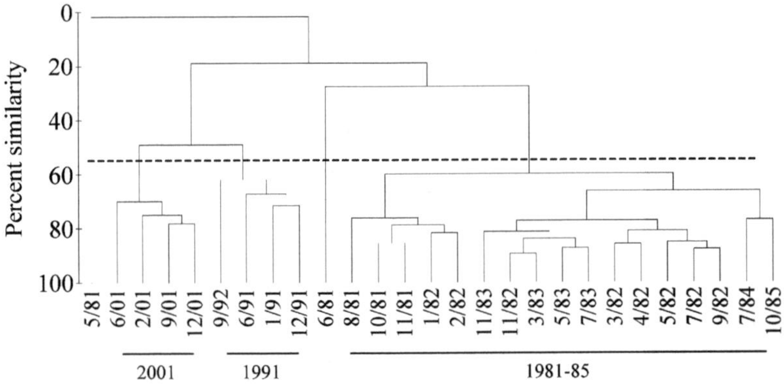

Fig. 3 Cluster diagram by group-average clustering based on the Bray-Curtis similarity index using the abundance of species

Table 1 Percentage contribution of typifying taxa to within-group similarity for the identified benthic assemblages by cluster analysis

Time	Typifying specie to within group similarity	Contribution %	Cumulative %
3.82–12.83	*Mytilus galloprovincialis*	75.66	75.66
	Balanus perforatus	18.97	94.64
1984–1985	*Balanus perforatus*	51.76	51.76
	Mytilus galloprovincialis	23.66	72.42
	Amphipoda	7.87	83.29
	Pomatoceros triqueter	4.44	87.73
	Pomatoceros lamarkii	2.22	89.94
	Hydroides pseudouncinatus pseudouncinatus	1.85	91.79
1991–1992	Amphipoda	53.17	53.17
	Polydora sp. *Pl.*	14.97	68.13
	Lysidice	5.68	73.81
	Hydroides pseudouncinatus pseudouncinatus	5.13	78.93
	Serpula concharum	3.78	82.71
	Syllis truncata cryptica	1.81	84.52
	Terebella lapidaria	1.25	85.77
	Sabellidae	1.25	87.02
	Pilumnus hirtellus	1.21	88.23
	Gastrochaena dubia	1.19	89.42
	Lumbrineris funchalensis	1.13	90.54
2001	Amphipoda	40.48	40.48
	Polydora sp. *Pl.*	16.69	57.17
	Sabellaria spinulosa	6.76	63.92
	Sabellidae	6.29	70.21
	Balanus perforatus	5.25	75.45
	Striarca lacteal	3.77	79.23
	Hydroides pseudouncinatus pseudouncinatus	2.03	81.26
	Aphelochaeta marioni	1.91	83.17
	Pilumnus hirtellus	1.54	84.71
	Serpula concharum	1.51	86.23
	Lumbrineris funchalensis	1.45	87.68
	Lysidice	1.36	89.04
	Syllis truncata cryptica	0.93	89.97
	Gastrochaena dubia	0.91	90.88

The monthly cumulative sum curve of monthly flow shows a clear upward trend between 1981 and 1985 and consequently an increase of quantity of sediment in the sea waters. Subsequently (1990–2001) this curve displays a reversed trend (Fig. 4) interpreted as a decrease of the terrigenous particles distributed from the Tevere to sea waters.

Discussion

Five different time-phases as determined by faunal similarity can be distinguished in the cluster analysis.

1. Pioneer species recruitment (May 1981–June 1981). During this phase the reef was colonised only by hydroids (*Obelia dichotoma*, *Bougainvillia ramosa*), serpulid polychaetes (*Pomatoceros triqueter*, *Hydroides elegans*), barnacles (*Balanus eburneus* and *B. perforatus*) and molluscs (*Anomia ephippium* and *Mytilus galloprovincialis*). Three months after artificial reef establishment, the mussel *M. galloprovincialis* was the most abundant species in the assemblage. Notably, macroalgae were absent.
2. *Mytilus galloprovincialis* dominance (August 1981–November 1983). The mussel *M. galloprovincialis* dominated the benthic assemblage

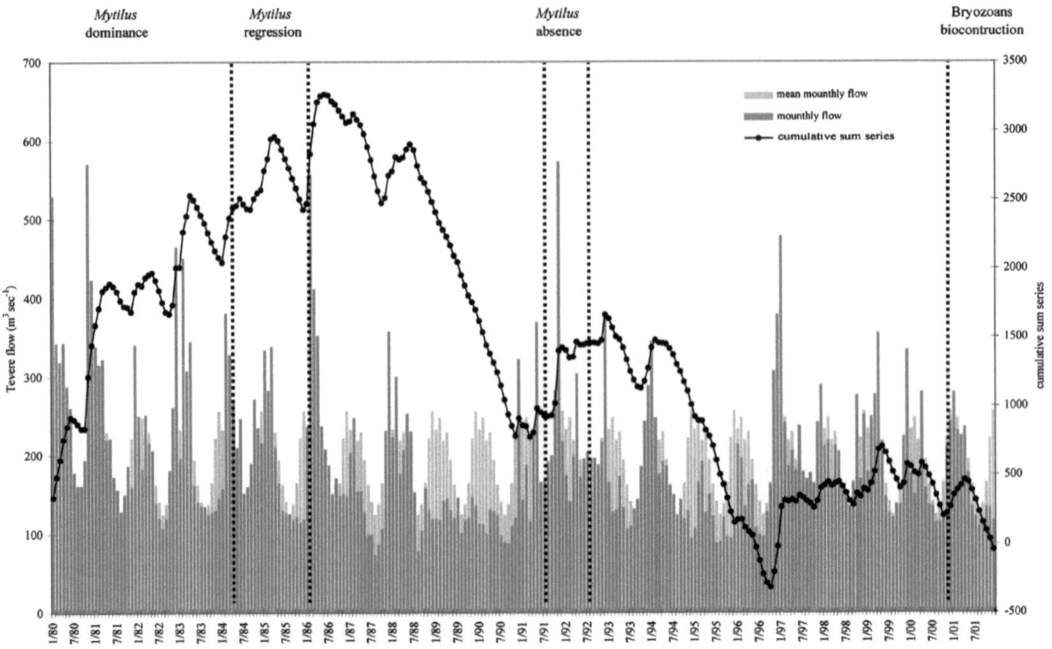

Fig. 4 Raw data and cumulative sum series of mean monthly flow of the Tevere River from 1980 to 2001. Raw data are compared to the average monthly flow calculated over the period 1980–2001. The reference value used to calculate the cumulative sum series is the long-term mean flow of the river (193 $m^3 s^{-1}$)

influencing the settlement of other epibenthic species. Mussel valves and barnacles altered the topography of substratum that became more heterogeneous, this gave origin to an increase in both number of individuals and species (Ardizzone et al., 1982a, b; Ardizzone & Chimenz, 1982).

3. *Mytilus galloprovincialis* regression (July 1984–October 1985). Gradual reduction of mussel density was observed and soft bottom species became frequent next to the hard bottom ones. Over-sedimentation and "mussel mud" production caused a decrease in the availability of hard substrata for epibenthic species colonization (Gravina et al., 1989).
4. *Mytilus galloprovincialis* absence (1991–1992). A new benthic assemblage, characterised by the absence of *M. galloprovincialis* and the numerical abundance of soft bottom species, established itself on the hard and very muddy artificial substratum after 10 years of artificial reef deployment. Bryozoans *Schizoporella errata* and *Scrupocellaria reptans* were the only sessile organism extensively distributed on the reef (Ardizzone et al., 1996).
5. Bryozoans bioconstruction dominance (2001). The benthic assemblage, 20 years after artificial reef deployment, is characterised by the presence of colonies of the bryozoans *S. errata* and *Turbicellepora magnicostata*. These bryozoans are strong space competitors and are able to build rigid frameworks. They are important "bioconstructors" *sensu* Bianchi (2001). The benthic assemblage present in 2001 is different from the ones observed in the previous periods when the bioconstructions were absent and there is a strong increase in the number of species typical of both hard and soft substrates.

The artificial reef at Fregene is located in a transitional zone, characterised by abiotic factors typical of the infralittoral zone (high hydrodynamism and eutrophic water) and the circalittoral zone (reduced illumination). During the second period of succession, a mussel dominated assemblage, characteristic of shallow exposed and eutrophic environments, was present. The colonisation of the artificial reef at Fregene was similar to the colonisation observed on other artificial reefs in the Adriatic Sea. These other artificial reefs in the Adriatic were also dominated by *M. galloprovincialis* (Fabi et al., 1986, 1989). The increased sediment recorded during 1981–1985 (Fig. 4) can be offered as one of the reasons that led to the decline in mussels. Other species, already present as a small percentage of the assemblage on the muddy substrate, became dominant. Later, from 1991 to 1992, the assemblage was characterised by the presence of sciaphilous species typical of muddy infralittoral hard bottoms. Subsequently, the ensuing dramatic decrease of the terrigenous particles from the Tevere (Fig. 4) may have favoured the high presence of suspension-feeding organisms, such as bryozoans. In 2001 these organisms, together with serpulids, barnacles and vermetids, gave rise to bioconstructions which changed the edaphic reef structure. Consequently, habitat heterogeneity increased providing new habitat and thus facilitating the increase in biological diversity (Bianchi & Morri, 1985; Ferdeghini et al., 2000, 2001; Cocito et al., 2000). The presence of these biocontructions allowed the presence of many species typical of "coralligenous" biocoenosis (i.e., polychaetes *Palola siciliensis*, *Serpula vermicularis*, *Eunice schizobranchia*, *Eunice oerstedi*, *Haplosyllis spongicola* and the crustacean *Leptochelia savignyi*). In fact, the bioconstruction present on the artificial reef is similar to a "coralligenous" bioconstruction in morphology and composition, thereby favouring the settlement of many typical coralligenous habitat species. The coexistence of these species with infralittoral species favours the maintenance of a high biodiversity. Due to the high heterogeneity of substratum the benthic settlement was also characterized by the presence of soft bottom species that occupied the interstices filled with mud in the bryozoan bioconstruction and the spaces left by dead organisms. A schematic representation of the cumulative sum series of the river juxtaposed with the different phases of the benthic assemblage succession is reported in Fig. 4.

Conclusion

This study underlines the importance of long-term investigations when attempting to understand benthic community succession. Long-term studies allow detection of the effects of rare events and slow acting processes, revealing subtle but consistent trends and environmental changes. Study results from short-term studies using static "snap-shots" of reality may not be able to detect these changes. As a consequence, short-term study results may lead investigators to arrive at incorrect conclusions regarding the dynamics of ecological systems. This 20 years study provided information on the succession of benthic assemblages in relation to the Tevere River flow variability. A benthic community during any point in time is determined by the interaction of biotic and abiotic factors. Among the abiotic factors, the Tevere River flow, and the subsequent sedimentation process, seemed to have influenced the benthic assemblage succession on the artificial reef at Fregene. The present analysis of a benthic community succession provides a contribution that is useful in understanding the dynamics of ecological systems. The analysis of historical data thereby allows a better understanding of how ecosystems behave and develop.

Acknowledgements In particular we would like to thank Dr. Andrea Belluscio for his support in conducting the field work and Dr. Paola La Valle (Molluscs), Dr. Loretta Lattanzi and Dr. Monica Targusi (Crustaceans) for their time and invaluable dedication to the laboratory activities.

References

Ardizzone, G. D., G. Bombace & P. Pelusi, 1982. Settlement and growth of *Mytilus galloprovincialis* Lamk on an artificial reef in the Tyrrhenian Sea. Journée Etud. Recifs Artif. et Maricult. suspend., Cannes, Rapp. CIESM: 59–61.

Ardizzone, G. D., & C. Chimenz, 1982a. Primi insediamenti bentonici della barriera artificiale di Fregene. Atti Conv. Ris. Biol. Inq. Mar. P. F. Oceanogr.: Fondi Marini. Roma, 10–11 Novembre 1981: 165–181.

Ardizzone, G. D., C. Chimenz & A. Belluscio, 1982b. Benthic community on the artificial reef of Fregene (Latium). Journée Etud. Récifs artif. Et Maricult. Suspend. – Cannes, C.I.E.S.M.: 55–57.

Ardizzone, G. D., M. F. Gravina & A. Belluscio, 1989. Temporal development of epibenthic communities on

artificial reefs in the central Mediterranean Sea. Bulletin of Marine Science 44(2): 592–608.
Ardizzone, G. D., A. Belluscio, M. F. Gravina & A. Somaschini, 1996. Colonization and disappearance of *Mytilus galloprovincialis* Lamk on an artificial habitat on the Mediterranean Sea. Estuarine, Coastal and Shelf Science 43: 665–676.
Bellotti, P. & P. Tortora, 1985. Il delta del Tevere: lineamenti batimetrici, morfologici e tessiturali della conoide sommersa e delle aree limitrofe. Bollettino della Società Geologica Italiana 104: 65–80.
Bellotti, P. & P. Tortora, 1996. I sedimenti sul fondale del delta del Fiume Tevere. Bollettino della Società Geologica Italiana 115: 449–458.
Bencivenga, M. & E. Ranieri, 1997. Il regime dei deflussi del Tevere a Roma. Acqua 3: 51–60.
Bianchi, C. N., 2001. La biocostruzione negli ecosistemi marini e la biologia marina Italiana. Biologia Marina Mediterranea 8(1): 112–130.
Bianchi, C. N. & C. Morri, 1985. I policheti come descrittori della struttura trofica degli ecosistemi marini. Oebalia 11: 203–214.
Bombace, G., 1977. Aspetti teorici e sperimentali concernenti le barriere artificiali. In Cinelli, F., E. Fresi & L. Mazzella (eds), Atti del IX Congresso della Società Italiana di Biologia Marina. Ischia, Naples: 29–41.
Bourget, E., J. DeGuise & G. Daigle, 1994. Scales of substratum heterogeneity, structural complexity, and the early establishment of a marine epibenthic community. Journal of Experimental Marine Biology and Ecology 181(1): 31–52.
Brown, K. M. & D. C. Swearingen, 1998. Effects of seasonality, length of immersion, locality and predation on an intertidal fouling assemblage in the Northern Gulf of Mexico. Journal of Experimental Marine Biology and Ecology 225: 107–121.
Clarke, K. R., 1993. Non-parametric multivariate analyses of changes in community structure. Australian Journal of Ecology 18: 117–143.
Cocito, S., F. Ferdeghini, C. Morri & C. N. Bianchi, 2000. Patterns of bioconstruction in the cheilostome *Schizoporella errata:* the influence of hydrodynamics and associated biota. Marine Ecology Progress Series 192: 153–161.
Fabi, G., L. Fiorentini & S. Giannini, 1986. Growth of *Mytilus galloprovincialis* Lamk on a suspended and immersed culture in the Bay of Portonovo (central Adriatic sea). FAO Fisheries Report 357: 144–154.
Fabi, G., L. Fiorentini & S. Giannini, 1989. Experimental shellfish culture on an artificial reef in the Adriatic Sea. Bulletin of Marine Science 44 (2): 923–933.
Ferdeghini, F., S. Cocito, C. Morri & C. N. Bianchi, 2000. Living Bryozoan Buildups: *Schizoporella errata* (Waters, 1848) (Cheilostomatida, Ascophorina) in the Northwestern Mediterranean (Preliminary Observations). Proceedings of the 11th International Bryozoology Association Conference: 238–244.
Ferdeghini, F., S. Cocito, L. Azzaro, S. Sgorbini & F. Cinelli, 2001. Bryozoan biocostructions in the coralligenous formations of S. M. Leuca (Apulia, Italy). Biologia Marina Mediterranea 8(1): 238–245.
Glasby, T. M. & S. D. Connell, 2001. Orientation and position of substrata have large effects on epibiotic assemblages. Marine Ecology Progress Series 214: 127–135.
Gray, J.S., 1981. The Ecology of Marine Sediments. Cambridge University Press, Cambridge, 185 pp.
Gray, J. S., K. R. Clarke, R. M. Warwick & G. Hobbs, 1990. Detection of initial effects of pollution on marine benthos: an example from the Ekofisk and Eldfisk oilfields, North Sea. Marine Ecology Progress Series 66: 285–299.
Gravina, M. F., G. D. Ardizzone & A. Belluscio, 1989. Polychaetes of an artificial reef in the Central Mediterranean Sea. Estuarine, Coastal and Shelf Science 28: 161–172.
Ibanez, F., J. M. Fromentin & J. Castel, 1993. Application de la méthode des sommes cumulées à l'analyse des séries chronologiques en océanographie. Comptes Rendus de l'Academie des Sciences Paris 316: 745–748.
La Monica, G. B. & R. Raffi, 1996. Morfologia e sedimentologia della spiaggia e della piattaforma continentale interna. In: Il Mare del Lazio. Università degli Studi di Roma "La Sapienza", Regione Lazio Assessorato Opere e Reti di Servizi e Mobilità, 62–105.
Kocak, F. & N. Zamboni, 1998. Settlement and seasonal changes of sessile macrobenthic communities on the panels in the Loano artificial reef (Ligurian sea, NW Mediterranean). Oebalia XXIV, 17–37.
Pearson, T. H. & R. Rosenberg, 1978. Macrobenthic succession in relation to organic enrichment and pollution of the marine environment. Oceanography and Marine Biology: An Annual Review 16: 229–311.
Riggio, S., 1995. Le barriere artificiali e l'uso conservativo della fascia costiera: risultati dei "Reefs" nella Sicilia N/O. Biologia Marina Mediterranea 2(1): 129–164.
Salen-Picard, C., D. Arlhac & E. Alliot, 2003. Responses of a Mediterranean soft bottom community to short-term (1993–1996) hydrological changes in the Rhone River. Marine Environmental Research 55: 409–427.
Smith, S. D. A. & M. J. Rule, 2002. Artificial substrata in a shallow sublittoral habitat: do they adequately represent natural habitats or the local species pool? Journal of Experimental Marine Biology and Ecology 277: 25–41.
Somaschini, A., G. D. Ardizzone & M. F. Gravina, 1997. Long-term changes in the structure of a polychaete community on artificial habitats. Bulletin of Marine Science 60(2): 460–466.
Warwick, R. M. & K. R. Clarke, 1991. A comparison of some methods for analysing changes in benthic community structure. Journal of Marine Biological Association of the United Kingdom 71: 225–244.
Wiens, J. A., 1997. Lengthy ecological studies. Trends in Ecology and Evolution 12(12): 499.

Hydrobiologia (2007) 580:241–244
DOI 10.1007/s10750-006-0449-9

Development of a transplantation technique of *Cystoseira amentacea* var. *stricta* and *Cystoseira compressa*

M. L. Susini · L. Mangialajo · T. Thibaut · A. Meinesz

Abstract On the North-Western Mediterranean coastline, the genus *Cystoseira* (Fucales) has a key role on upper infralittoral community structure. *Cystoseira amentacea* var. *stricta* and *C. compressa* are the most common superficial *Cystoseira* species. In this area, the coastline is subjected to a strong anthropogenic pressure which has caused severe *Cystoseira* population decreases. The present study consists in the development of a cheap and easy to use transplantation technique of *C. amentacea* var. *stricta* and *C. compressa* in order to either restore former natural populations or help their settlement on artificial substrates. The thalli were fixed with epoxy glue in holes made by a drill. The success of the technique was assessed by counting the number of remaining transplanted thalli after 3 weeks, 3 months and 6 months. Results were found encouraging with 75% of survival for both species after 6 months. Moreover fertile fronds were observed on the transplanted thalli showing the harmlessness of the technique.

Keywords *Cystoseira amentacea* var. *stricta* · *Cystoseira compressa* · Transplantation technique

In the Western Mediterranean Basin, *Cystoseira amentacea* (C. Agardh) Bory var. *stricta* Montagne is located on the upper infralittoral zone, on wave-exposed rocky shores, forming monospecific belts in the most exposed parts, whereas the ubiquitous *Cystoseira compressa* (Esper) Gerloff & Nizamuddin is present in mixed populations (Ollivier, 1929). *C. amentacea* var. *stricta* and *C. compressa* form complex communities providing habitats for numerous epiphytic species and shelter for many shade-loving organisms (Ollivier, 1929). Due to its structuring role on superficial Mediterranean communities and to its endemic status in the Mediterranean Sea, *C. amentacea* var. *stricta* has been strictly protected in Appendix I of the Bern Convention on the Conservation of European Wildlife and Natural Habitats (Council of Europe, 1979). For decades, the Région Provence-Alpes-Côte d'Azur's coastline (France) has been exposed to a strong anthropogenic pressure (Meinesz et al., 1990) and many studies have reported severe degradations or even disappearance of *Cystoseira* popula-

Guest editors: G. Relini & J. Ryland
Biodiversity in Enclosed Seas and Artificial Marine Habitats

M. L. Susini (✉) · T. Thibaut · A. Meinesz
E.A 3156 "Gestion de la Biodiversité", Laboratoire Environnement Marin Littoral, Faculté des Sciences, Université de Nice-Sophia Antipolis, Parc Valrose, Nice Cedex 2 06108, France
e-mail: susini@unice.fr

L. Mangialajo
Dipartimento per lo Studio del Territorio e delle sue Risorse, University of Genoa, C.so Europa 26, 16132 Genoa, Italy

tions (Arnoux & Bellan-Santini, 1972; Bellan-Santini & Desrosiers, 1977; Soltan, 2001). Despite the setting up of a wastewater treatment plant in the vicinity of the Marseille sewage outfall 8 years ago, *C. amentacea* var. *stricta* recolonization has still not recolonized its former habitat (Soltan et al., 2001). One can hypothesize that the lack of recolonization in disturbed areas could be explained by a small dispersal range and therefore by recruitment being linked to the presence of reproductive individuals in the neighborhood, as reported by Clayton (1990) for the order Fucales. Since natural restoration has never occurred, we propose to develop a technique for recolonization of impacted areas. To this aim, as a preliminary work, we tested the success of the technique in a natural undisturbed population.

The experiment started in March 2004, at the exposed rocky cape of the "Pointe du Colombier", St Jean-Cap Ferrat (France). Three stations, 20 m apart, (A, B and C) were randomly selected along the shore in belts of *C. compressa* and *C. amentacea* var. *stricta*. At each station and for both species, 15 holes were drilled on the bare rock. Holes were 40 mm deep and 30 mm large. *Cystoseira* thalli were then collected from nearby natural populations. One thallus was fixed per hole with epoxy putty. The harmlessness of the glue was tested by Sant Funk (2003) on *Cystoseira* spp. A total number of 45 thalli per species were set up. The success of the technique was assessed by counting the remaining transplanted thalli after 3 weeks (April 2004), 3 months (June 2004) and 6 months (September 2004). In addition, we compared the fitness of transplanted thalli to the one of individuals, growing in the natural population, called neighboring controls. After checking the assumption of the normality of frequencies, a χ^2 test ($\nu = 2$; $\alpha = 0.05$) was performed to test the hypothesis that transplantation success is equivalent for both species at each observed date and among the three stations (A, B, C). Additionally, χ^2 tests ($\nu = 4$; $\alpha = 0.05$) were done to test whether the transplantation success differed between seasons (three dates) and stations for each of the two species of *Cystoseira*.

Transplantation success remained high during the 6 months experiment for both species (from 75% to 96%, Table 1). There was no significant difference in the transplantation success between the two species ($\chi^2_{\text{obs}} < \chi^2_{(2,0.05)}$) whatever the date. Looking at each species separately, there was no significant variation between the numbers of remaining transplanted thalli when comparing the three stations at a fixed date and the three dates at a fixed station (Table 1). After 3 weeks of experiment (April 2004), transplantations were successful for both species, with >93% of the transplanted thalli remaining. All of them were alive and showed no sign of degradation. We observed growing parts of fronds on the transplanted thalli. Neighboring controls were also growing. After 3 months (June 2004), >84% of the thalli were still in place and 11 or more thalli out of 15 were still present at each station (Table 1). In June 2004, all the transplanted thalli of both species were fertile, like the neighboring controls. This observation clearly shows that transplantation did not affect *Cystoseira* development. After 6 months (September 2004), only 75% of transplanted thalli were still in place for both species (Table 1). However, these losses could be explained by the strong waves observed during August 2004 or by a high fish-grazing pressure by *Sarpa salpa* L. We observed grazed apices on the transplanted thalli and on natural populations. Indeed, *S. salpa* eats *Cystoseira* fronds (Verlaque, 1990) and exerts a strong traction on the thallus which can be torn off. Although the success of transplantation did not significantly differ between both species, the morphology of *C. amentacea* seems to be better adapted to fit the transplantation hole. Both species are caespitose but *C. amentacea* var. *stricta* has long and thin erect axes which can be easily enclosed in the epoxy glue and then inserted in the holes. *C. compressa* fronds, on the contrary, are wide and flat, and grow directly from the base, giving the thallus a shape of rosette (Cormaci et al., 1992). Falace et al. (2006) encountered similar problems when trying to transplant *C. compressa* individuals on artificial reefs. Their transplantation technique was successful for the species *Cystoseira barbata* (Stackhouse) C. Agardh but not for *C. compressa*. They explained this result by the morphology of the thalli, since *C. barbata* has a single axis, contrary to *C. compressa*. During the whole experiment,

Table 1 Total transplantation success and number of remaining transplanted thalli of *C. compressa* and *C. amentacea* at each sampling date and at each station

	C. compressa					*C. amentacea* var. *stricta*				
	Total transplantation success (%)	Station A	Station B	Station C	χ^2_{obs}	Total transplantation success (%)	Station A	Station B	Station C	χ^2_{obs}
April 2004	93	15	13	14	0.03	96	15	14	14	0.08
June 2004	84	13	11	14	0.11	93	14	14	14	0.10
September 2004	75	13	10	11	0.10	75	12	13	9	0.38
χ^2_{obs}		0.06	0.04	0.11			0.02	0.02	0.32	

Results of the χ^2 ($\chi^2_{(4,0.05)} = 9.49$)

most of the plants remained glued, but they did not generate new attachment structures. These preliminary results show that transplantation has no inhibiting or toxic effect on *Cystoseira* survival and development (fertility). This innovative technique is cheap and efficient, although strongly dependent on wave conditions during and after transplantation. Moreover, such a technique is easy to reproduce and could be used on other natural or artificial hard substrates. It would be interesting to test this technique on *Cystoseira* species with single axis, and also on species living in sheltered environments. Indeed creating or re-creating *Cystoseira* populations will increase biodiversity of coastal marine environments.

Acknowledgements We are very grateful for field assistance from Jean-Michel Cottalorda and Fabrice Javel. This work was supported by the Région Provence-Alpes-Côte d'Azur (PhD grant for ML Susini) and by the EGIDE and the CRUI, within the framework of the Galileo exchange Programme.

References

Arnoux, A. & D. Bellan-Santini, 1972. Relations entre la pollution du secteur de Cortiou par les détergents anioniques et les modifications des peuplements de *Cystoseira stricta*. Tethys 4: 583–586.

Bellan-Santini, D. & G. Desrosiers, 1977. Distribution du benthos de substrat dur dans un golfe soumis à de multiples pollutions (golfe de Fos). IIIèmes Journées Etudes Pollutions, Split, CIESM, 153–157.

Clayton, M., 1990. The adaptive significance of life history characters in selected orders of marine brown macroalgae. Australian Journal of Ecology, Oxford 15: 439–452.

Cormaci, M., G. Furnari, G. Giaccone, B. Scammacca & D. Serio, 1992. Observations taxonomiques et biogéographiques sur quelques espèces du genre *Cystoseira* C. Agardh. Bulletin de l'Institut Océanographique, Monaco 9: 21–35.

Council of Europe, 1979. Convention on the Conservation of European Wildlife and Natural Habitats, Appendix I, Strictly protected Flora species. Bern, Switzerland. http://www.conventions.coe.int/Treaty/FR/Treaties/Html/104-1.htm.

Falace, A., E. Zanelli & G. Bressan, 2006. Algal transplantation as a potential tool for artificial reef management and environmental mitigation. Bulletin of Marine Science 78(1): 161–166.

Meinesz, A., E. Bellone, J. Lefevre & P. Vitiello, 1990. Impact des aménagements construits sur le domaine maritime de la région Provence-Alpes-Côte d'Azur. Rapport au Conseil Régional PACA – Délégation Régionale à l'Architecture et à l'Environnement.

Ollivier, G., 1929. Etude de la flore marine de la Côte d'Azur. Annales de l'Institut Océanographique, Tome VII, Fasc. III, 173 pp.
Sant Funk, N., 2003. Algues bentòniques mediterrànies: comparació de mètodes de mostreig, estructura de comunitats i variació en la resposta fotosintètica. Ph.D. Thesis, Departement d'Ecologia, Universitat de Barcelona, 249 pp.
Soltan, D., 2001. Etude de l'incidence de rejets urbains sur les peuplements superficiels de macroalgues en Méditerranée nord-occidentale. Ph.D. Thesis, Université de la Méditerranée, Centre d'Océanologie de Marseille, 137 pp.
Soltan, D., M. Verlaque, C. F. Boudouresque & P. Francour, 2001. Changes in macroalgal communities in the vicinity of a Mediterranean sewage outfall after the setting up of a treatment plant. Marine Pollution Bulletin 42: 59–70.
Verlaque, M., 1990. Relationships between *Sarpa salpa* (L.) (Teleostei, Sparidae), other browser fishes, and the Mediterranean algal phytobenthos. Oceanologica Acta 13: 373–388.

Hydrobiologia (2007) 580:245–254
DOI 10.1007/s10750-006-0448-x

BIODIVERSITY IN ENCLOSED SEAS

Gametogenesis and maturity stages scale of *Raja asterias* Delaroche, 1809 (Chondrichthyes, Raijdae) from the South Ligurian Sea

M. Barone · S. De Ranieri · O. Fabiani · A. Pirone · F. Serena

Abstract This stu\dy was aimed at the acquisition of basic life-history information on *Raja asterias* through an objective approach in the assignment of maturity stage, using both histological techniques and a multivariate analysis aimed at macroscopic evaluations of the reproductive system. The samples examined were collected from landings of commercial trawl vessels from the port of Viareggio. From July 2001 to August 2002 a total of 351 specimens, 166 females and 155 males, were purchased. A description of the main stages of oogenesis and spermatogenesis is presented here. Moreover, on the basis of the maturity stage definition, maturity ogives were constructed. Length at 50% maturity was estimated to be 51.7 cm and 56.1 cm total length for males and females respectively. Finally an informal table for the maturity stage assignment, which can be useful for a quick and easier definition both on board and in the laboratory, is formulated.

Keywords Shark fisheries · Sexual maturity · Oogenesis · Spermatogenesis

Guest editors: G. Relini & J. Ryland
Biodiversity in Enclosed Seas and Artificial Marine Habitats

M. Barone · F. Serena (✉)
Agenzia Regionale per la Protezione Ambientale della Toscana, Via Marradi, 114, 56100 Livorno, Italy
e-mail: f.serena@arpat.toscana.it

S. De Ranieri
Centro Interuniversitario di Biologia Marina, Livorno, Italy

O. Fabiani · A. Pirone
Sezione di anatomia – Dipartimento di produzioni animali, Università di Pisa, Pisa, Italy

Introduction

The starry ray *Raja asterias* Delaroche, 1809 is common in the whole Mediterranean excluding probably the Black Sea (Fischer et al., 1987; Relini et al., 2000). Although along the Atlantic coasts of northern Morocco and in southern Portugal it has been recorded (Whitehead et al., 1984), *R. asterias* may be considered in general terms a Mediterranean endemic species. It is a benthic species generally found in shallow waters on muddy or sandy soft bottoms, mostly concentrated at depths up to 150 m (Serena et al., 1988).

In the southern Ligurian Sea port of Viareggio, *R. asterias* is among the most important components of the fish assemblages caught all year round with beam-trawling. Elasmobranchs are particularly vulnerable to fishing pressure, because of their biological characteristics: low growth rates, late sexual maturation, low fecundity, and a long reproductive cycle. The decline of many species as a consequence of removals that went beyond the sustainable levels has been stressed by many authors (Musick, 1999; Stevens et al., 2000). However, it seems that this did not occur for *R. asterias*. On the grounds where the fishery operates, data on catch trends for the last 15 years suggest that the biomass of the stock is in a steady state (Serena et al., 1988; Abella & Serena, 2005).

Almost nothing is known about the reproductive biology of the starry ray in the South Ligurian Sea. Lo Bianco (1908) reported observation of specimens from the Golfo di Napoli, Tortonese (1956) of specimens from markets of Livorno, Napoli and the Adriatic Sea, Capapé (1977) also studied the biology of *R. asterias* in the Tunisian waters. This species is oviparous, mature females being present throughout the year (Capapé, 1977) and the egg-cases are usually deposited at a depth of about 30–40 m. The only available data on rates of egg laying and incubation were obtained under experimental conditions. The rate of egg lying estimated by Lo Bianco (1908) was one egg capsule every 4 days; the incubation period lasted about 5–6 months (Capapé, 1977). As newborn juveniles of about 8 cm TL have been found distributed primarily in the northern Tyrrhenian Sea at depth of 8–12 m, this area is very likely to be a nursery (Abella et al., 1997).

This study was aimed at the acquisition of basic life-history information on *R. asterias* through an objective approach in the assignment of maturity stage, using both histological techniques and a multivariate analysis aimed at macroscopic evaluations of the reproductive system. A description of the main stages of oogenesis and spermatogenesis, the definition of a maturity scale and an estimate of length at first maturity are showed here.

Materials and methods

The samples examined were collected from landings of commercial vessels utilizing a variant of the beam-trawl called *rapido* and operating from the estuary of the Arno River to the Isle of Tino. From July 2001 to August 2002, monthly purchases amounted to a total of 351 specimens, 166 females and 155 males. At the laboratory, total length (nearest 0.5 cm below), from tip of snout to terminal point of the caudal fin, disc width and sex were recorded for each specimen. In addition, in females the maximum nidamental gland width (nearest 0.1 mm) was measured.

Methodology for definition of maturation stages

In order to determine maturity stages, for females, the size of the ovarian eggs, the condition of the oviducts, the measurements of the nidamental gland, and the presence of egg-cases were considered; for males the measurements and the consistency of the claspers, the development of the sperm ducts and the gonads (testes) were observed. Each characteristic was associated with a score. As a result, each specimen was allocated a code made of four scores. In order to group together specimens with the same macroscopic features, data of reproductive apparatus were analyzed with multivariate techniques carried out through the option "agglomerative hierarchical clustering" of S-Plus program (MathSoft, 1999).

Determination of the length at maturity

The length at 50% maturity (L_{50}) for both sexes was estimated. Females were classified as mature when they were in stages up to III (vitellogenic follicles, well differentiated oviducts, nidamental gland of large size, including specimens with egg capsules), conferred by the previous method; males when they were in stage IV (claspers longer than tips of posterior pelvic fin lobes and with skeleton hardened, sperm ducts filled, gonads wide, rose-coloured, with seminiferous follicles filling the whole volume).

The proportion of mature females and males at length was described through a logistic function as follows:

$$Y = 1/1 + e^{a+bX}$$

where Y is the estimated mature proportion, a and b are the estimated coefficients of the logistic equation and X is the total length. The length at maturity can be estimated as the ratio of the coefficients (a/b) by substituting $Y = 0.5$ in equation. The curve fitting was done using the specific tool included in Microsoft Excel.

Histological techniques

The ovaries of 34 females and the testes of 7 males, representing all stages of morphological maturity, were extracted in their entirely and fixed in 10% formaldehyde 0.1 M phosphate buffer, pH 7.4. Small fragments of each gonad were extensively washed in fresh water for one day, dehydrated in a graded ethanol series (30 min at 30%, 50%, 70%; 1 h at 80%; 2 h at 95%), then embedded in glycol methacrylate resin (JB-4, Polysciences Inc.). Microtomed sections of 5 μm were collected on glass slides coated with 0.5% gelatine containing 0.05% chrome alum. Sections were subsequently stained with toluidine blue, methylene blue and PAS reaction, cleared in xylene, mounted with DPX and examined with a light microscope (×10–×1000 magnification).

Methodology for ovarian egg counts

One of the ovaries of 166 female specimens was fixed in modified Gilson's solution (Simpson, 1951), which composition was: 100 ml ethanol 60%, 15 ml nitric acid 80%, 18 ml acetic acid, 20 g mercury chloride, 800 ml distilled water. The total number of eggs in each ovary was counted and their diameter was measured with the eyepiece micrometer of a stereoscope.

Results

Developing follicles of various sizes were present in the ovary of *R. asterias*. In small previtellogenic follicles ~50 μm in diameter, lying immediately below the ciliated peritoneal epithelium, follicular and theca layers were very thin, both composed by a single layer of squamous cells (Figs. 1, 2). In follicles ~150 μm in diameter the follicular epithelium showed two kinds of cells: small and large cells (Fig. 3). Small cells were localized both under the basal lamina, surrounding the whole follicle, and along the vitelline envelope. Large cells were cubic and occupied the centre of the follicular epithelium. A zona pellucida was evident between the oocyte and the granulosa cells. Large follicles, up to 1500 μm, characterized by the presence of a third follicle cell type, pyriform cells, spanned the whole follicular epithelium (Fig. 4). They showed an elongated apex pointing toward the oocyte surface. Cell protrusions, linking pyriform cells with oocyte, were often observed inside the zona pellucida (Fig. 5). The nucleus of an oocyte of ~1600 μm contained chromosomes showing a lampbrush appearance (Fig. 6). Yolk droplets, positive to the PAS reaction, first appeared in follicles up to 3000 μm. Initially, very small yolk droplets were evident in the cortical regions of the oocyte; then their size increased and they filled the entire ooplasm (Figs. 7, 8). Atretic previtellogenic and postovulatory follicles were also found (Figs. 9,

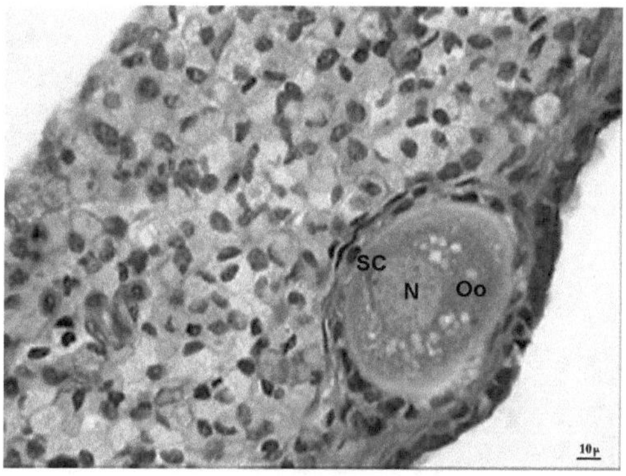

Fig. 1 Small previtellogenic follicles ~50 μm in diameter. The follicular epithelium is single layered and made up to small squamous cells (SC). TC, theca cells; N, nucleus; Oo, oocyte

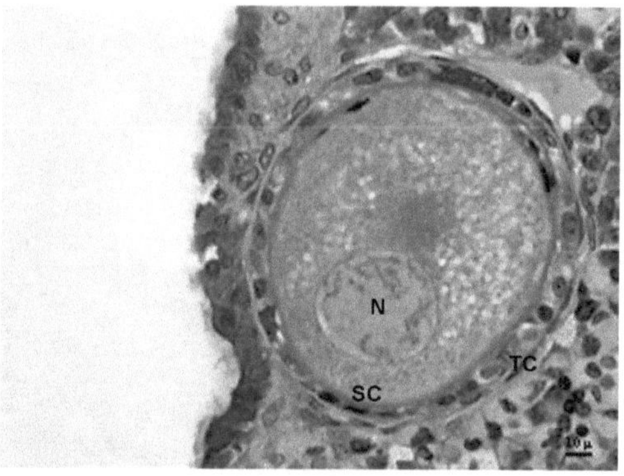

Fig. 2 Small previtellogenic follicles ~50 μm in diameter. The follicular epithelium is single layered and made up to small squamous cells (SC). TC, theca cells; N, nucleus; Oo, oocyte

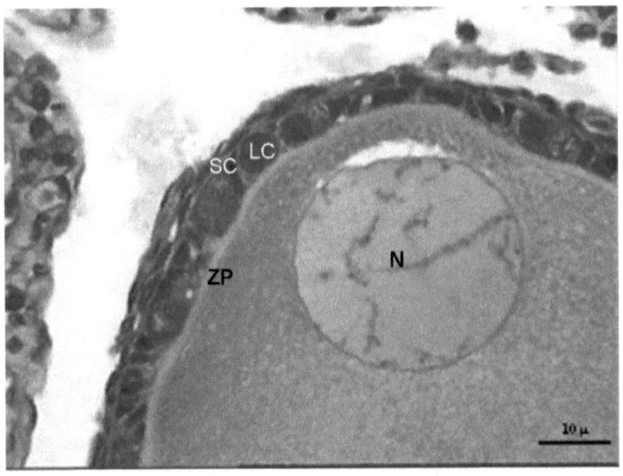

Fig. 3 Previtellogenic follicle ~100 µm in diameter. The follicular epithelium is double layered and made up to small (SC) and large (LC) cells. ZP, zona pellucida

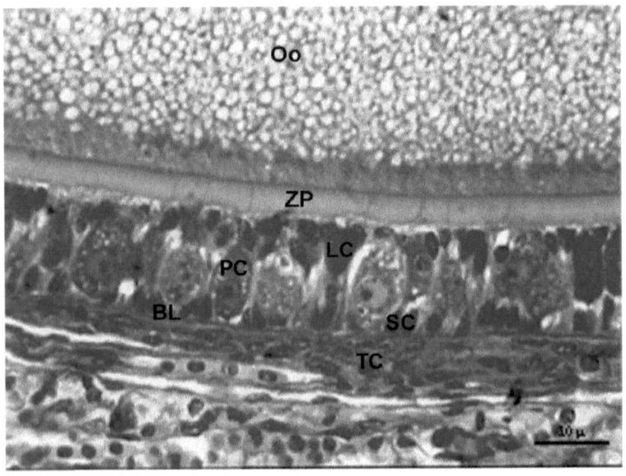

Fig. 4 Large previtellogenic follicles, up to 1500 µm in diameter. The follicular epithelium is multilayered and made up to small (SC), large (LC) and pyriform (PC) cells. Oo, oocyte; ZP, zona pellucida; BL, basal lamina

10). The former were characterized by follicle and theca cells that later become hypertrophic. In postovulatory follicles the basal lamina, positive to the PAS reaction, appeared collapsed and invaded the central lumen.

In the testes, six stages of spermatocyst development were observed. At the beginning Sertoli cells lined the lumen and a lesser number of spermatogonia were in a peripherical position in the spermatocyst (Fig. 11). In the second stage the Sertoli cells were also found in peripherical position just inside the basement membrane (Fig. 12). Successively, the primary spermatocytes showed large nuclei and in the fourth stage secondary spermatocytes appeared with nuclei

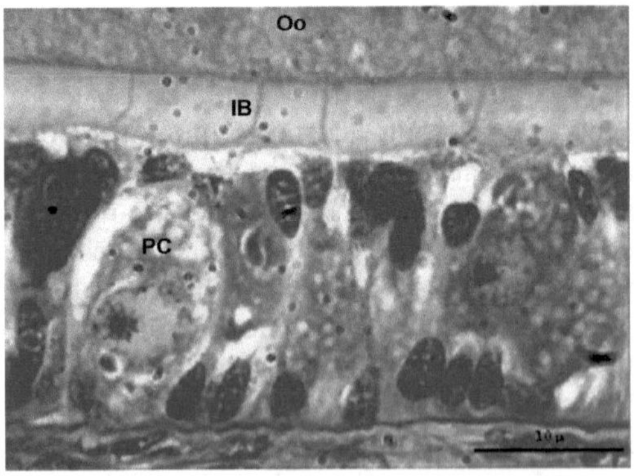

Fig. 5 Intercellular bridges (IB) link pyriform (PC) cells with oocyte (Oo)

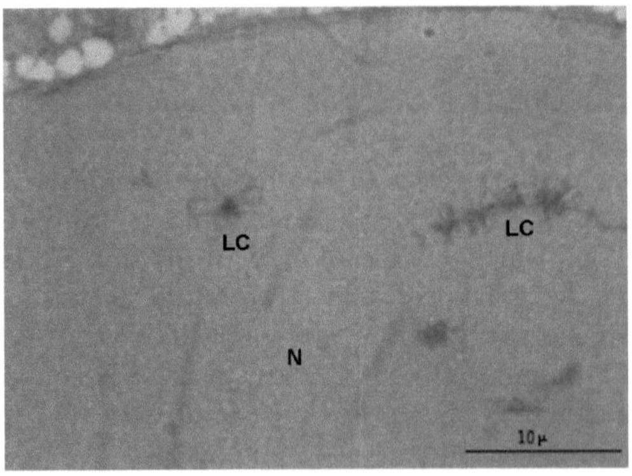

Fig. 6 Nucleus (N) of oocyte 1600 µm in diameter contained lampbrush chromosomes (LC)

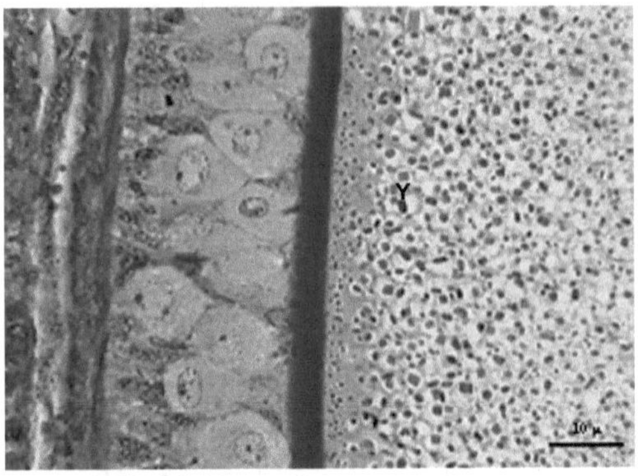

Fig. 7 Follicles up to 3000 µm filled with yolk platelets (Y)

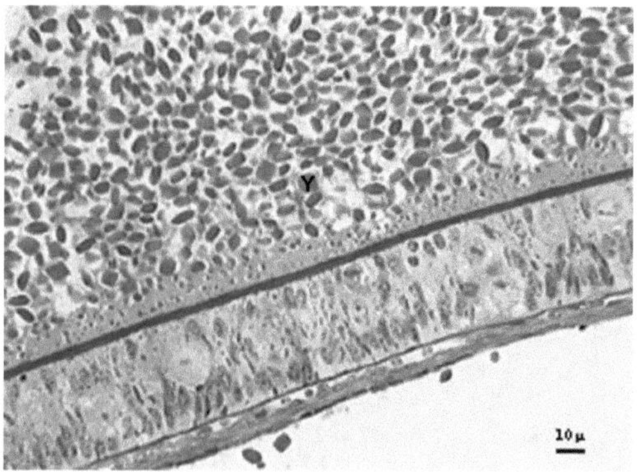

Fig. 8 Follicles up to 3000 μm filled with yolk platelets (Y)

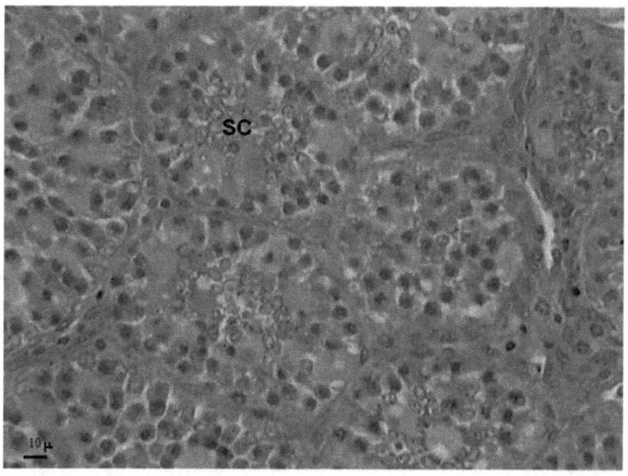

Fig. 11 Sertoli cell nuclei (Sc) migrating toward the periphery

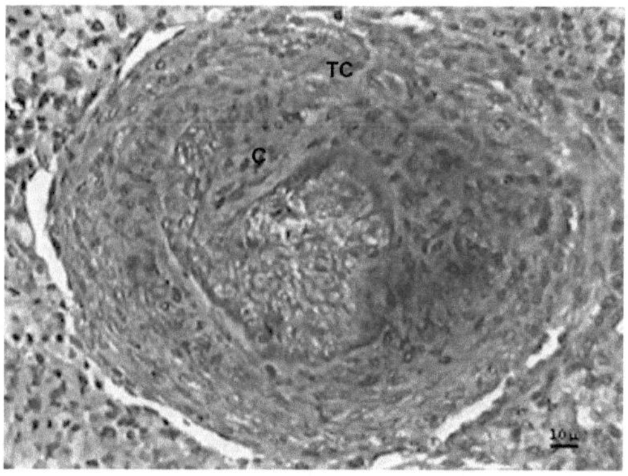

Fig. 9 Atretic previtellogenic follicle. Granulosa cells (C) and theca cells (TC) are hypertrophic

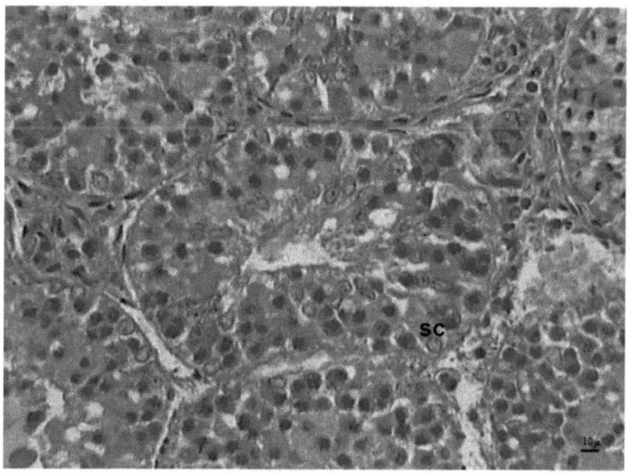

Fig. 12 Sertoli cells (Sc) in peripherical position just inside the basement membrane

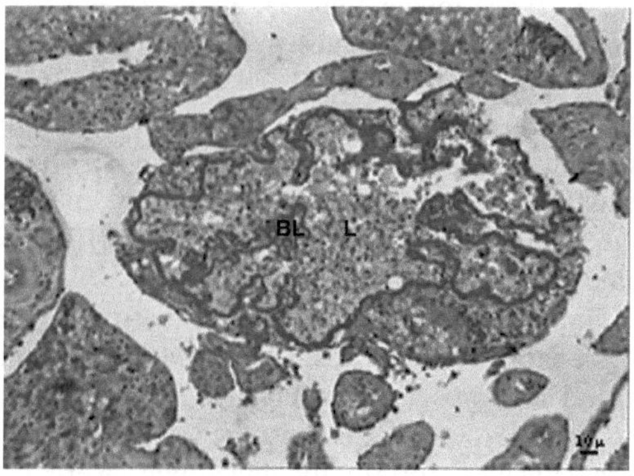

Fig. 10 Postovulatory follicle. The basal lamina (BL) appears collapsed and invades the central lumen (L)

containing condensed chromosomes (Figs. 13, 14). The fifth stage was characterized by spermatids with ellipsoidal nuclei and emerging flagella. In this stage, spermatozoa began to aggregate around the periphery of the spermatocyst to form tight bundles (Fig. 15). In the terminal stage of spermatogenesis, spermatozoa were present in the lumen (Fig. 16).

The diameter of oocytes contained in the ovaries of 118 specimens were measured and distributed in classes of 150 μm. The diameters were correlated to the maturity stage previously assigned. Table 1 shows the presence of oocytes up to 3 mm in diameter, in which it was supposed that vitellogenesis had begun. The mean number

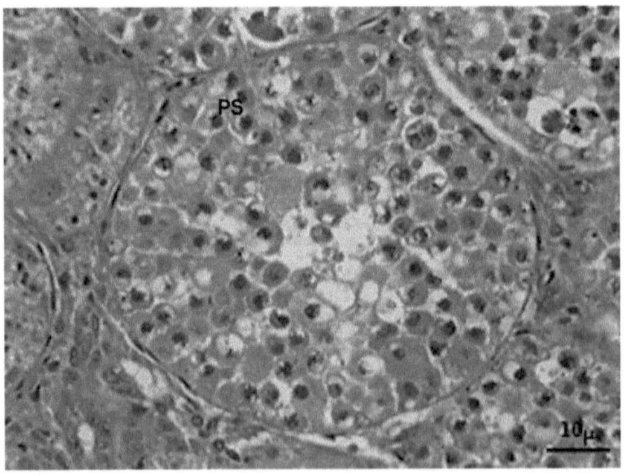

Fig. 13 Primary spermatocytes (Ps) showing large nuclei

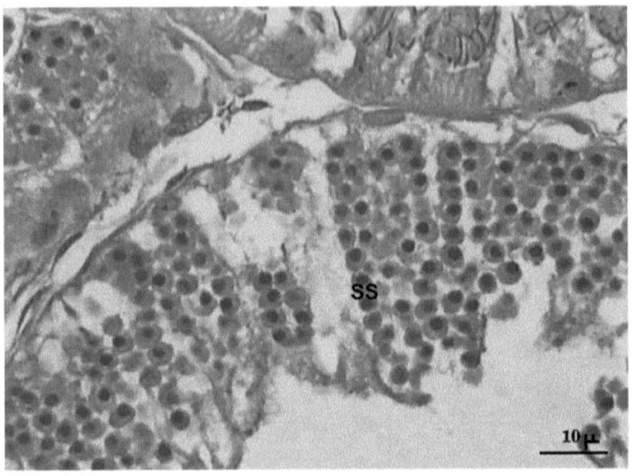

Fig. 14 Secondary spermatocytes (Ss) with nuclei containing condensed chromosomes

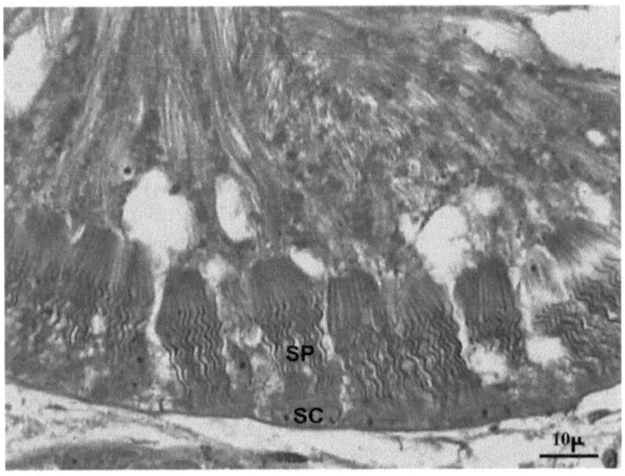

Fig. 15 Spermatozoa (sp) associated in linear arrays in the Sertoli cells (Sc)

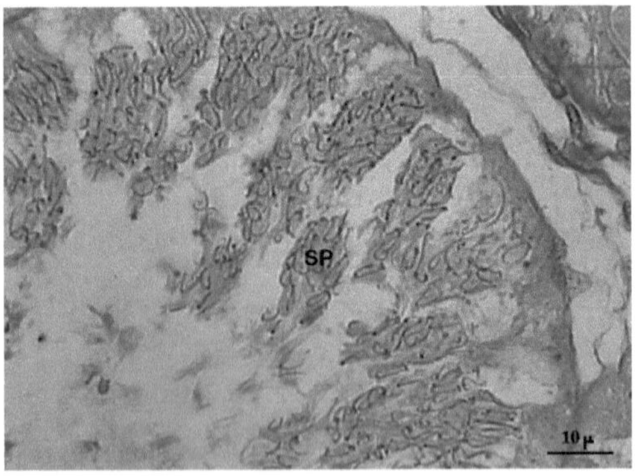

Fig. 16 Spermatozoa (sp) within the lumen

of eggs up to 3 mm in each ovary increased in stage II (Maturing) and reached a maximum in stage III (Mature) (Table 1).

The cluster analysis in the case of the males (n = 121) showed 18 groups and for females (n = 150) 12 groups of specimens with the same features. These groups were further aggregated on the basis of their proximity in the cluster and, for females, considering also the results of microscopic observations. Finally, to each of the five groups, for both males and females, a maturity stage was assigned as shown in Tables 2 and 3.

On the basis of the morphological determination of maturity, maturity ogives were constructed for males (n = 136) and females (n = 165) (Fig. 17). Males mature by 45 cm TL. Males smaller than 45 cm TL (n = 49) were immature, specimens 46–55 cm TL (n = 85) were maturing and all those longer than 55 cm TL (n = 4) were sexually mature. Females may mature by 53 cm TL. Females 21–53 cm TL (n = 94) were immature, those 54–59 cm TL (n = 50) were maturing. All females longer than 59 cm TL (n = 13) were sexually mature. Total length at 50% maturity was estimated to be 51.7 cm and 56.1 cm for males and females respectively.

The monthly percentages of mature female rays containing egg capsules during the sampling period are presented in Fig. 18. Most (40%) of the females with egg capsules, including those with capsules not totally developed, were captured in May.

Table 1 Number of oocytes in females of *R. asterias* at different maturity stages

Stage	No. of females	No. of oocytes	Mean number oocytes <3 mm/female	Mean number oocytes >3 mm/female
I	71	6371	86	6
II	19	1794	74	21
III	11	1852	135	33
IV	11	1599	117	27
V	6	637	84	16

Table 2 Maturity scale for females of *R. asterias*

Stage	State	Description	
		Macroscopic	Microscopic
I	Immature	Oocytes uniformly small and white. Ovary very small. Oviducts thread-like and nidamental gland absent or little differentiated.	The anterior region of the ovary containing numerous follicles in first phase of previtellogenesis, up to 150 μm in diameter. Atretic follicles present.
II	Maturing	Ovary small, oocytes differentiated to various small sizes but all white. Oviducts and nidamental gland well developed but small (<2 mm maximum width).	Sections appear filled with previtellogenic follicles of various sizes.
III	Mature	Ovary walls more transparent. Ovary filled with yellow Oocytes, some large. Oviducts well developed and nidamental gland large (>2 mm maximum width).	Sections of oocytes of 3–7 mm show yolk droplets.
IV	Extruding	Ovary filled with large yellow oocytes. Egg capsules more or less formed in one or both oviducts.	Only the follicular walls were embedded because of large dimension of oocytes. Yolk droplets evident in the cortical region.
V	Resting	Ovary walls transparent. Oocytes of different sizes, white or yellow. Oviducts empty but much enlarged and vascularized. Nidamental gland small (<2 mm).	Postovulatory follicles observed.

Table 3 Maturity stages for males of *R. asterias*

Stage	State	Description
I	Immature	Clasper shorter or as long as the extreme tips of posterior pelvic fin lobes. Ducti deferentes not differentiated or narrow. Testes small and white occupying up to the half of abdominal cavity.
II	Virgin	Clasper longer than tips of posterior pelvic fin lobes but skeleton still flexible. Ducti deferentes well developed. Testes occupying over the half of abdominal cavity with walls transparent but seminiferous follicles not fill the whole testes.
III	Maturing	Clasper longer than tips of posterior pelvic fin lobes, skeleton hardened but axial cartilages still soft. Ducti deferentes whitish. Testes enlarged, not entirely filled with seminiferous follicles.
IV	Mature	Clasper longer than tips of posterior pelvic fin lobes, skeleton hardened with axial cartilages hardened and pointed. Ducti deferentes full. Testes wide, rose-coloured, with seminiferous follicles filling whole volume.
V	Resting	Clasper longer than tips of posterior pelvic fin lobes, skeleton hardened with axial cartilages hardened and pointed. Testes scarcely occupying half of abdominal cavity.

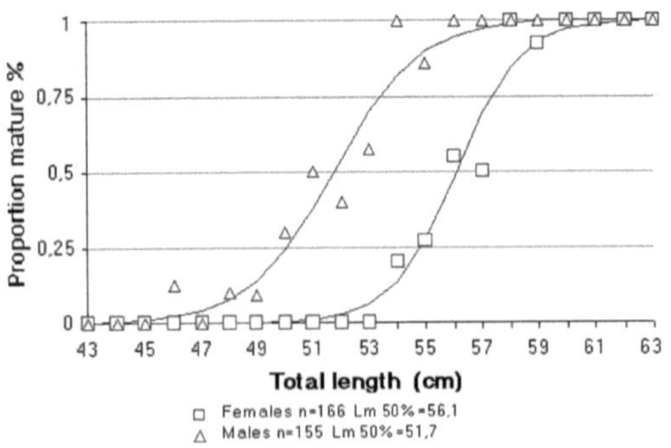

Fig. 17 Maturity ogives of *R. asterias*

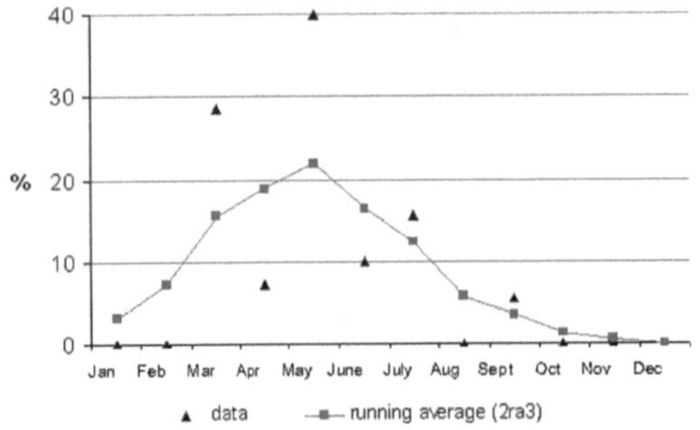

Fig.18 Percentage and running average (2ra3) of egg-capsules in females of *R. asterias*

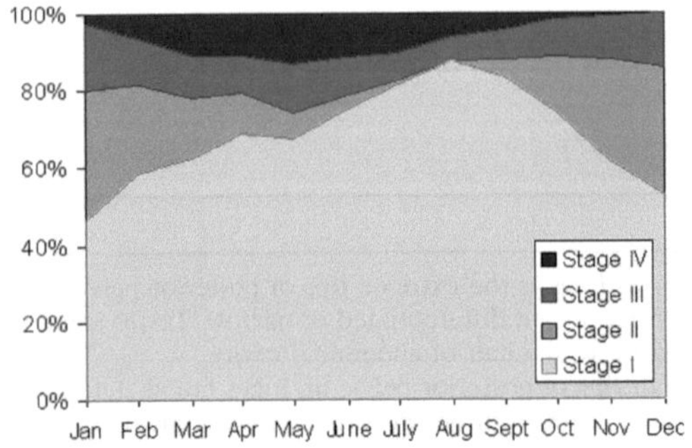

Fig. 19 Running average (2ra3) of maturity stages in *R. asterias*

The monthly abundances of females of *R. asterias* at different stages of maturity is presented in Fig. 19. The greatest number (80%) of females at stage I (Immature) was observed in August. This percentage diminished from September to December, when the specimens at stage II increased until the 50%. Stage III (Mature) stays almost constant throughout the year, but stage IV (Extruding) were mainly observed in May and June, when it reached the 20 %, decreasing in autumn.

Discussion

The analysis of histological maturity in female of *R. asterias* indicates that both ovaries contain developing follicles of various sizes. In all the ovaries examined, the great majority of follicles was in the phase of previtellogenesis, characterized by the progressive growth of the oocytes and the increase in complexity of the follicular epithelium. The change observed in the follicular epithelium, during the early previtellogenic stage, corresponds to that described by Andreuccetti et al. (1999) in the same species. In addition, an evidence of the phase of previtellogenesis is the presence of chromosome with lampbrush appearance in the nucleus of oocytes 50–1500 μm in diameter. References to lampbrush chromosomes in elasmobranch oocytes have not been found, although the lampbrush chromosomes were described by Guraya (1986) in the ovary of several teleosts, in which they appear during the first meiotic prophase. Furthermore, vitellogenis starts when the follicle measures about 3000 μm in diameter and is indicated by the appearance of minute platelets of yolk in the cortical region. We have been unable to describe the successive phases of oogenesis because of the large size of the oocytes. Instead, the count of oocytes allowed us to observe that only few follicles continue with vitellogenesis; therefore, the ovary of *R. asterias* seems to contain a reserve of oocytes, present both at every maturity stage and throughout the year.

A brief analysis of the histology of testes indicates that spermatogenesis of *R. asterias* is similar to that described by Hamlett (1999) in *Urolophus jamaicensis*. In the testes spermatocysts at different stages of development were present, in accordance with the statement that the gonads of rajids continuously produce reproductive gametes (Dodd, 1983).

Gross morphology of the reproductive tracts of females at any time, supported by the histology, facilitate the assignment of maturity stages, and indicate that the process of maturation is both progressive and gradual. In males only qualitative analysis of the gonads were done; however, the assignment of maturity stages was based on accurate observations not only of claspers, as was done by Capapé (1977), but also of gonads and sperm ducts.

The maturity stages both for females and males resulting from the cluster analysis were numerous. However, this method was used to eliminate the subjectivity of the current maturity scales (Stehmann, 2002) postponing the classification into different stages until separate examination of each feature of the reproductive tracts had been carried out. The first three groups resulting from the clustering were joined together in stage I (Immature) both in males and females. Between females classified as immature, the differences consisted in the appearance of oviducts and nidamental gland during growth; in all cases the oocytes are small and white. In males, the appearance of sperm ducts and the start of the development of the gonads were the reasons determining the different groups in the cluster. In addition, in females, stage III (Mature) is characterized by two phases in which the oocytes grow and become yellow, contemporary the nidamental gland reaches a large size. Another finding, resulting from the application of cluster analysis, was the stage of resting hypothesized for both males and females and rarely described for elasmobranchs.

In females, at the onset of maturity, the ovaries contain vitellogenic oocytes and the nidamental gland is of medium dimension; in males claspers are longer than the tips of pelvic fins, axial cartilages are pointed, and gonads contain seminiferous ampullae filling the whole volume. The estimation of size at first sexual maturity (L_m) for males (51.7 cm) is somewhat discordant from that estimated by Capapé (1977) (54 cm); also in the case of females the estimated value of L_m (56.1 cm) is lower than that calculated by Capapé (1977) and Tortonese (1956) (60 cm). These differences could be related to environmental and/or geographical aspects (Frisk et al., 2001).

In evaluating the reproductive cycle, the low number of specimens reaching lengths at 50% maturity, 36 females and 46 males, was insufficient to estimate the gonado-somatic indexes for each month. From the occurrence of mature females and females with egg-capsules during the sampling period, it is concluded that the main period of spawning activity occurs from March to July. On the basis of these results, the tables of maturity stages provide a more reliable tool for describing rajid life history data.

Acknowledgements We wish to express our thanks to Prof. Giorgio Mancino (Sezione di Biologia cellulare e dello sviluppo del Dipartimento di Fisiologia e Biochimica—Università di Pisa) for his advice on histological investigations. We are very grateful to Alvaro Abella and Romano Baino (Agenzia Regionale per la Protezione Ambientale della Toscana) for their useful support in processing the data. Many thanks also to Cecilia Mancusi for her help and encouragement. We owe special thanks to Caroline Bennett (Fao-AdriaMed) for her assistance with the translation of this paper.

References

Abella, A. J. & F. Serena, 2005. Comparison of elasmobranch catches from Research Trawl Surveys and commercial landings at port of Viareggio, Italy, in the last decade. e-Journal of Northwest Atlantic Fishery Science V35: art. 23.

Abella, A., R. Auteri, R. Baino, A. Lazzeretti, P. Righini, F. Serena, R. Silvestri, A. Voliani & A. Zucchi, 1997. Reclutamento di forme giovanili nella fascia costiera toscana. Biologia Marina del Mediterraneo 4(1): 172–181.

Andreuccetti, P., M. Iodice, M. Prisco & R. Gualtieri, 1999. Intercellular bridges between granulosa cells and the oocyte in the elasmobranch *R. asterias*. Anatomical Record 255(2): 180–187.

Capapé, C., 1977. Contribution à la biologie des Rajidae des côtes tunisiennes. 4. *Raja asterias* Delaroche, 1809: répartition geographique et bathymétrique, sexualité, reprodution et fecondité. Bulletin du Museum d'Histoire Naturelle, Paris, 3e ser., 435, Zool., 305: 305–326.

Dodd, J. M., 1983. Reproduction in cartilagineus fishes (Chondrichthyes). In Hoar, W. S., D. J. Randall & E/M/ Donaldson (eds), Fish Physiology, Vol. 9. Academic Press, San Diego, pt A, 31–95.

Fischer, W., M. L. Bouchot & M. Schneider, (rédacteurs) 1987. Fishes FAO d'identification des espéces pour les besoins de la pêche. (Revision 1). Méditerranée et Mer Noire. Zone de peche 37, Vol. II. Rome, FAO, Vértébres, 847–876.

Frisk, G. M., J. T. Miller & M. J. Fogarty, 2001. Estimation and analysis of biological parameters in elasmobranch fishes: a comparative life history study. Canadian Journal of Fisheries and Aquatic Science 58(5): 969–981.

Guraya, S. S., 1986. The Cell and Molecular Biology of Fish Oogenesis. Monographs in Developmental Biology, Vol. 18. Karger, Basel, 223 p.

Hamlett, W. C., 1999. Sharks, Skates and Rays: The Biology of the Elasmobranchs Fishes. John Hopkins University Press, Maryland, 515 pp.

Lo Bianco, S., 1908. Notizie biologiche riguardanti specialmente il periodo di maturità sessuale degli animali del golfo di Napoli. Pubblicazione della Stazione Zoologica di Napoli 19: 513–761.

MathSoft, 1999. S-PLUS User's Guide. Data Analysis Products Division, MathSoft, Seattle, WA, 620 pp.

Musick, A. J., 1999. Criteria to define extinction risk in marine fishes. The American Fisheries Society initiative. Fisheries 24(12): 6–14.

Relini, G., F. Biagi, F. Serena, A. Belluscio, M. T. Spedicato, P. Rinelli, M. C. Follesa, C. Piccinetti, N. Ungaro, L. Sion & D. Levi, 2000. I selaci pescati con lo strascico nei mari italiani. Biologia Marina del Mediterranaeo 7(1): 347–384.

Serena, F., R. Baino & P. Righini, 1988. Geographical and depth distribution of Rays in Northern Tyrrenian Sea. Commission Internationale pour l'Exploration Scientifique de la Mer Méditerranée 31(II): 277.

Simpson, A. C., 1951. The fecundity of the plaice. Fisheries Investigation, Ministry of Agriculture, Fisheries and Food (G.B.), Ser. 11 17(5): 1–27.

Stehmann, M. F. W., 2002. Proposal of a maturity stages scale for oviparous and viviparous cartilaginous fishes (Pisces, Chondrichthyes). Archive of Fishery and Marine Research 50(1): 23–48.

Stevens, J. D., R. Bonfil, N. K. Dulvy & P. A. Walker, 2000. The effects of fishing on shark, rays and chimaeras (chondrichthyans), and implications for marine ecosystem. ICES Journal of Marine Science 57: 476–494.

Tortonese, E., 1956. Leptocardia, Cyclostomata, Selachii. Fauna d'Italia, Vol. II. Calderini (ed.), Bologna, 545 pp.

Whitehead, P. J. P., M. L. Bauchot, J. C. Hureau, J. Nielsen & E. Tortonese, 1984. Fishes of the North-Eastern Atlantic and Mediterranean (FNAM), Vol. I. UNESCO, Paris, 510 p.

Hydrobiologia (2007) 580:255–257
DOI 10.1007/s10750-006-0447-y

Feeding strategy of the sacoglossan opisthobranch *Oxynoe olivacea* on the tropical green alga *Caulerpa taxifolia*

Paola Gianguzza · Franco Andaloro · Silvano Riggio

Abstract The feeding behaviour of the shelled sacoglossan *Oxynoe olivacea* was investigated to better understand the role and importance of this species in influencing encroachments of the alien alga *Caulerpa taxifolia* in the Mediterranean sea. We tested whether this slug preferred, as preliminary field observations suggested, an aggregative feeding behaviour and which part of the algal thallus, phylloid *vs* rhizoid, it preferred. Results showed that *O. olivacea* fed in groups and actively selected phylloid. This outcome poses important questions regarding the possibility that this species, fragmenting the alga thallus, could enhance dispersion and regeneration of *C. taxifolia*.

Keywords *Oxynoe olivacea* · *Caulerpa* spp. · Feeding strategy · Introduced species

Guest editors: G. Relini & J. Ryland
Biodiversity in Enclosed Seas and Artificial Marine Habitats

P. Gianguzza (✉) · F. Andaloro · S. Riggio
Dipartimento di Biologia Animale,
Università degli Studi di Palermo,
via Archirafi 18, 90123 Palermo, Italy
e-mail: pgiangu@unipa.it

F. Andaloro
ICRAM, Via Emerico Amari 124,
90134 Palermo, Italy

In recent ecological studies about coastal marine invasions much attention has been paid to the potential role of native consumers and their competitors influencing the invasiveness of introduced algae (Trowbridge, 2004).

Caulerpa taxifolia (Vahl) C. Agardh (Ulvophyceae: Caulerpales) is a green alga commonly occurring in tropical and subtropical areas (Meinesz et al., 2001). Over the last two decades a cold-tolerant strain of this species, known as "aquarium strain" (Jousson et al., 2000), has been undergoing prolific spread along the coastline of the Mediterranean Sea (Meinesz et al., 2001) and recently it has also been discovered on the coasts of California and Australia (Jousson et al., 2000; Millar; 2004). Following this Mediterranean invasion, *Oxynoe olivacea* (Rafinesque, 1819), a native sacoglossan species that previously lived and foraged exclusively on the autochthonous *Caulerpa prolifera* (Forskål) Lamouroux, has adapted its diet to include the invasive alga to its repertoire of potential food for consumption and spawning (Gianguzza et al., 2002; Gianguzza et al., 2005).

This phenomenon, known as host range expansion (Secord, 2003), is well documented for native sacoglossans that frequently modify their diet and/or host-plant as introduced plants become available (Trowbridge, 2004). Therefore, it is reasonable to expect that *O. olivacea* will increase its feeding activity on this new Mediterranean host.

Preliminary field observations on the feeding preferences of *O. olivacea* among *Caulerpa* species showed that this species often occurred in groups of three or more animals on the upper surface of *C. taxifolia* phylloids (Gianguzza, unpublished data). The feeding behaviour of *O. olivacea* was investigated to better understand its influence on encroachments of *C. taxifolia*. Specifically, we tested whether this sea slug selected an aggregative feeding behaviour and which part of the algal thallus, phylloid *vs* rhizoid, was preferred.

One hundred *O. olivacea* were collected by scuba-diving, between March 2001 and November 2002, from a stand of *C. taxifolia* in the Straits of Messina at the Torre Faro Marina (38°15.95' N; 15°39.10' E Italy, Sicily, western Mediterranean). When not used, the animals were maintained in 35 l aquaria with circulated natural seawater, at ambient light and temperature regimes (22°C). Lighting inside the laboratory was programmed to an alternating cycle of 12 h light and 12 h dark. The animals were not starved before starting the experiment. *C. taxifolia* was supplied daily *ad libitum* to avoid artificially stimulating them to discriminate less between portion of the algae (Haven & Scrosati, 2004). All experiments involved slugs ranging from 30 to 35 mm long: the precision of the measuring protocol has been previously evaluated as ca. 2.8% (Gianguzza et al., 2005). To prevent changes in slug's natural feeding responses, consequent to collecting and transferring stress, all the experiments were run after an adaptation period of 24 h.

The laboratory experiment consisted in offering a thallus of *C. taxifolia* (~15 g wet weight) comprising of the phylloid and a rhizoid portion to three sea slugs placed simultaneously in a 30 cm glass beaker filled with filtered, still sea water. Then the feeding duration of each possible *O. olivacea* aggregation arbitrarily classified as: isolate = 1 animal feeding, pair = 2 animals feeding, and group = 3 animals feeding, was independently recorded either on the algal phylloid or rhizoid. All feeding experiments were observed through a Leitz M3C dissection microscope for two hours.

The duration of the feeding events was analysed by using an orthogonal two-way ANOVA model considering the following fixed factors: Feeding Aggregation with 3 levels (isolate, pair, group) and Slug Attack with 2 levels (phylloid and rhizoid). Data were independently collected for each combination of factors and there were five replicates *per* level and a total of 90 specimens have been observed. All data were tested for homogeneity of variance with Cochran's C-test (Winer, 1971) and when the assumption was violated, data were transformed prior to analysis. After ANOVA, means were compared (at $\alpha = 0{\cdot}05$) with Student-Newman-Keuls (SNK) test (Underwood, 1997). The GMAV 5.0 software (University of Sydney, Australia) was used to perform the statistical tests.

Results from ANOVA showed that "Feeding Aggregation", "Slug Attack" and their interaction remarkably affected feeding duration in *O. olivacea* (Table 1a). SNK test results also highlighted that, in experimental conditions, *O. olivacea* preferentially selected feeding in a group and thus, apparently, slug feeding stimulated other conspecifics to consume the alga (Table 1b). In particular, the feeding events of groups were notably the longest in duration. Furthermore, the groups significantly attacked and consumed phylloids, the softest part of *C. taxifolia* thallus, rather than the thicker rhizoids (Table 1b).

Feeding strategy is a determinant factor to consider in the evolution of sacoglossan opisthobranchs because it is the result of important biological and ecological interactions between these molluscs and their food algae (Williams & Walker, 1999).

The sea slug *O. olivacea* feeds by piercing the cell wall of *Caulerpa* spp. with a specialized blade-shaped tooth, and then extracting the cell-sap with a muscular pharyngeal pump (Jensen, 1991). Both in field and laboratory conditions *O. olivacea* exhibited an aggregative feeding behaviour and laboratory observations indicated that most *O. olivacea* groups were found exclusively on phylloids of *C. taxifolia* and thus, apparently, slug feeding stimulated other conspecifics to feed upon these softer portions of the alga. This behaviour has two known advantages. Gregarious feeding is one way in which small consumers can successfully feed on large, sedentary items and may reduce individual risk of predation. Also, as in the case of suctorial consumers such as the sacoglos-

Table 1 Outcome of the ANOVA on the effects of Feeding Aggregation (isolate *vs* pair *vs* group), Slug Attack (phylloid *vs* rhizome) on feeding choice duration in *Oxynoe olivacea* (a) and results of SNK test (b) ($N = 5$)

	SS	df	MS	*F*
(a) Source of variation				
Feeding Aggregation (FA)	93.28	2	46.64	5.47**
Slug Attack (SA)	170.32	1	170.32	19.98***
FA × SA	89.85	2	44.93	5.27**
RES	204.60	24	8.5	
TOT	558.06	29		
Transformation	Sqrt (X + 1)			
Cochran's Test	C = 0.39 (ns)			

(b) SNK Test			
Feeding Aggregation		Slug Attack	
Isolate	ns	phylloid	rhizoid
Pair	ns	isolate < pair < group	ns
Group	rhizoid < phylloid		

* denotes $P < 0{\cdot}05$; ** $P < 0{\cdot}01$; *** $P < 0{\cdot}001$; ns not significant

san *Placida dendritica* A. & H. and *Stiliger aureomarginatus* (Jensen, 1993), this behaviour may reduce the viscosity of algal prey thus resulting in chemosensory stimulated slug feeding and enhancing the food uptake (Trowbridge, 1991; Jensen, 1999).

This study poses important questions regarding the possible use of *O. olivacea* as a biocontrol agent of *C. taxifolia*. This aggregative grazing could have the effect, as discovered in *L. serradifalci* (Žuljevic et al., 2001), of fragmenting the alga thallus and thus enhancing its dispersion and regeneration into new individuals: a hypothesis that in any case remains to be experimentally tested.

This study is part of PhD thesis by Paola Gianguzza and was funded by the M.U.R.S.T. ex 60% research project to Prof. Silvano Riggio.

References

Gianguzza, P., L. Airoldi, R. Chemello, C. D. Todd & S. Riggio, 2002. Feeding preferences of *Oxynoe olivacea* (Mollusca, Opisthobranchia, Sacoglossa) among three *Caulerpa* species. Journal of Molluscan Studies 68: 315–316.

Gianguzza, P., F. Badalamenti, K. R. Jensen & S. Riggio, 2005. Relationship between egg size and maternal body size in the simultaneous hermaphrodite *Oxynoe olivacea* (Mollusca, Opisthobranchia, Sacoglossa). Marine Biology 148: 117–122.

Haven, C. & R. Scrosati, 2004. Feeding preference of *Littorina* snails (Gastropoda) for bleached and photosynthetic tissues of the seaweed *Mazzaella parksii* (Rhodophyta). Hydrobiologia 513: 239–243.

Jensen, K. R., 1991. Comparison of alimentary system in shelled and non-shelled Sacoglossa (Mollusca, Opisthobranchia). Acta Zoologica 72: 143–150.

Jensen, K. R., 1999. Aggregative Behaviour in *Stiliger aureomarginatus* Jensen, 1993 (Mollusca, Opisthobranchia) in Rottnest Island, Western Australia. In Walker D. I., & F. E. Wells (eds), The Seagrass Flora and Fauna of Rottnest Island, Western Australia Western Australian Museum, Perth, 275–280.

Jousson, O., J. Pawlowski, L. Zaninetti, F. W. Zechman, F. Dini, G. Di Gujiseppe, R. Woodfield, A. Millar & A Meinesz, 2000. Invasive alga reaches California. Nature 408: 157–158.

Meinesz, A., T. Belsher, T. Thibaut, B. Antolic, K. Ben Mustapha, C.F. Boudouresque, D. Chiaverini, F. Cinelli, J. M. Cottalorda, A. Djellouli, A. El Abed, C. Orestano, A. M. Grau, L. Ivesa, A. H. Jaklin Langar, E. Massuti-Pascual, A. Peirano, A. Tunesi, J. De Vaugelas, N. Zavodnik & A. Zuljevic, 2001. The introduced green alga *Caulerpa taxifolia* continues to spread in the Mediterranean. Biological Invasions 3: 201–210.

Millar, A., 2004. New records of marine benthic algae from New South Wales, eastern Australia. Phycological Research 5: 117–128.

Trowbridge, C. D., 1991. Group membership facilitates feeding of the herbivorous sea slug *Placida dendritica*. Ecology 72: 2193–2203.

Trowbridge, C. D., 2004. Emerging associations on marine rocky shores: specialist herbivores on introduced macroalgae. Journal of Animal Ecology 73: 294–308.

Underwood, A. J., 1997. Experiments in Ecology: Their Logical Design and Interpretation Using Analysis of Variance. Cambridge University Press, Cambridge.

Williams, S. I. & D. I. Walker, 1999. Mesoherbivore-macroalgal interactions: Feeding ecology of sacoglosan sea slugs (Mollusca, Opisthobranchia) and their effects on their food algae. Oceanography and Marine Biology Annual Revie, 37: 87–128.

Winer, B. J., 1971. Statistical principles in experimental designs, 2nd edn. McGraw-Hill, New York.

Žuljevic, A., T. Thibaut, H. Elloukal & A. Meinesz, 2001. Sea slug disperses the invasive *Caulerpa taxifolia*. Journal of the Marine Biological Association of the United Kingdom 81: 343–344.

Hydrobiologia (2007) 580:259–263
DOI 10.1007/s10750-006-0446-z

BIODIVERSITY IN ENCLOSED SEAS

Species-specific probe, based on 18S rDNA sequence, could be used for identification of the mucilage producer microalga *Gonyaulax fragilis* (Dinophyta)

F. Tinti · L. Boni · R. Pistocchi · M. Riccardi · F. Guerrini

Abstract In the Adriatic Sea, the correlation between mucilage phenomena and the presence of *Gonyaulax fragilis* (Schütt) Kofoid (Dinophyta) has been recently demonstrated. The application of PCR-based methods and the development of species-specific molecular probes might represent powerful technologies for rapid and specific monitoring of microalgal species in seawater samples. Here, we report sequencing of the small subunit (SSU) ribosomal RNA gene (18S rDNA) of *G. fragilis* and its comparative analysis within the Dinophyta. Total DNAs were extracted and amplified from cultured cells of *G. fragilis*, which were isolated from natural phytoplanktonic association in the northern Adriatic Sea. Total 18S rDNA gene was amplified using 16S1N and 16S2N primers and sequenced using *ad hoc* designed internal primers. The primers amplified a product of expected size (length 1700/1800 bp). The phylogenetic analysis carried out by comparing *G. fragilis* sequence to homologous sequences of *Lingulodinium polyedrum* (Stein) Dodge, *Gonyaulax spinifera* (Claparède et Lachmann) Diesing, *Protoceratium reticulatum* (Claparède et Lachmann) Bütschli revealed a great nucleotide divergence of *G. fragilis* SSU sequence. Therefore, the SSU sequence could be used as species-specific marker for the identification of this mucilage producer microalga. In addition, such sequence could be used as target to design oligonucleotide probes for the construction of DNA microchips as diagnostic tool for the routine monitoring of harmful algae in seawater.

Keywords *Gonyaulax fragilis* · 18S rDNA · Mucilage · Adriatic Sea

Introduction

Mucilage formation represents a particular situation of uncertain origin appearing in marine environments with heavy consequences on human activities such as fishing, tourism, aquaculture etc. In the Adriatic Sea, the presence of gelatinous masses has been reported since 1729 and many accounts of them have appeared in both newspapers and scientific journals (Molin et al., 1992). In the summers of 1988 and 1989 (July and August respectively) the phenomenon affected a very extended area (Rinaldi et al., 1995), and in

Guest editors: G. Relini & J. Ryland
Biodiversity in Enclosed Seas and Artificial Marine Habitats

F. Tinti · L. Boni · R. Pistocchi · M. Riccardi (✉) · F. Guerrini
Centro Interdipartimentale di Ricerche sulle Scienze Ambientali (CIRSA), Alma Mater Studiorum University of Bologna , Via S. Alberto 163, 48100 Ravenna, Italy
e-mail: manuela.riccardi@unibo.it

summers 1991, 1997, 2000, 2001, 2002 similar, but less intense events have been observed. Observing the historical data (Molin et al., 1992), the phenomenon appeared time by time, probably linked to some particular environmental condition, especially phytoplankton distribution and abundance. It is anyway quite impossible to understand the phenomenon periodicity because only in recent years mucilage has become an object of complete research programmes, and field observations are now much more frequent and accurate than in the past. During many years, naturalists and fishermen attributed the production of mucilage to many different causes. Only at the end of XIX–beginning of XX century did the production of mucus in the sea start to be related to phytoplankton.

The mucilage phenomenon of the Adriatic Sea had been usually associated with extracellular organic matter production by phytoplankton, mainly diatoms, which are abundant in mucilage (Rinaldi et al., 1995) and known to produce extracellular polysaccharide substances (Myklestad, 1995; Myklestad et al., 1989); a relevant role of bacteria was also often underlined (Azam et al., 1999; Decho & Herndl, 1995; Herndl et al., 1999).

Alldredge et al. (1998) in California and Mackenzie et al. (2002) in New Zealand demonstrated the relevance of dinoflagellates in the extracellular production of mucilaginous matter. Pompei et al. (2003) and Pistocchi et al. (2005) showed that the mucilage of the Adriatic Sea could be related to the presence of the dinoflagellate *G. fragilis* in fact, the role played by this species was evidenced by in situ observations and laboratory experiments during many years. The correlation between the presence of *G. fragilis* and the appearance of the mucilage was difficult to demonstrate because cells in the samples are often broken, decomposed or so strongly aggregated that their microscope identification was very hard. On the contrary, the diatoms initially entrapped in the mucilage could survive and proliferate, appearing as mucilage producers. However it can be supposed that the diatoms are only growing in the mucus as on a bottom.

The application of PCR-based methods and the development of species-specific molecular probes might be powerful technologies for rapid and specific monitoring of microalgal species in mucilage and seawater samples (Edvardsen et al., 2003). In this report, we have sequenced a large part of the small subunit (SSU) ribosomal RNA gene (18S rDNA) of *G. fragilis.* The nucleotide sequence of *G. fragilis* was compared with those of closely related species of Dinophyta.

Materials and methods

Cultures

G. fragilis (Fig. 1) was isolated in June–July 2002 from mucilage samples collected in coastal waters of the north–western Adriatic Sea facing Emilia-Romagna, Italy. For the experimental work, we used seven monospecific cultures, obtained by isolating single cells with the micropipette method (McLachlan, 1973). The cultures were grown into sterile flasks with the medium GP (Loeblich & Smith, 1968), at 20°C under a 16:8 light-dark period at about 90 μmol m^{-2} s^{-1} from cool white lamps.

DNA extraction

20 ml of monospecific cultures were centrifuged at 6000 × *g* for 10 min and the pellet was resuspended in 420 μl of Nuclease-Free water; the DNA was extracted using the protocol described

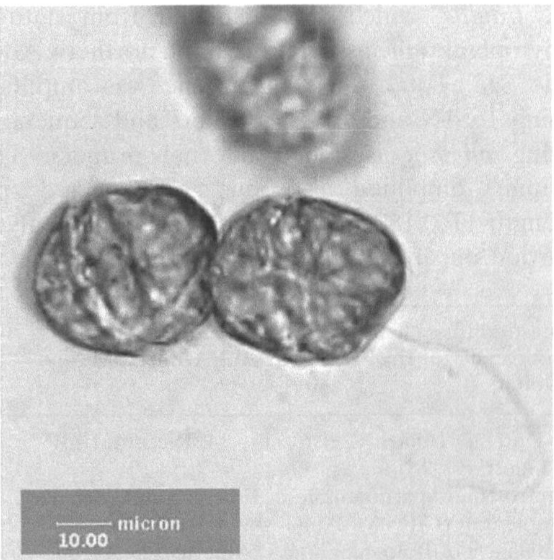

Fig. 1 *Gonyaulax fragilis* (Schütt) Kofoid

by Godhe et al., 2001. The quality and amount of the genomic DNA was controlled on an agarose TAE gel before PCR reaction.

Polymerase chain reaction amplification

Almost the entire SSU gene was specifically amplified using the 16S1N-16S2N primer pair (Grzebyk et al., 1998) and applying PCR conditions reported by Grzebyk et al. (1998), Godhe et al. (2001) and Edvardsen et al. (2003).

DNA sequencing

The PCR products were purified using the Exo-SAP-IT kit and sequenced with BigDye Terminator v1.1 cycle sequencing kit (Applied Biosystems). Nucleotide sequences were resolved on an ABI PRISM 310 Genetic Analyzer. The full-length sequence of the *G. fragilis* SSU was obtained using the PCR and internal primers. The internal forward and reverse primers were designed ad hoc with program Primer3 (Rozen & Skaletsky, 2000).

Sequence analysis

The SSU sequences of *G. fragilis* were analysed with ProSeq v2.91 (Filatov, 2002). A multiple alignment of the SSU sequences of *G. fragilis* and other species of Dinophyta (see below) was obtained using the ClustalX 1.83 (Thompson et al., 1997) program. Molecular systematics and phylogenetic analyses were carried out comparing the SSU sequence of *G. fragilis* to those of *L. polyedrum* AF274269, *G. spinifera* AF022155, *P. reticulatum* AF274273 retrieved from the GenBank. *L. polyedrum*, *G. spinifera* and *P. reticulatum* were often found together with *G. fragilis* in the phytoplanktonic associations in the Adriatic Sea. In these analyses we applied the Tamura–Nei distance model and the Neighbor-Joining method of reconstruction implemented in Mega 2.1 program (Kumar et al., 2001).

Results and discussion

The primer pair used (16S1N/16S2N) amplified DNA fragments of expected size (1,700 bp) from all cultured cells of *G. fragilis* (Fig. 2).

Results from BLAST searches into the GenBank databases confirmed that obtained sequences were homologous to ribosomal genes of dinoflagellates. The sequence of SSU of *G. fragilis* was deposited into the GenBank with the accession number AY672702.

The sequences obtained from seven monoculture isolates of *G. fragilis* showed very low nucleotide variation (0.41%).

The comparative sequence analysis within the selected species of Dinophyta revealed that seven polymorphic regions at the between-species level could be identified. These regions were of variable length (70/200 bp) and their genetic variation ranged from 43.5% to 60%. Within each region, the *G. fragilis* SSU sequence showed from 2 to 17 species-specific nucleotide positions (Fig. 3). This species-specific variation allowed consideration of such fragments as powerful regions to identify *G. fragilis* species-specific markers.

Although the Tamura–Nei distances (Table 1) showed a great and comparable divergence of the SSU sequences among the species, the Neighbor-Joining phylogenetic tree (Fig. 4) revealed a clustering of species highly similar to previously presented phylogenies inferred with SSU sequence or LSU partial sequence (Daugbjerg et al., 2000; Saldarriaga et al., 2001; Edvardsen et al., 2003). Now new phylogenetic data are available for *G. fragilis*, which formed with *G. spinifera* a clade highly supported by bootstrap analysis (92%). These species then clustered to *P. reticulatum* into a common clade with 90% bootstrap support.

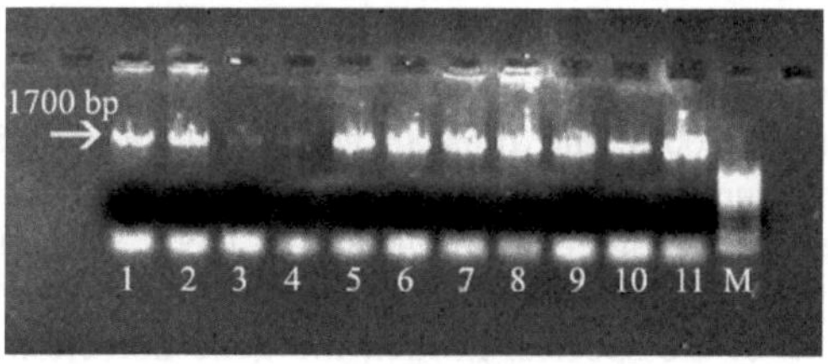

Fig. 2 Electrophoresis of the amplified SSU gene fragment (~1700 bp) from monocultural isolates of *G. fragilis* (lanes 1–11) and molecular marker (lane M)

Fig. 3 Diagram representing the variable SSU gene regions with Range, Genetic Variation (GV) and Species-Specific Nucleotide Position (SSNP) among the Dinophyta species here analysed (*G. fragilis*)

Region	Range	G.V.	SSNP
1	61-134	54.8%	6
2	174-310	56%	13
3	650-763	51.3%	17
4	862-920	60%	16
5	1065-1111	43.5%	2
6	1374-1411	51.3%	4
7	1513-1571	56%	7

Table 1 Tamura–Nei distances (below diagonal) and S.E. (above diagonal) among Dinophyta species

		1	2	3	4	5
[1] *Sarcocystis muris*	1		0.013	0.011	0.012	0.014
[2] *Gonyaulax fragilis*	2	0,221		0.011	0.011	0.012
[3] *Lingulodinium polyedrum*	3	0,180	0,168		0.009	0.013
[4] *Protoceratium reticulatum*	4	0,186	0,156	0,131		0.012
[5] *Gonyaulax spinifera*	5	0,269	0,214	0,209	0,199	

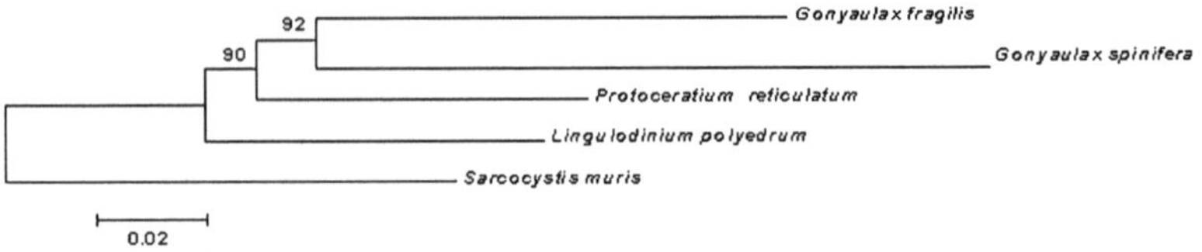

Fig. 4 Phylogenetic Neighbor-Joining tree illustrating the relationships and genetic divergence among the Dinophyta species; *S. muris* was used as outgroup species

Conclusion

Several studies (Godhe et al., 2001; Galluzzi et al., 2004) showed that PCR-based methods would be very useful for detection of microalgae in natural environments and are particular important when dealing with harmful algae.

The successful DNA sequencing of *G. fragilis* strains could be very interesting for its future application. As we wrote in the introduction, this species is hardly recognizable in the mucilage due to its fragility, but the presence of low cell numbers was usually noticed about 20 days before the appearance of mucilage (Pompei et al., 2003) and thus an early detection of its presence in the water has great importance in respect of practical applications. DNA sequencing may allow a more certain and easy observation of the species. In fact, the gene detection is the first step to develop species-specific probes which could be a very useful tool to identify *G. fragilis* cells in natural samples, especially when their number is very low. In such a way it could be possible to notice the presence of this harmful species in the sea as an early warning of the appearance of the mucilage phenomenon.

In addition, such sequences could be used as targets to design oligonucleotide probes for the construction of DNA microchips (Metfies & Medlin, 2004) as diagnostic tool for the routine monitoring of harmful algae in seawater.

References

Alldredge, A. L, U. Passow & S. H. D. Haddock, 1998. The characteristics and transparent exopolymer (TEP) content of marine snow from thecate dinoflagellates. Journal of Plankton Research 20: 393–406.

Azam, F., S. Fonda Umani & E. Funari, 1999. Significance of bacteria in the mucilage phenomenon in the Northern Adriatic Sea. Annali Istituto Superiore Sanità 35: 411–419.

Daugbjerg, N., G. Hansen J. Larsen & Ø. Moestrup, 2000. Phylogeny of some of the major genera of dinoflagellates based on ultrastructure and partial LSU rDNA sequence data, including the erection of three new genera of unarmoured dinoflagellates. Phycologia 39: 302–317.

Decho, A. W. & G. J. Herndl, 1995. Microbial activities and the transformation of organic matter within mucilaginous material. Science of the Total Environment 165: 33–42.

Edvardsen, B., K. Shalchian-Tabrizi, K. S. Jakobsen, L. K. Medlin, E. Dahl, S. Brubak & E. Paasche, 2003. Genetic variability and molecular phylogeny of *Dinophysis* species (Dinophyceae) from norwegian waters inferred from single cell analyses of rDNA. Journal of Phycology 39: 395–408.

Filatov, D. A., 2002. ProSeq: a software for preparation and evolutionary analysis of DNA sequence data sets. Molecular Ecology Notes 2: 621–624.

Galluzzi, L., A. Penna, E. Bertozzini, M. Vila, E. Garces & M. Magnani, 2004. Development of a real-time PCR assay for rapid detection and quantification of *Alexandrium minutum* (a Dinoflagellate). Applied Environmental Microbiology 70: 1199–1206.

Godhe, A., S. K. Otta, A. S. Rehnstam-Holm, I. Karunasagar & I. Karunasagar, 2001. Polymerase chain reaction in detection of *Gymnodinium mikimotoi* and *Alexandrium minutum* in field samples from Southwest India. Marine Biotechnology 3: 152–162.

Grzebyk, D., Y. Sako & B. Berland, 1998. Phylogenetic analysis of nine species of *Prorocentrum* (Dinophyceae) inferred from 18S ribosomal DNA sequences, morphological comparisons, and description of *Prorocentrum panamensis*, sp.nov. Journal of Phycology 34: 1055–1068.

Herndl, G. J., J. M. Arrietta & K. Stoderegger, 1999. Interaction between specific hydrological and microbial activity leading to extensive mucilage formation in the Northern Adriatic Sea. Annali Istituto Superiore Sanità 35: 405–409.

Kumar, S., K. Tamura, I. B. Jakobsen & M. Nei, 2001. MEGA2: molecular Evolutionary Genetics Analysis software. Bioinformatics 17(12): 1244–1245.

Loeblich, A. R. & V. E. Smith, 1968. Chloroplast pigments of the marine dinoflagellate *Gyrodinium resplendens*. Lipids 3: 5–13.

MacKenzie, L., I. Sims, V. Beuzenberg & P. Gillespie, 2002. Mass accumulation of mucilage caused by dinoflagellate polysaccharide exudates in Tasman Bay, New Zealand. Harmful Algae 1: 69–83.

McLachlan, J., 1973. Culture methods and growth measurements. In Stein, J. R. (ed.), Handbook of Phycological Methods. Cambridge University Press, New York, 25–51.

Metfies, K. & L. K. Medlin, 2004. DNA microchips for phytoplankton: the fluorescent wave of the future. Nova Hedwigia 79: 321–327.

Molin, D., E. Guidoboni & A. Lodovisi, 1992. Mucilage and the phenomena of algae in the history of the Adriatic: periodicization and the anthropic context (17–20th centuries). In Vollenweider, R. A., R. Marchetti & R. Viviani (eds), Marine Coastal Eutrophication. Elsevier, Amsterdam, 511–524.

Myklestad, S. M., 1995. Release of extracellular products by phytoplankton with special emphasis on polysaccharides. Science of the Total Environment 165: 155–164.

Myklestad, S. M., O. Holm-Hansen, K. M. Vårum & B. E. Volcani, 1989. Rate of release of extracellular amino acids and carbohydrates from the marine diatom *Chaetoceros affinis*. Journal of Plankton Research 11: 763–73.

Pistocchi R., M. Cangini, C. Totti, R. Urbani, F. Guerrini, T. Romagnoli, P. Sist, S. Palamidesi, L. Boni & M. Pompei, 2005. Relevance of the dinoflagellate *Gonyaulax fragilis* in mucilage formations of the Adriatic Sea. Science of the Total Environment 353: 307–316.

Pompei, M., C. Mazziotti, F. Guerrini, M. Cangini, S. Pigozzi, M. Benzi, S. Palamidesi, L. Boni & R. Pistocchi, 2003. Correlation between the presence of *Gonyaulax fragilis* (Dinophyceae) and the mucilage phenomena of the Emilia-Romagna coast (northern Adriatic Sea). Harmful Algae 2: 301–316.

Rinaldi, A., R. A. Vollenweider, G. Montanari, C. R. Ferrari & A. Ghetti, 1995. Mucilages in Italian Seas: the Adriatic and Tyrrhenian Seas, 1988–1991. Science of the Total Environment 165: 165–183.

Rozen, S. & H. J. Skaletsky, 2000. Primer3 on the WWW for general users and for biologist programmers. In Krawetz, S. & S. Misener (eds), Bioinformatics Methods and Protocols: Methods in Molecular Biology. Humana Press, Totowa, NJ, 365–386.

Saldarriaga, J. F., F. J. R. Taylor, P. J. Keeling & T. Cavalier-Smith, 2001. Dinoflagellate nuclear SSU rRNA phylogeny suggests multiple plastid losses and replacements. Journal of Molecular Evolution 53: 204–213.

Thompson, J. D., T. J. Gibson, F. Plewniak, F. Jeanmougin & D. G. Higgins, 1997. The ClustalX windows interface: flexible strategies for multiple sequence alignment aided by quality analysis tools. Nucleic Acids Research 25: 4876–4882.

Hydrobiologia (2007) 580:265–271
DOI 10.1007/s10750-006-0445-0

Assemblages in a submarine canyon: influence of depth and time

A. Sabatini · M. C. Follesa · I. Locci · A. A. Pendugiu · P. Pesci · A. Cau

Abstract In this work we have studied the assemblages from Quirra canyon (Sardinia). We analysed data from 25 trawl samples from the canyon, made at different times and depth. A total of 71 demersal species (38 teleosts, 5 cartilaginous fishes, 13 molluscs and 15 crustaceans) were examined. We found four groups using cluster analysis; each group is characterised by a certain depth or time. In particular we have shown that the time of the day appears to have a role in the movement of shrimps and other species. This phenomenon seems to be linked to trophic need and by the consequent different food availability into the Canyon.

Keywords Cluster analysis · Multidimensional scaling · Submarine canyon · Fish assemblages · Depth · Mediterranean sea

Guest editors: G. Relini & J. Ryland
Biodiversity in Enclosed Seas and Artificial Marine Habitats

Electronic supplementary material Supplementary material is available for this article at http://dx.doi.org/10.1007/s10750-006-0445-0 and accessible for authorised users

A. Sabatini (✉) · M. C. Follesa · I. Locci · A. A. Pendugiu · P. Pesci · A. Cau
D.B.A.E., University of Cagliari, viale Poetto, 1-09126 Cagliari, Italy
e-mail: asabati@unica.it

Introduction

Submarine canyons are areas rich in nutrients, due to both their geomorphology and their material and hydrographic flows; they are areas with strong turbidity currents that cause morphology changes (Shepard et al., 1974).

Although less deep than in the Atlantic Ocean, submarine canyons are also present in the Mediterranean Sea. In the seas around Sardinia a number of submarine canyons with highly variable morphology divide the upper slope, mainly along the eastern coast (Western part of Central Tyrrhenian Sea). One of these is Quirra Canyon; it slopes steeply down from a depth of 120 m to 1400 m in only 5 nautical miles, perpendicular to the coast. Fishing grounds are very irregular in this canyon, because of its morphology.

Submarine canyons are complex environments, described as unstable (Thorne-Miller & Catena, 1991) and species that live there have greater mobility than those of typical deep-sea assemblages (Rowe, 1971). Their influence on the movements of a number of species has been widely studied (Tudela et al., 2003; Sardà et al., 1997; Tursi et al., 1996, among many), but information concerning day-night movements in submarine canyons is scarce.

In this work we have studied the structure of demersal assemblages in Quirra Canyon. This canyon has not been widely studied till now and there are no works about its macro-faunal communities. This paper represents the first description of its demersal assemblages. We have also determined the diversity of assemblages related to depth and time and highlined day-night movements of the species which live in the canyon.

Materials and methods

The present work is based on data collected between April 1996 and October 1997 in the Quirra Canyon. This canyon is located in South-Eastern Sardinia (Western part of Central Tyrrhenian Sea) (Fig. 1) and is surrounded by a continental slope characterized by a marked variation in depth.

We analysed 25 samples, taken at different hours, between depths of 160 and 600 m with an 'Italian type' traw used by local fishermen; the hauls lasted between 30 and 100 min. We classified each haul according to time and depth; those made during daylight hours (from about 7 a.m. to 6 p.m.) were classified 'L' ('light'), those during the first part of the night (about 6 p.m. to 1 a.m.) were 'N1' ('night 1') while those made in the second part of the night (1 to 7 a.m.) were classified 'N2' ('night 2'). In terms of depth, 'C' is the code for hauls made down to 350 m, 'D' for those between 350 m and 500 m, and 'E' for hauls beyond 500 m.

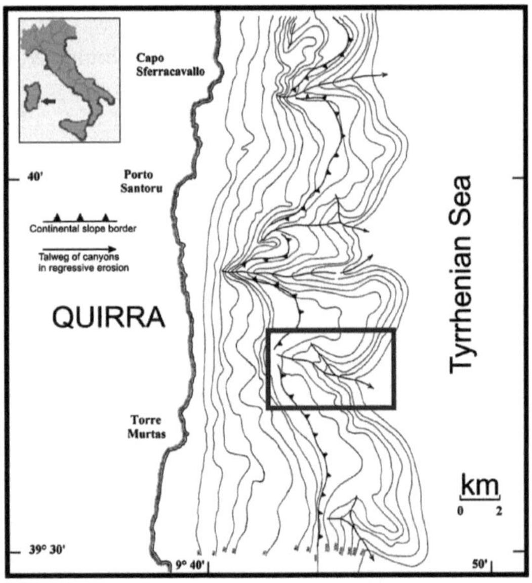

Fig. 1 Map of study area: rectangle indicates Quirra Canyon; lines represent depth curves; arrow in upper left rectangle indicates the geographical position of Quirra Canyon

All species caught in each haul were counted and weighed; these data were used to obtain biomass indexes standardised per hours of haul. Data were checked to reduce the potentially confusing effects of rare species and occasional catches of pelagic ones (Biagi et al., 2002). We accepted for the analysis only the species with an abundance of at least 10 individuals and that had occurred in a minimum of 2 hauls. In consequence, 71 species were selected for the cluster analysis.

Cluster analysis was carried out using biomass data for each species caught, according to depth (C, D and E) and time of day (L, N1 and N2). The data was 'root-root' transformed (Field et al., 1982) and subsequently the similarity between hauls was calculated using the Bray-Curtis measure (Bray & Curtis, 1957; Field et al., 1982). Samples were classified by hierarchical agglomerative cluster analysis using the group average linking method. To show the percentage contribution of each taxon to the average dissimilarity between samples of the various groups' pair combinations, we used the SIMPER (similarity percentage) analysis (Clarke, 1993; Clarke & Warwick, 1994). This procedure indicates the average contribution (%) of each species to the similarity (typifying species) and dissimilarity (discriminating species) between groups of samples. Ordination of samples was then made using multidimensional scaling techniques (MDS) (Kruskal, 1964). MDS was applied to the distances matrix generated by clustering, in order to obtain a 2-dimensional representation of hauls.

Statistical analyses were all performed using the PRIMER package (release 4.0) (Clarke & Warwick, 1994).

Results

About 12 of the 25 analysed hauls were performed in daylight, 10 during the first part of the night (N1) and 3 in the second (N2). In terms of depth range, the hauls were made between 166 m and 639 m. Specifically, 14 hauls were made in the C stratum (<350 m), 2 in the D (between 350and 500 m deep) and 9 in the E stratum (>500 m) (Table 1).

We analysed 71 species, 38 of which were teleosts, 5 cartilaginous fishes, 13 molluscs and 15 crustaceans. Most species were caught at only one particular depths, both during the day and at night, revealing their typical depth range in the Quirra Canyon (see Electronic Supplementary Material). Species such as *Aspitrigla cuculus* or *Mullus barbatus* were found in the canyon only at a depth of less than 350 m throughout the whole day. Another group is formed by those species captured beyond 500 m; they were never, at any time, found at different depth ranges (for example *Paromola cuvieri* and *Polycheles typhlops*). The other two groups identified are characterized by a wider extension: they were caught at depth up to 500 m (such as *Capros aper* or *Sepia orbignyana*) or from 350 to over 500 m (like *Stomias boa* or *Epigonus telescopus*). None of these were ever found at different depth ranges.

Moreover, there are also ubiquitous species; these were caught at all the depths examined, both during the day and at night, for example *Phycis blennoides* and *Helicolenus dactylopterus*.

More interesting, however, are those species which, although spending their lives at a specific depth range, move at a certain time of the day. In Canyon Quirra we found 9 species (including the giant red and rose shrimps, *Aristaeomorpha foliacea* and *Aristeus antennatus*) which, during the first part of the night only, are found in the C stratum, moving from a depth of more than 500 m; in the second part of the night, we find these species at depths of more than 500 m, therefore returning to their usual depth range. Furthermore, we observed that two species (*Lophius budegassa* and *Pasiphaea multidentata*) are found down to 350 m all through the day and go deeper only in the night, to all appearances moving in the opposite direction to the previous group.

To evaluate what relation there was, in terms of depth and time of day, between the analysed hauls, we have used cluster analysis. We found four main groups (Fig. 2). The first two are composed of the hauls made in the Canyon at depths of less than 350 m: the first group (1) comprises the hauls made during the first part of the night only (N1), in which the absence of hauls made during daylight or in the second part of the night should be noted; the second (2) is made up of hauls made during the second part of the night and in daylight. The third group (3) brings together the hauls made at depths of over 500 m irrespective of time of day, while the fourth groups four hauls (4), two made in daylight, in the D and E strata (G-D-9 and G-E-21), and two others made during the first part of the night (N1) at lesser depths (N1-C-17 and N1-D-14). The four groups gather together hauls made in different seasons, so that seasonality doesn't seem to be relevant in this analysis. The ordination of the 25 hauls with MDS was in agreement with the clustering, showing the same aggregation of samplings previously observed.

SIMPER analysis revealed that the main indicator species for the first group is *Merluccius merluccius* (7.24%), for the second *Glossanodon leioglossus* (13.44%), for the third *A. antennatus* (8.84%) and for the fourth *Chlorophthalmus agassizi* (10.24%). The discriminating species between group 2 and all the others is *G. leioglossus* (with% variable from 5.76, group 2 versus group 1, to 7.24, group 2 versus group 4) (Table 2). For group 4, the discriminating species is *C. agassizi*, with 5.99% versus group 1 and 6.53% versus group 3. For the third group versus group 1 the discriminating species is *G. leioglossus* (4.86%); for this group only, the discriminating species does not coincide with that of the indicator. The analysis also shows that about 50% of the contribution to the total dissimilarity is accounted for by a minimum of 16 to a maximum of 20 species.

Discussion and conclusion

Submarine canyons are very different areas from the typical continental slope (Haedrich et al., 1975, 1980). They generally have higher productivity and a regular influx of organic matter that influences the life of species living in. It has been noted that submarine canyons in the Mediterranean

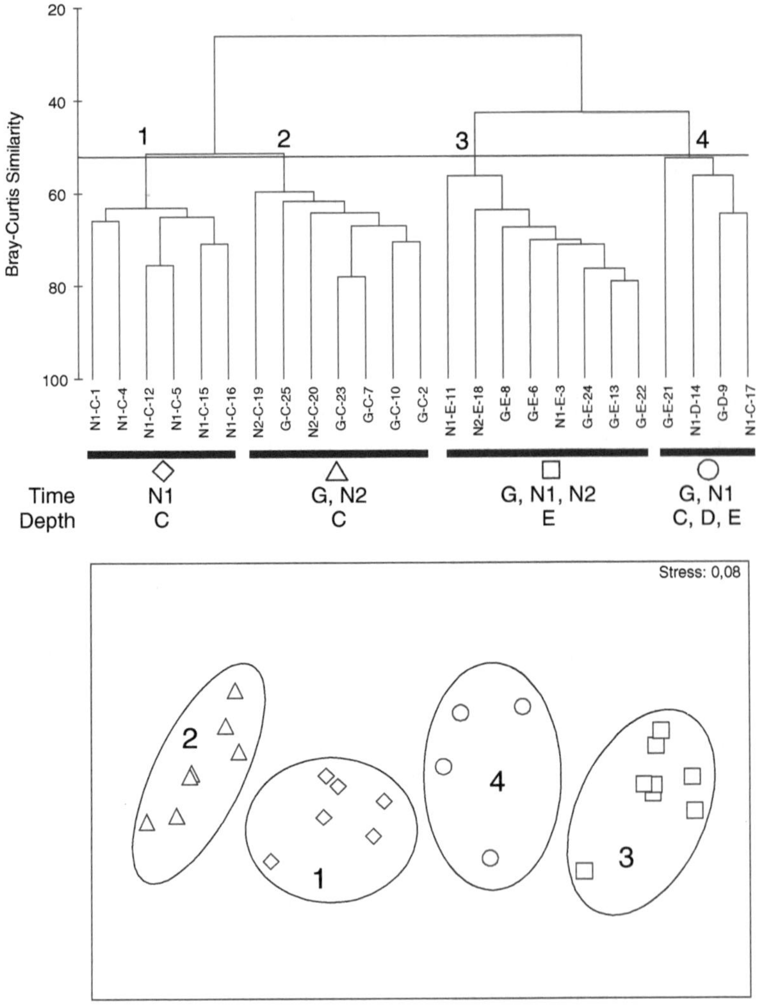

Fig. 2 Classification (cluster analysis) and ordination (multidimensional scaling ordination analysis) of species assemblages in Quirra Canyon

Sea (Stefanescu et al., 1994) exert an influence on fish assemblages, just as in the Atlantic Ocean (Haedrich et al., 1975); we studied this phenomenon in Quirra Canyon.

The analysis carried out in this paper clearly distinguishes between the fauna of the deeper waters (E stratum) and that living at less depths (C stratum). Each association appears to be characterized by an indicator species which typifies the group. The first group is characterised by hauls made in the C stratum during the first part of the night only; indicator species for this group appeared to be *M. merluccius*. In fact, we found this species at all the depths analysed but, above all, in the C stratum. As regards the second group, the indicator species is *G. leioglossus*. This group brings together the hauls made at depths of <350 m during the day only or in the second part of the night. This species was found in the C stratum at all hours of testing. The third group gathers together the hauls made at greater depths, in the E stratum. The species characterising this group are those which typically live at greater depths, hence the indicator species for this group is *A. antennatus*. The fourth group collects together different kind of hauls, some made in the E stratum during the day and others made at lesser depths during the night only. The indicator species for this group is *C. agassizi*. Other species that characterize this group are *A. antennatus* and

Table 1 Code, date, season, start and end time (solar time), last and depth for each hauls studied

Code	Date	Season	Start sampling	End sampling	Last (minutes)	Depth (m)	Belonging group	
N1-C-1	22/04/1996	Spring	21,48	22,28	40	198	N1	C
G-C-2	08/05/1996	Spring	16,05	17,30	85	206	G	C
N1-E-3	08/05/1996	Spring	18,50	19,50	60	595	N1	E
N1-C-4	08/05/1996	Spring	22,00	23,18	78	212	N1	C
N1-C-5	29/07/1996	Summer	0,01	1,03	62	200	N1	C
G-E-6	29/07/1996	Summer	17,30	18,28	58	590	G	E
G-C-7	29/07/1996	Summer	19,36	20,07	31	170	G	C
G-E-8	27/08/1996	Summer	13,55	15,10	75	586	G	E
G-D-9	27/08/1996	Summer	16,15	17,10	55	420	G	D
G-C-10	27/08/1996	Summer	18,05	19,20	75	231	G	C
N1-E-11	27/08/1996	Summer	20,30	21,45	75	587	N1	E
N1-C-12	27/08/1996	Summer	22,50	23,59	69	208	N1	C
G-E-13	05/11/1996	Autumn	13,35	14,34	59	583	G	E
N1-D-14	11/02/1997	Winter	19,55	20,31	36	411	N1	D
N1-C-15	11/02/1997	Winter	21,24	22,30	66	205	N1	C
N1-C-16	11/02/1997	Winter	23,26	24,10	44	166	N1	C
N1-C-17	20/05/1997	Spring	22,34	24,00	86	308	N1	C
N2-E-18	21/05/1997	Spring	0,56	1,50	54	599	N2	E
N2-C-19	21/05/1997	Spring	3,18	4,05	47	173	N2	C
N2-C-20	21/05/1997	Spring	4,50	6,30	100	198	N2	C
G-E-21	21/05/1997	Spring	7,42	9,00	78	519	G	E
G-E-22	24/06/1997	Summer	8,25	9,25	60	580	G	E
G-C-23	24/06/1997	Summer	10,38	11,08	30	188	G	C
G-E-24	13/10/1997	Autumn	8,50	10,02	72	639	G	E
G-C-25	13/10/1997	Autumn	11,25	12,05	40	247	G	C

A. foliacea. Movement patterns for these species have been examined by other authors (Tudela et al., 2003; Sardà et al., 1997; Tursi et al., 1996) who have noted the ability of deep-water shrimps to adapt their life cycle to the morphology of a canyon. In this work we have shown that shrimps and other species inside Quirra Canyon move during a day-night cycle, probably with light as a parameter indirectly related to their activity, as proposed by Tobar & Sardà (1992). These species are mainly predators (Gristina et al., 1992; Fischer et al., 1987) moving during the night to catch prey. They find the canyon a suitable place for finding food, in which they can reach different depths in a relative short distance; this is what probably happens in Quirra Canyon.

In fact, this Canyon is characterised by a regressive erosion (Palomba & Ulzega, 1984) that is gradually modifying its morphology, bringing the limit of its borders closer and closer to the coast. A great movement of materials inside Quirra Canyon makes it, in a certain way, an "active canyon". This probably causes transportation of sediments rich in organic matter from the shelf to deeper regions, and even diel migratory plankton from upper levels, in the same way seen for other canyons (Macquart-Moulin & Patriti, 1993).

This phenomenon, together with the fact that in a canyon it is possible to reach different depths with minimum distances, determines the variations in species assemblages with time and depth observed in this paper. Observed movements do not seem to be primarily linked to seasonality in Quirra Canyon, although a number of nocturnal movements have been attributed primarily to hydrographical phenomenon due to seasons (Sardà et al., 1997; Cartes et al., 1994; Sardà et al., 1994; Cau & Deiana, 1982; Bombace, 1975, Maurin, 1960). The results of our work show that movements of species inside the Canyon, and so the different species' compositions observed, seems to be linked to a day-night cycle, and probably by the consequent different food availability, as suggested by Cau & Deiana (1982) and Stefanescu et al. (1994).

Table 2 Between group comparison, results of SIMPER analysis

Species	Group	Group	Cum. %	Species	Group	Group	Cum. %
Av. diss. = 48.53	2	1		Av. diss = 52.17	4	1	
Glossanodon leioglossus	2.73	1.58	5.76	*Chlorophthalmus agassizi*	1.88	0.30	5.99
Aristeus antennatus	0.00	1.17	10.67	*Raja clavata*	0.11	1.16	9.91
Aristaeomorpha foliacea	0.00	1.16	15.48	*Glossanodon leioglossus*	0.77	1.58	13.47
Zeus faber	0.99	0.29	19.09	*Galeus melastomus*	1.03	0.33	16.77
Plesionika edwardsii	0.00	0.8	22.46	*Aspitrigla cuculus*	0.00	0.83	19.92
Gnathophis mystax	0.18	0.77	25.5	*Gnathophis mystax*	0.00	0.77	22.82
Mullus barbatus	0.74	0.18	28.24	*Capros aper*	0.76	0.63	25.69
Raja clavata	1.69	1.16	30.93	*Merluccius merluccius*	0.66	1.36	28.45
Calappa granulata	0.00	0.61	33.49	*Plesionika martia*	0.80	0.10	31.11
Dipturus oxyrinchus	0.37	0.57	35.97	*Scyliorhinus canicula*	0.94	0.93	33.67
Scyliorhinus canicula	1.49	0.93	38.29	*Nephrops norvegicus*	0.75	0.06	36.22
Trigla lyra	0.74	0.65	40.47	*Dipturus oxyrinchus*	0.48	0.57	38.65
Lepidorhombus boscii	0.21	0.57	42.53	*Trigla lyra*	0.16	0.65	40.91
Argentina sphyraena	0.32	0.61	44.53	*Plesionika edwardsii*	0.51	0.80	43.12
Loligo forbesi	0.19	0.42	46.52	*Gadiculus argenteus*	0.91	0.47	45.20
Scaeurgus unicirrhus	0.52	0.07	48.49	*Micromesistius poutassou*	0.69	0.41	47.25
Lepidotrigla cavillone	0.41	0.37	50.45	*Macrorhamphosus scolopax*	0.00	0.54	49.29
Av. diss = 88.1	2	3		Av. diss = 69.34	2	4	
Glossanodon leioglossus	2.73	0.09	7.24	*Glossanodon leioglossus*	2.73	0.77	5.83
Raja clavata	1.69	0.05	11.79	*Chlorophthalmus agassizi*	0.09	1.88	11.20
Scyliorhinus canicula	1.49	0.00	15.89	*Raja clavata*	1.69	0.11	15.95
Galeus melastomus	0.00	1.30	19.49	*Galeus melastomus*	0.00	1.03	19.13
Aristeus antennatus	0.00	1.28	23.04	*Zeus faber*	0.99	0.00	22.10
Phycis blennoides	0.08	1.12	25.90	*Aspitrigla cuculus*	0.93	0.00	24.92
Zeus faber	0.99	0.00	28.63	*Gadiculus argenteus*	0.00	0.91	27.64
Aristaeomorpha foliacea	0.00	0.97	31.34	*Aristaeomorpha foliacea*	0.00	0.90	30.35
Aspitrigla cuculus	0.93	0.00	33.92	*Scyliorhinus canicula*	1.49	0.94	32.92
Lophius budegassa	1.02	0.11	36.45	*Aristeus antennatus*	0.00	0.85	35.49
Illex coindetii	1.00	0.22	38.69	*Coelorhynchus coelorhynchus*	0.00	0.84	38.02
Capros aper	0.81	0.00	40.91	*Merluccius merluccius*	1.46	0.66	40.47
Macrorhamphosus scolopax	0.80	0.00	43.11	*Macrorhamphosus scolopax*	0.80	0.00	42.87
Lampanyctus crocodilus	0.00	0.75	45.18	*Plesionika martia*	0.00	0.80	45.25
Sepia orbignyana	0.75	0.00	47.23	*Phycis blennoides*	0.08	0.85	47.56
Eledone cirrhosa	0.79	0.07	49.27	*Capros aper*	0.81	0.76	49.84
Av. diss = 57.25	4	3		Av. diss = 70.76	3	1	
Chlorophthalmus agassizi	1.88	0.33	6.53	*Glossanodon leioglossus*	0.09	1.58	4.86
Scyliorhinus canicula	0.94	0.00	10.29	*Raja clavata*	0.05	1.16	8.40
Gadiculus argenteus	0.91	0.00	14.02	*Galeus melastomus*	1.30	0.33	11.78
Capros aper	0.76	0.00	17.23	*Scyliorhinus canicula*	0.00	0.93	14.78
Glossanodon leioglossus	0.77	0.09	20.32	*Aspitrigla cuculus*	0.00	0.83	17.46
Lampanyctus crocodilus	0.00	0.75	23.41	*Plesionika edwardsii*	0.00	0.80	20.06
Lepidorhombus whiffiagonis	0.73	0.07	26.22	*Lophius budegassa*	0.11	0.89	22.62
Galeus melastomus	1.03	1.30	28.95	*Gnathophis mystax*	0.00	0.77	25.09
Eledone cirrhosa	0.73	0.07	31.67	*Phycis blennoides*	1.12	0.37	27.48
Argentina sphyraena	0.59	0.00	34.18	*Etmopterus spinax*	0.72	0.00	29.79
Etmopterus spinax	0.16	0.72	36.63	*Lepidorhombus whiffiagonis*	0.07	0.79	32.09
Illex coindetii	0.78	0.22	39.04	*Lampanyctus crocodilus*	0.75	0.07	34.29
Merluccius merluccius	0.66	0.84	41.35	*Trigla lyra*	0.00	0.65	36.38
Helicolenus dactylopterus	0.66	0.64	43.55	*Capros aper*	0.00	0.63	38.40
Micromesistius poutassou	0.69	0.46	45.74	*Calappa granulata*	0.00	0.61	40.37
Plesionika edwardsii	0.51	0.00	47.91	*Argentina sphyraena*	0.00	0.61	42.34
Paromola cuvieri	0.00	0.52	50.01	*Plesionika martia*	0.66	0.10	44.23
				Nephrops norvegicus	0.63	0.06	46.12
				Hoplostethus mediterraneus	0.59	0.00	48.01
				Merluccius merluccius	0.84	1.36	49.89

Acknowledgements The authors wish to thank prof. F. Bertolino for his suggestions and help with statistical analyses, and the two anonymous referee for their valuable comments.

References

Biagi, F., P. Sartor, G. D. Ardizzone, P. Belcari, A. Belluscio & F. Serena, 2002. Analysis of demersal assemblages off the Tuscany and Latium Coasts (nort-western Mediterranean). Scientia Marina 66: 233–242.

Bombace, G., 1975. Considerazioni sulla distribuzione delle popolazioni di livello batiale con particolare riferimento a quelle bentonectoniche. Pubblicazioni della Stazione Zoologica di Napoli 39: 7–21.

Bray, J. R. & J. T. Curtis, 1957. An ordination of the upland forest communities of southern Wisconsin. Ecological Monography 27: 325–349.

Cartes, J. E., J. B. Company & F. Maynou, 1994. Deep-water decapod crustacean communities in the North-western Mediterranean: influence of submarine canyons and season. Marine Biology 120: 221–229.

Cau, A. & Deiana A. M., 1982. Sulle variazioni di cattura del gambero rosso "*Aristaeomorpha foliacea*" in relazione alla sua eco-etologia. Bollettino dei Musei e degli Istituti Biologici dell'Università di Genova 50: 145–150.

Clarke, K. R., 1993. Non-parametric multivariate analyses of change in community structure. Australian Journal of Ecology 18: 117–143.

Clarke, K. R. & R. M. Warwick, 1994. Change in marine communities: an approach to statistical analysis and interpretation. Natural Environmental Research Council - Plymouth Marine Laboratory, UK.

Field, J. G., K. R. Clarke & R. M. Warwick, 1982. A practical strategy for analysing multispecies distribution patterns. Marine Ecology Progress Series 8: 37–52.

Fischer, W., M. L. Bauchot, M. Schneider (rédacteurs) 1987 Fiches FAO d'identification des espèces pour les besoins de la pêche (Révision 1). Méditerranée et mer Noire. Zone de pêche 37. Vol. II. Rome, FAO, 2: 761–1530.

Gristina, M., F. Badalamenti, G. Barbera, G. D'Anna & C. Pipitone, 1992. Preliminary data on feeling habitus of Aristaeomorpha foliacea (Risso) in the Sicilian Channel. Oebalia suppl XVII: 143–144.

Haedric, R. L., G. T. Rowe & P. T. Polloni, 1975. Zonation and faunal composition of epibenthic populations on the continental slope south of New England. Journal of Marine Research 33: 191–212.

Haedric, R. L., G. T. Rowe & P. T. Polloni, 1980. The megabenthic fauna in the deep sea south of New England, USA. Marine Biology 57: 165–179.

Kruskal, J. B., 1964. Multidimensional scaling by optimising goodness of fit to a non-metric hypothesis. Psychometrike 29: 1–27.

Macquart-Moulin, C. & G. Patriti, 1993. Canyons sous-marins et advection vers le talus continental du plankton néritique. Oceanologica Acta 16: 179–189.

Maurin, C., 1960. Les crevettes profondes du littoral français de la Mèditerranée répartition selon la profondeur. Notes biométriques. Rapports et Proces-Verbaux des Reunions CIESM 5: 147–154.

Palomba, M. & A. Ulzega, 1984. Geomorfologia dei depositi quaternari del Rio Quirra e della piattaforma continentale antistante (Sardegna Orientale). Rendiconti del Seminario della Facoltà di Scienze 54: 109–121.

Rowe, G. T., 1971. Observations in bottom currents and epibenthic populations in Hatteras Submarine Canyon. Deep-Sea Research 18: 569–581.

Sardà, F., J. E. Cartes & J. B. Company, 1994. Spatio-temporal variations in megabenthos abundance in three different habitats of the Catalan deep-sea (Western Mediterranean). Marine Biology 120: 211–219.

Sardà, F., F. Maynou & Ll. Tallò, 1997. Seasonal and spatial mobility patterns of rose shrimp (*Aristeus antennatus* Risso, 1816) in the western Mediterranean: results of a long-term study. Marine Ecology Progress. Series 159: 133–141.

Shepard, F. P., N. F. Marshall & P. A. McLoughlin, 1974. Current in submarine canyons. Deep-Sea Research 21: 691–706.

Stefanescu, C., B. Morales-Nin & E. Massuti, 1994. Fish assemblages on the slope in the Catalan Sea (Western Mediterranean): influence of a submarine canyon. Journal of the Marine Biological Association of the United Kingdom 74: 499–512.

Thorne-Miller, B. & J. Catena, 1991. The living ocean. Understanding and protecting marine biodiversity. Island Press, Washington DC.

Tobar, R. & F. Sardà, 1992. Annual and diel cycle as a predictive factor in deep-water fisheries for the prawn *Aristaeus antennatus* Risso, 1816. Fisheries Research 15: 169–179.

Tortonese, E., 1970. Osteichthyes. Pesci ossei. Vol X. Edizioni Calderini, Bologna.

Tortonese, E., 1975. Osteichthyes Pesci ossei. Vol XI. Edizioni Calderini, Bologna.

Tudela, S., F. Sardà, F. Maynou & M. Demestre, 2003. Influence of submarine canyons on the distribution of the deep-water shrimp, *Aristeus antennatus* (Risso, 1816) in the NW Mediterranean. Crustaceana 76: 217–225.

Tursi, A., A. Matarrese, G. D'Onghia, M. Panza, P. Maiorano, M. Basanisi, F. Perri, C. A. Marano, F. Casamassima 1996. Density, abundance and structure of population of red shrimps, Aristeus antennatus and Aristaeomorpha foliacea, in the Ionian Sea (southern Italy). EC Final Report Contract MED92.015 DG XIV 1-264.

MIX
Papier aus verantwortungsvollen Quellen
Paper from responsible sources
FSC® C105338

If you have any concerns about our products,
you can contact us on
ProductSafety@springernature.com

In case Publisher is established outside the EU,
the EU authorized representative is:
Springer Nature Customer Service Center GmbH
Europaplatz 3, 69115 Heidelberg, Germany

Printed by Libri Plureos GmbH
in Hamburg, Germany